*To Lin Youjin and Liu Xin*

# MS&A

Weijiu Liu

# Introduction to Modeling Biological Cellular Control Systems

 Springer

**Weijiu Liu**
Department of Mathematics
University of Central Arkansas
Conway, AR, USA

MS&A – Modeling, Simulation & Applications
ISSN print version: 2037-5255          ISSN electronic version: 2037-5263

ISBN 978-88-470-5636-7          ISBN 978-88-470-2490-8 (eBook)
DOI 10.1007/978-88-470-2490-8

Springer Milan Dordrecht Heidelberg London New York

Cover design: Beatrice ꓷ, Milan

Typesetting with LaTeX: PTP-Berlin, Protago TeX-Production GmbH, Germany (www.ptp-berlin.eu)

Springer-Verlag Italia S.r.l., Via Decembrio 28, I-20137 Milano
Springer is a part of Springer Science+Business Media (www.springer.com)

# Preface

When I tried to design output feedback controllers for nonlinear control systems, I found that design methods are limited. I thought that a cellular process in living organisms, such as the blood glucose control system, would be a perfect feedback control system because the cellular process should be tightly controlled so that cells in the living organism can carry out numerous tasks to survive. Thus I was wondering whether I can get inspiration from cells to design feedback controllers and started to read biology books. As I read more and more in biology, I found that there are numerous perfect feedback control mechanisms in life, for instance, enzyme feedback inhibition, blood glucose regulation, and store-operated calcium entry in cells, to mention a few. This motivated me to open a class to teach mathematical modeling in biology with a focus on cellular processes and motivated me to write this textbook for my class.

Because the general idea of establishing a model for every cellular control system is similar, I have selected a number of cellular control systems that I have understood most, such as the blood glucose control system and intracellular calcium control system, to demonstrate how to model them mathematically in the setting of control theory. Once one masters the methods of modeling these selected cellular systems, one will be able to use them to handle other cellular systems.

This textbook contains the essential knowledge in modeling, simulation, analysis, and applications in dealing with biological cellular control systems. In particular, the book shows how to use the law of mass balance and the law of mass action to derive an enzyme kinetic model - the Michaelis-Menten function or the Hill function, how to use a current-voltage relation, Nernst potential equilibrium equation, and Hodgkin and Huxley's models to model an ionic channel or pump, and how to use the law of mass balance to integrate these enzyme or channel models into a complete feedback control system. The book also illustrates how to use data to estimate parameters in a model, how to use MATLAB to solve a model numerically, how to do computer simulations, and how to provide model predictions. Furthermore, the book demonstrates how to conduct a stability and sensitivity analysis on a model.

This textbook is self-contained and easy to read. Modeling, simulation, and applications are explained in details. Whenever possible, a schematic diagram is drawn to help understand the biology in a cellular process. A good background in ordinary differential equations and molecular biology is sufficient to understand the materials in this textbook.

This text is designed as a textbook for a one-semester course in mathematical modeling in biology with a focus on cellular processes in living organisms. There are exercises in the end of each chapter and preliminary MATLAB is introduced in Appendix A.

Although all models in this textbook are ordinary differential equation (ODE) models, a diffusion term can be easily added to these ODE models for some cellular systems such as the intracellular calcium control system to lead to partial differential equation (PDE) models. However, the PDE models will greatly increase the complexity and difficulty of computer simulation.

Although I tried hard to make the text error-free, there are certainly still numerous errors and mistakes of different types, such as typos, grammatical errors, and even scientific mistakes. I apologize for making these mistakes and you are welcome to send your comments and criticisms to me at weijiul@uca.edu or liuweijiu@hotmail.com.

I thank Dr. Fusheng Tang for collaborative work on mathematical biology, constant inspirational discussions on biology, and insightful comments on this manuscript. I thank the reviewer for evaluating this text and giving constructive comments. I thank my students for using this text and correcting mistakes.

Conway (Arkansas), July 2011                                          *Weijiu Liu*

**Acknowledgement**
The MS&A Editorial Board thanks Prof. Juan Jose Lopez Velazquez for the suggestions and comments given to the Author on a preliminary version of the manuscript.

# Contents

**1 Overview** . . . . . . . . . . . . . . . . . . . . . . . . . . . . . . . . . . . . . . . . . . . . . . . . . . 1
    1.1   Examples of Biological Cellular Control Systems . . . . . . . . . . . . . 1
    1.2   Modeling Methodology . . . . . . . . . . . . . . . . . . . . . . . . . . . . . . . . . . . 3
    1.3   Computer Simulation . . . . . . . . . . . . . . . . . . . . . . . . . . . . . . . . . . . . . 5
    1.4   Impact . . . . . . . . . . . . . . . . . . . . . . . . . . . . . . . . . . . . . . . . . . . . . . . . 6
    1.5   Audience . . . . . . . . . . . . . . . . . . . . . . . . . . . . . . . . . . . . . . . . . . . . . . 6
    References . . . . . . . . . . . . . . . . . . . . . . . . . . . . . . . . . . . . . . . . . . . . . . . . 7

**2 Enzyme Kinetics** . . . . . . . . . . . . . . . . . . . . . . . . . . . . . . . . . . . . . . . . . . . . 11
    2.1   The Law of Mass Balance . . . . . . . . . . . . . . . . . . . . . . . . . . . . . . . . 11
    2.2   The Law of Mass Action . . . . . . . . . . . . . . . . . . . . . . . . . . . . . . . . . 12
    2.3   The Michaelis-Menten Equation . . . . . . . . . . . . . . . . . . . . . . . . . . 14
    2.4   Bi-substrate Enzymes . . . . . . . . . . . . . . . . . . . . . . . . . . . . . . . . . . . 18
    2.5   Inhibitors . . . . . . . . . . . . . . . . . . . . . . . . . . . . . . . . . . . . . . . . . . . . . 19
        2.5.1   Competitive Inhibition . . . . . . . . . . . . . . . . . . . . . . . . . . . 20
        2.5.2   Uncompetitive Inhibition . . . . . . . . . . . . . . . . . . . . . . . . . 22
        2.5.3   Noncompetitive Inhibition . . . . . . . . . . . . . . . . . . . . . . . . 23
    2.6   Cooperativity . . . . . . . . . . . . . . . . . . . . . . . . . . . . . . . . . . . . . . . . . 24
    2.7   Chemical Potential . . . . . . . . . . . . . . . . . . . . . . . . . . . . . . . . . . . . . 27
    2.8   The Arrhenius Formula . . . . . . . . . . . . . . . . . . . . . . . . . . . . . . . . . 28
    2.9   Effects of Energy . . . . . . . . . . . . . . . . . . . . . . . . . . . . . . . . . . . . . . 28
    2.10  Effects of pH . . . . . . . . . . . . . . . . . . . . . . . . . . . . . . . . . . . . . . . . . 29
    2.11  The Ion Hopping Model . . . . . . . . . . . . . . . . . . . . . . . . . . . . . . . . 30
    Exercises . . . . . . . . . . . . . . . . . . . . . . . . . . . . . . . . . . . . . . . . . . . . . . . . . 31
    References . . . . . . . . . . . . . . . . . . . . . . . . . . . . . . . . . . . . . . . . . . . . . . . . 35

**3 Preliminary Systems Theory** . . . . . . . . . . . . . . . . . . . . . . . . . . . . . . . . . . 37
    3.1   Elementary Matrix Algebra . . . . . . . . . . . . . . . . . . . . . . . . . . . . . . 37
        3.1.1   Matrix Sums . . . . . . . . . . . . . . . . . . . . . . . . . . . . . . . . . . . 38
        3.1.2   Scalar Multiple . . . . . . . . . . . . . . . . . . . . . . . . . . . . . . . . . 38
        3.1.3   Matrix Multiplication . . . . . . . . . . . . . . . . . . . . . . . . . . . . 39

3.1.4   Powers of a Matrix .............................. 39
3.1.5   Transpose of a Matrix ........................... 39
3.1.6   The Determinant ................................ 40
3.1.7   Eigenvalues .................................... 40
3.1.8   Rank of a Matrix ............................... 42
3.2   Stability of Equilibrium Points ........................... 43
3.2.1   Definition of Stability .......................... 44
3.2.2   Lyapunov's Stability Theorem .................... 45
3.2.3   Lyapunov's Indirect Method ...................... 48
3.2.4   Invariance Principle ............................ 51
3.2.5   Input-output Stability .......................... 53
3.3   Controllability and Observability ......................... 55
3.4   Feedback Control ...................................... 57
3.5   Parametric Sensitivity ................................. 62
Exercises ..................................................... 65
References .................................................... 68

4  Control of Blood Glucose ....................................... 69
4.1   A Control System of Blood Glucose ....................... 70
4.2   Design of Feedback Controllers .......................... 71
4.2.1   Insulin and Glucagon Transition .................. 71
4.2.2   Glucagon Signaling Pathway ..................... 72
4.2.3   Insulin Signaling Pathway ....................... 73
4.2.3.1   Insulin Receptor Recycling Subsystem ........ 74
4.2.3.2   Postreceptor Signaling Subsystem ........... 75
4.2.3.3   GLUT4 Transport Subsystem .............. 78
4.2.4   Dynamical Feedback Controllers .................. 79
4.3   Estimation of Parameters ................................ 80
4.4   Simulation of Glucose and Insulin Dynamics ................ 81
4.5   Model Prediction of Parametrical Sensitivity ............... 84
4.6   Simulation of Glucose and Insulin Oscillations ............. 86
Exercises ..................................................... 90
References .................................................... 91

5  Control of Calcium in Yeast Cells .............................. 95
5.1   A Model of Aging Process .............................. 95
5.2   Calcium Uptake from Environment ........................ 96
5.3   Calcium Movement across the Vacuolar Membrane ........... 98
5.4   Calcium Movement across the Golgi Membrane .............. 99
5.5   Calcium Movement across the Endoplasmic Reticulum Membrane 100
5.6   A Calcium Control System .............................. 101
5.7   Design of Feedback Controllers .......................... 102
5.7.1   Control of Calcium Uptake from Environment ........ 102
5.7.2   Control of $Ca^{2+}/H^+$ Exchanger Vcx1p ............. 103
5.7.3   Control of Calcium Pumps Pmc1p and Pmr1p .......... 103

5.7.4    Control of Channel Yvc1p ........................ 107
5.7.5    Control of X-induced Calcium Channel on the Vacuolar
         Membrane .......................................... 107
5.7.6    Control of Calcium Homeostasis in Golgi and ER ...... 109
5.8    Simulation of Calcium Shocks ............................ 109
5.9    Simulation of Calcium Accumulations ..................... 112
5.10   Prediction of Cell Cycle-dependent Oscillations of Calcium ..... 113
5.11   Prediction of an Upper-limit of Cytosolic Calcium Tolerance for
       Cell Survival ........................................... 114
5.12   Model Limitation ........................................ 117
Exercises ..................................................... 118
References .................................................... 119

6  Kinetics of Ion Pumps and Channels ........................... 123
6.1    The Nernst-Planck Equation ............................ 124
6.2    The Nernst Equilibrium Potential ....................... 125
6.3    Current-voltage Relations .............................. 125
6.4    The Potassium Channel ................................. 127
       6.4.1    The Voltage-Gated Potassium Channel............. 127
       6.4.2    The Calcium-Activated Potassium Channel ......... 131
       6.4.3    The ATP-Sensitive Potassium Channel............. 132
6.5    The Voltage-Gated Sodium Channel ...................... 137
6.6    The Voltage-Gated Calcium Channel ..................... 141
6.7    The IP$_3$ Receptor ..................................... 144
6.8    The Ryanodine Receptor ................................ 150
6.9    The Sarcoplasmic or Endoplasmic Reticulum Calcium ATPase... 154
6.10   The Plasma Membrane Calcium ATPase ................... 156
6.11   The Sodium/Potassium ATPase .......................... 157
6.12   The Sodium/Calcium Exchanger ......................... 158
6.13   Membrane Potential Models ............................. 160
6.14   The Hodgkin-Huxley Model ............................. 161
Exercises ..................................................... 164
References .................................................... 168

7  Control of Intracellular Calcium Oscillations ................. 173
7.1    The Chay-Keizer Feedback Control System ............... 175
7.2    The BSSMAMS Feedback Control System ................. 177
7.3    The FTMP Feedback Control System ..................... 179
Exercises ..................................................... 186
References .................................................... 187

8  Store-Operated Calcium Entry ............................... 189
8.1    A Calcium Control System ............................. 190
8.2    Design of an Output Feedback Controller ............... 191
8.3    SOCE Computer Simulation ............................ 193

8.4    Simulation of Rejection of Agonist Disturbances ............... 197
8.5    Stability Analysis ...................................... 198
8.6    Remarks ............................................. 200
Exercises ................................................... 201
References ................................................. 202

**9    Control of Mitochondrial Calcium** ............................. 207
9.1    Respiration-driven Proton Ejection ......................... 208
9.2    ATP Synthesis and Proton Uptake by the $F_1F_0$-ATPase ........ 212
9.3    ATP and ADP Transport by the Adenine Nucleotide Translocator  214
9.4    Calcium Uptake by the Uniporter ........................... 217
9.5    Calcium Efflux via the Sodium/Calcium Exchanger ............ 217
9.6    Governing Equations of Calcium Dynamics ................... 218
Exercises ................................................... 220
References ................................................. 220

**10   Control of Phosphoinositide Synthesis** ......................... 223
10.1   PIP Synthesis from PI ................................... 223
10.2   $PIP_2$ Synthesis from PIP .............................. 224
10.3   $PIP_2$ Hydrolysis ..................................... 224
10.4   A Control System for Phosphoinositide Synthesis .............. 225
Exercises ................................................... 228
References ................................................. 228

**Appendix A   Preliminary MATLAB** ............................. 229
A.1    MATLAB Desktop ...................................... 229
       A.1.1    Command Window ............................... 229
       A.1.2    Help Browser .................................. 230
       A.1.3    Editor / Debugger .............................. 230
A.2    Creating, Writing, and Saving a MATLAB File ................ 231
A.3    Simple Mathematics .................................... 232
       A.3.1    Variables ..................................... 232
       A.3.2    Operators ..................................... 232
       A.3.3    Built-in Functions ............................. 232
       A.3.4    Mathematical Expressions ....................... 233
A.4    Vectors and Matrices ................................... 234
       A.4.1    Generating vectors ............................. 234
       A.4.2    Generating matrices ............................ 235
       A.4.3    Array Addressing or Indexing .................... 236
       A.4.4    Arithmetic Operations on Arrays ................. 237
A.5    M-Files .............................................. 239
       A.5.1    Scripts ....................................... 239
       A.5.2    Functions ..................................... 240
A.6    Basic Plotting ........................................ 241
A.7    Relational Operators ................................... 242

A.8    Flow Control ............................................ 243
       A.8.1    If-Else-End Constructions ....................... 243
       A.8.2    For Loops ....................................... 244
       A.8.3    While Loops ..................................... 245
A.9    Logical Operators ....................................... 246
A.10   Solving Symbolic Equations .............................. 247
A.11   Solving Ordinary Differential Equations ................. 248
A.12   Data Fitting ............................................ 251
A.13   Parameter Estimation .................................... 255
Exercises ...................................................... 258
References ..................................................... 260

**Appendix B   Units, Physical Constants and Formulas** ........ 261
B.1    Physical Formulas ....................................... 261
B.2    Units, Unit Scale Factors and Physical Constants......... 262
References ..................................................... 264

**Index** ...................................................... 265

# 1

# Overview

## 1.1 Examples of Biological Cellular Control Systems

There are numerous cellular control systems in living organisms. These cellular systems are tightly controlled through numerous feedback control mechanisms so that cells in the living organisms can carry out numerous functions to survive. In this textbook, I select a number of cellular control systems that I have understood most to demonstrate how to model them mathematically in the setting of control theory.

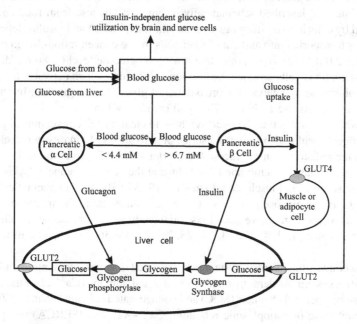

**Fig. 1.1.** A schematic description of a simplified blood glucose control system. Molecular control mechanisms of blood glucose sketched in this figure are described in the text

Liu W.: Introduction to Modeling Biological Cellular Control Systems.
DOI 10.1007/978-88-470-2490-8_1, © Springer-Verlag Italia 2012

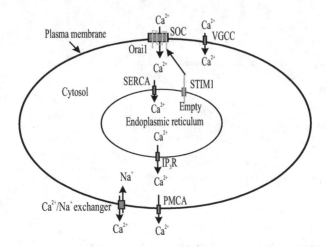

**Fig. 1.2.** Schematic diagram of an intracellular calcium control system. The abbreviations used are: SOC, store-operated channel; VGCC, voltage-gated calcium channel; PMCA, plasma membrane $Ca^{2+}$ ATPase; SERCA, sarcoplasmic or endoplasmic reticulum $Ca^{2+}$-ATPase; $IP_3R$, inositol (1,4,5)-trisphosphate receptor. The diagram is explained in the text

The first cellular control system to be discussed in this textbook is the important well-studied blood glucose control system. A simplified version of the blood glucose control system is described schematically in Fig. 1.1. Glucose from food and liver is utilized by cells in two different ways: insulin-independent and insulin-dependent. Glucose is transported into and out of liver cells by the concentration-driven glucose transporter 2 (GLUT2). In response to a low blood glucose level ($< 80$ mg/dl or 4.4 mM), pancreatic $\alpha$ cells produce the hormone glucagon, which activates the glycogen phosphorylase to catalyze the breakdown of glycogen into glucose. In response to a high blood glucose level ($> 120$ mg/dl or 6.7 mM), pancreatic $\beta$ cells secrete insulin, which activates the glycogen synthase to catalyze the conversion of glucose into glycogen. Insulin also initiates a series of activations of kinases in muscle cells to lead to the redistribution of glucose transporter 4 (GLUT4) from intracellular storage sites to the plasma membrane (PM). Once at the cell membrane, GLUT4 transports glucose into the muscle or fat cells [6, 19, 35]. Because mathematical modeling of the blood glucose control system could contribute to the treatment of diabetes, it has received intensive attentions and there have been numerous publications on it, for instance [1, 2, 4, 12, 18, 24, 25, 26, 30, 35, 36, 37, 38, 41], to mention a few.

The second cellular control system to be modeled is an important intracellular calcium control system. As demonstrated in Fig. 1.2, calcium ions $Ca^{2+}$ enter the cytosol through store-operated channels (SOC) and voltage-gated calcium channels (VGCC). The sarcoplasmic or endoplasmic reticulum $Ca^{2+}$-ATPases (SERCA) pump $Ca^{2+}$ from the cytosol into the endoplasmic reticulum (ER) and $Ca^{2+}$ in ER are released to the cytosol through the inositol (1,4,5)-trisphosphate ($IP_3$)- and $Ca^{2+}$-mediated inositol (1,4,5)-trisphosphate receptors (IP$_3$R). Ca$^{2+}$ exit the cytosol through plasma membrane Ca$^{2+}$-ATPases (PMCA) and Ca$^{2+}$/Na$^+$ exchangers. Depletion of ER Ca$^{2+}$ stores causes STIM1 to move to ER-PM junctions, bind to Orai1, and activate store-operated channels for Ca$^{2+}$ entry [11, 23, 28, 29, 31, 32, 39, 46]. Because rises in cytoplasmic Ca$^{2+}$ concentration are used by all cells as a signalling mechanism for numerous cellular processes [5, 8], the cytoplasmic Ca$^{2+}$ concentration must be tightly controlled within a narrow range by the above mentioned calcium pumps or channels.

Other cellular control systems to be considered include a mitochondrial calcium control system and a phosphoinositide synthesis control system. Mitochondrial Ca$^{2+}$ uptake has profound consequences for physiological cell functions. Intramitochondrial Ca$^{2+}$ stimulates oxidative phosphorylation and controls the rate of adenosine triphosphate (ATP) production. Mitochondrial Ca$^{2+}$ uptake modifies the shape of cytosolic Ca$^{2+}$ pulses or transients [9, 13, 14] and regulates store-operated calcium entry [10, 33]. Furthermore, for any repetitive physiological process dependent on intramitochondrial free Ca$^{2+}$ concentration, a kind of intramitochondrial Ca$^{2+}$ homeostasis must exist and be controlled dynamically to avoid either Ca$^{2+}$ buildup or depletion in mitochondria [3, 13].

Phosphatidylinositol 4,5-bisphosphate (PIP$_2$) is the predominant (99%) phosphoinositide in mammalian cells [43]. PIP$_2$ is synthesized from phosphatidylinositol-4-phosphate (PIP) by PIP$_2$ synthases while PIP is synthesized from Phosphatidylinositol (PI) by PIP synthases [19]. PIP$_2$ in cells is normally hydrolyzed by phospholipase C (PLC) to generate inositol 1,4,5-trisphosphate (IP$_3$) and diacylglycerol (DAG), which serve as second messengers for intracellular Ca$^{2+}$ mobilization and PKC (protein kinase C) activation, respectively [31, 43]. Thus, PIP$_2$ plays important roles in PLC-mediated cellular processes, such as glucose-stimulated insulin secretion [15], store-operated calcium entry [22, 42], and sterol trafficking [34, 44]. Hence phosphoinositide synthesis must be delicately controlled by molecular feedback mechanisms.

## 1.2 Modeling Methodology

To simplify computer simulation, we assume that concentrations of molecules involved in a cellular control system are uniform in space. So they can be modeled by ordinary differential equations (ODEs). However, a diffusion term can be easily added to these ODE models for some cellular systems such as the intracellular calcium control system, leading to partial differential equation (PDE) models. The PDE models will greatly increase the complexity and difficulty of computer simulation.

In modeling a cellular control system, we use the law of mass balance and the law of mass action. The *law of mass balance* states that the input rate of mass into a system is equal to the sum of the rate of change of mass in the system in time and the rate of output from the system. Mathematically the law can be expressed as

$$\frac{dx}{dt} = \text{input rate} - \text{output rate},$$

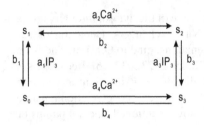

**Fig. 1.3.** A four state model of the IP$_3$-receptor proposed in [20]

where $x$ denotes the amount of mass in the system. The *law of mass action* is a mathematical model that explains and predicts the kinetics of a chemical reaction. It states that each individual forward reaction rate or backward reaction rate is proportional to the product of the concentrations of the participating molecules.

In order to use the law of mass balance to establish a mathematical model for a cellular control system, we first need to model enzymes, ionic pumps, ionic channels, and glucose transporters involved in the cellular system. A general idea of deriving such a model is to propose a chemical reaction model of multiple states of an enzyme, a channel, a pump, or a transporter. As an example, we consider the IP$_3$-receptor, which is a tetramer and has four subunits. Keizer *et al* [20] proposed a four state model for the IP$_3$-receptor, as demonstrated in Fig. 1.3. Each subunit is endowed with an IP$_3$ and a Ca$^{2+}$ binding site that interact with each other, such that when the Ca$^{2+}$ site is occupied, the affinity for binding of IP$_3$ is increased. Thus a subunit can exist in four states: state $s_0$ consists of a subunit with neither IP$_3$ nor Ca$^{2+}$ bound; $s_1$ has only IP$_3$ bound; $s_2$ has both IP$_3$ and Ca$^{2+}$ bound; $s_3$ has only Ca$^{2+}$ bound. An open channel is assumed to result only when each one of the four subunits is in the state $s_1$. All other states of the tetramer are assumed to be closed. Thus, the open probability of the receptor is equal to $x_1^4$, where $x_1$ denotes the fraction of subunits in state $s_1$. By an equilibrium analysis, $x_1$ can be expressed in terms of the concentrations of IP$_3$ and Ca$^{2+}$. The parameters in this model can be determined by fitting the model into experimental data.

In some cases, experimental data show that the kinetics of an enzyme or channel follows the Michaelis-Menten function or the Hill function. In this case, a chemical reaction model of multiple states is not needed. Instead, the model can be obtained by fitting the function into data.

Usually, an enzyme can bind more than one substrate molecule and the binding of one substrate molecule promotes the binding of subsequent ones. This results in a positive cooperativity. The *Hill function* is a mathematical model to describe such a positive cooperativity:

$$V = \frac{V_{max}[S]^n}{K_m^n + [S]^n},$$

where $V$ is the velocity of converting substrate to product by an enzyme, $V_{max}$ is the maximal velocity, $K_m$ is a positive constant called *Michaelis-Menten constant*, $n$ is a positive constant called the *Hill exponent*, and $[S]$ denotes the concentration of the substrate. If $n = 1$, the Hill function is called the *Michaelis-Menten function*.

Because many ionic pumps or channels are electrogenic and are gated or controlled by the membrane potential, current-voltage relations and the Nernst equilibrium potential equation are needed to model them. In many cases, the following linear current-voltage relation

$$I_{ion} = g(V - V_I),$$

which is derived from Ohm's law, is sufficient to approximate the current generated by the movement of ions through a pump or channel across the plasma membrane. In this current-voltage equation, $I_{ion}$ denotes the current, $V$ denotes the electrical potential difference across the plasma membrane, $V_I$ is the equilibrium potential of the ion, and $g$ is a membrane conductance. The equilibrium potential $V_I$ is given by the Nernst equation

$$V_I = \frac{RT}{zF} \ln\left(\frac{c_o}{c_i}\right),$$

where $z$ is the valence of the ion, $R$ is the universal gas constant, $T$ is the absolute temperature, $F$ is Faraday's constant, and $c_o$ ($c_i$) is the concentration of the ion outside (inside) the membrane. This potential is called the *Nernst potential*.

In many cases, experiments showed that the conductance $g$ is not constant and varies with time $t$ and voltage $V$ [16]. Distinct models of the conductance $g$ for different ionic channels or pumps were established and many of them were based on Hodgkin and Huxley's models [16] such as the model for the voltage-gated potassium channel of the squid giant axon:

$$g = Gn^4,$$
$$\frac{dn}{dt} = \frac{n_\infty - n}{\tau},$$

where $G$ is the maximal conductance, $n_\infty = n_\infty(V)$ is a steady state of activation probability, and $\tau = \tau(V)$ is a time constant. $n_\infty$ and $\tau$ are determined by experimental data.

Due to the complexity of cellular control systems, Lyapunov stability analysis in many cases is difficult. So linear stability analysis will be used.

## 1.3 Computer Simulation

Mathematical models will be solved numerically by using the function ode15s from MATLAB, the Mathworks, Inc. The function ode15s is a variable order solver based on the numerical differentiation formulas (NDFs). MATLAB does not recommend to reduce MaxStep in ode15s for the accuracy of solutions since this can significantly slow down solution time. Instead the relative error tolerance and the absolute error tolerance will be used.

A mathematical model of a biological system usually contains numerous parameters. Estimation of these parameters is challenging because no universal methods can be used and in some cases no complete sets of data are available for fitting a model into data. Hence we usually have to appeal to different means, such as adoption of existing parameters in the literature, numerical simulations, use of optimization methods like the simulated annealing optimization [7, 21] and nonlinear least squares, and use of built-in parameter estimation functions like "SBparameterestimation" from the SBToolbox [47] and "sbioparamestim" from the SimBiology toolbox of MATLAB.

## 1.4 Impact

Mathematical models for cellular control systems are of potential interest to disease treatments. Mathematical models are required for integrating a glucose monitoring system into insulin pump technology to form a closed-loop insulin delivery system on the feedback of blood glucose levels, the so-called "artificial pancreas" [17, 27, 37]. Calcium ions play a central role in the process of insulin secretion. Release of $Ca^{2+}$ from intracellular stores is essential for the amplification of insulin secretion by promoting the replenishment of the readily releasable pool of secretory granules, while voltage-dependent $Ca^{2+}$ entry is directed to the sites of exocytosis via the binding of the L-type $Ca^{2+}$ channels to SNARE proteins [39, 40]. Therefore, mathematical models for intracellular calcium control systems would provide an important component in building an artificial pancreas. It has been discovered that the proteins Orai1 and STIM1 control calcium influx from the extracellular environment to the inside of cells [11, 23, 28, 29, 31, 32, 39, 46] and they are required for tumor cell migration, invasion, and metastasis, being potential targets for therapeutic intervention to inhibit tumor metastasis [45]. So a mathematical model for the calcium entry control system would provide a theory for such therapeutic intervention.

Mathematical models for cellular control systems might serve as a tool for extracting information from experimental data to obtain new insights into molecular mechanisms in living organisms and then provide guidelines for wet experiments. On the other hand, since the design of output feedback controllers for nonlinear systems is challenging, exploring mathematically cellular control systems in living organisms will inspire the mathematical control design and advance mathematical control theory.

## 1.5 Audience

The potential audience are applied mathematics students who are interested in applying mathematics to life sciences, biology students who like to use mathematics to study biology, and researchers in mathematical biology. If you have a good background in ordinary differential equations and molecular biology, you will not have any difficulties to read through this textbook.

This text is designed as a textbook for a one-semester course in mathematical modeling in biology with a focus on cellular processes in living organisms. There are exercises in the end of each chapter and preliminary MATLAB is introduced in Appendix A.

# References

1. Ackerman E., Gatewood L.C., Rosevear J.W., Molnar G.D.: Model studies of blood glucose regulation. Bull. Math. Biophys. **27**, 21-27 (1965).
2. Albisser A.M., Leibel B.S., Ewart T.G., Davidovac Z., Botz C.K., Zingg W.: An artificial endocrine pancreas. Diabetes **23**, 389-396 (1974) .
3. Balaban, R.S.: Cardiac energy metabolism homeostasis: role of cytosolic calcium. J. Mol. Cell. Cardiol. **34**, 1259-1271 (2002).
4. Bergman R.N., Finegood D.T., Ader M.: Assessment of insulin sensitivity in vivo. Endocrine Reviews. **6**, 45-86 (1985) .
5. Berridge M.J., Bootman M.D., Roderick H.L.: Calcium signalling: dynamics, homeostasis and remodelling. Nat. Rev. Mol. Cell Biol. **4**, 517-29 (2003).
6. Bevan P.: Insulin signaling. J. Cell Sci. **114**, 1429-1430 (2001).
7. Cerny V.: A thermodynamical approach to the travelling salesman problem: an efficient simulation algorithm. J. Optim. Theory and Appl. **45**, 41-51 (1985).
8. DeHaven W., Jones B., Petranka J., Smyth J., Tomita T., Bird G., Putney J.: TRPC channels function independently of STIM1 and Orai1. J. Physiol. **587**, 2275-2298 (2009).
9. Duchen M.R.: Mitochondria and calcium: from cell signalling to cell death. J. Physiol. **529**, 57-68 (2000).
10. Feldman B., Fedida-Metula S., Nita J., Sekler I., Fishman D.: Coupling of mitochondria to store-operated $Ca^{2+}$-signaling sustains constitutive activation of protein kinase B/Akt and augments survival of malignant melanoma cells. Cell Calcium **47**, 525-537 (2010).
11. Feske S., Gwack Y., Prakriya M., Srikanth S., Puppel S.H., Tanasa B., Hogan P.G., Lewis R.S., Daly M., Rao A.: A mutation in Orai1 causes immune deficiency by abrogating CRAC channel function. Nature **441**, 179-185 (2006).
12. Fridlyand L.E., Philipson L.H.: Glucose sensing in the pancreatic beta cell: a computational systems analysis. Theor. Biol. Med. Model. **24**, 7-15 (2010).
13. Gunter T.E., Yule D.I., Gunter K.K., Eliseev R.A., Salter J.D.: Calcium and mitochondria. FEBS Lett. **567**, 96-102 (2004).
14. Gunter T.E., Sheu S.-S.: Characteristics and possible functions of mitochondrial $Ca^{2+}$ transport mechanisms. Biochim. Biophys. Acta. **1787**, 1291-1308 (2009).
15. Hao M., Bogan J.S.: Cholesterol regulates glucose-stimulated insulin secretion through phosphatidylinositol 4,5-bisphosphate. J. Biol. Chem. **284**, 29489-29498 (2009).
16. Hodgkin A.L., Huxley A.F.: A quantitativfe description of membrane current and its application to conduction and excitation in nerve. J. Physiol. **117**, 500-544 (1952).
17. Hovorka R.: Continuous glucose monitoring and closed-loop systems. Diabetic Medicine **23**, 1-12 (2006).
18. Jauslin P.M., Frey N., Karlsson M.O.: Modeling of 24-hour glucose and insulin profiles of patients with type 2 diabetes. J. Clin. Pharmacol. **51**, 153-164 (2011).
19. Karp G.: Cell and Molecular Biology, 4th edition. John Wiley & Sons, Inc., Hoboken, NJ (2005).
20. Keizer J., De Young G.W.: Two roles for $Ca^{2+}$ in agonist stimulated $Ca^{2+}$ oscillations. Biophys. J. **61**, 649-660 (1992).
21. Kirkpatrick S., Gelatt C.D., Vecchi M.P.: Optimization by Simulated Annealing. Science, New Series **220**, 671-680 (1983).
22. Korzeniowski M.K., Popovic M.A., Szentpetery Z., Varnai P., Stojilkovic S.S., Balla T.: Dependence of STIM1/Orai1-mediated calcium entry on plasma membrane phosphoinositides. J. Biol. Chem. **284**, 21027-21035 (2009).
23. Lewis R.S: The molecular choreography of a store-operated calcium channel. Nature **446**, 284-287 (2007).

24. Liu W., Hsin C., Tang F.: A molecular mathematical model of glucose mobilization and uptake. Math. Biosciences **221**, 121-129 (2009).
25. Liu W., Tang F.: Modeling a simplified regulatory system of blood glucose at molecular levels. J. Theor. Biol. **252**, 608-620 (2008).
26. Man C.D., Rizza R.A., Cobelli C.: Meal simulation model of the glucose-insulin system. IEEE Tran. Biomed. Eng. **54**, 1740-1749 (2007).
27. Panteleon A.E., Loutseiko M., Steil G.M., Rebrin K.: Evaluation of the effect of gain on the meal response of an automated closed-loop insulin delivery system. Diabetes **55**, 1995-2000 (2006).
28. Parekh A.B.: On the activation mechanism of store-operated calcium channels. Pflugers Arch. - Eur. J. Physiol. **453**, 303-311 (2006).
29. Park C.Y., Hoover P.J., Mullins F.M., Bachhawat P., Covington E.D., Raunser S., Walz T., Garcia K.C., Dolmetsch R.E., Lewis R.S.: STIM1 clusters and activates CRAC channels via direct binding of a cytosolic domain to Orai1. Cell **136**, 876-890 (2009).
30. Percival M.W., Bevier W.C., Wang Y., Dassau E., Zisser H.C., Jovanovič L., Doyle F.J. 3rd.: Modeling the effects of subcutaneous insulin administration and carbohydrate consumption on blood glucose. J. Diabetes Sci. Technol. **4**, 1214-1228 (2010).
31. Potier M., Trebak M.: New developments in the signaling mechanisms of the store-operated calcium entry pathway. Pflugers Arch. **457**, 405-415 (2008).
32. Putney J.W. Jr.: Recent breakthroughs in the molecular mechanism of capacitative calcium entry (with thoughts on how we got here), Cell Calcium **42**, 103-110 ( 2007).
33. Ryu S.Y., Peixoto P.M., Won J.H., Yule D.I., Kinnally K.W.: Extracellular ATP and P2Y2 receptors mediate intercellular $Ca^{2+}$ waves induced by mechanical stimulation in submandibular gland cells: Role of mitochondrial regulation of store operated $Ca^{2+}$ entry. Cell Calcium **47**, 65-76 (2010).
34. Schulz T.A., Choi M.G., Raychaudhuri S., Mears J.A., Ghirlando R., Hinshaw J.E., Prinz W.A.: Lipid-regulated sterol transfer between closely apposed membranes by oxysterol-binding protein homologues. J. Cell. Biol. **187**, 889-903 (2009).
35. Sedaghat A.R., Sherman A., Quon M.J.: A mathematical model of metabolic insulin signaling pathways. Am. J. Physiol. Endocrinol. Metab. **283**, E1084-E1101 (2002).
36. Sherman A.: Lessons from models of pancreatic beta cells for engineering glucose-sensing cells. Math. Biosci. **227**, 12-19 (2010).
37. Steil G.M., Rebrin K., Darwin C., Hariri F., Saad M.F.: Feasibility of automating insulin delivery for the treatment of type 1 diabetes. Diabetes **55**, 3344-3350 (2006).
38. Sturis J., Polonsky K.S., Mosekilde E., Cauter E.V.: Computer model for mechanisms underlying ultradian oscillations of insulin and glucose. Am. J. Physiol. Endocrinol. Metab. **260**, E801-E809 (1991).
39. Tamarina N.A., Kuznetsov A., Philipson L.H.: Reversible translocation of EYFP-tagged STIM1 is coupled to calcium infux in insulin secreting $\beta$-cells. Cell Calcium **44**, 533-544 (2008).
40. Tengholm A., Gylfe E.: Oscillatory control of insulin secretion. Molecular and Cellular Endocrinology. **297**, 58-72 (2009).
41. Toffolo G., Campioni M., Basu R., Rizz R.A., Cobelli C.: A minimal model of insulin secretion and kinetics to assess hepatic insulin extraction. Am. J. Physiol. Endocrinol. Metab. **290**, E169-E176 (2006).
42. Walsh C.M., Chvanov M., Haynes L.P., Petersen O.H., Tepikin A.V., Burgoyne R.D.: Role of phosphoinositides in STIM1 dynamics and store-operated calcium entry. Biochem. J. **425**, 159-168 (2010).
43. Xu C., Watras J., Loew L.M.: Kinetic analysis of receptor-activated phosphoinositide turnover. J. Cell Biology **161**, 779-791 (2003).

44. Xu J., Dang Y., Ren Y.R., Liu J.O.: Cholesterol trafficking is required for mTOR activation in endothelial cells. Proc. Natl. Acad. Sci. USA **107**, 4764-4769 (2010).

45. Yang S., Zhang J.J., Huang X.-Y.: Orai1 and STIM1 are critical for breast tumor cell migration and metastasis. Cancer Cell **15**, 124-134 (2009).

46. Yuan J.P., Zeng W., Dorwart M.R., Choi Y.-J., Worley P.F., Muallem S.: SOAR and the polybasic STIM1 domains gate and regulate Orai channels. Nature Cell Biology **11** 337-343 (2009).

47. Systems Biology Toolbox for MATLAB. http://www.sbtoolbox.org/ (2011). Accessed 18 June 2011.

**2**

# Enzyme Kinetics

Enzymes catalyze biochemical reactions by lowering the free energy of activation of the reactions. In the reactions, other molecules called *substrates* are converted into products, but the enzymes themselves are not changed. An enzyme is usually a large protein, considerably larger than its substrate molecules. An enzyme protein has one or more active sites, to which its substrates can bind to form a complex. Enzymes are highly specific, usually catalyzing a reaction of only one particular substrate. They are regulated by complex feedback control mechanisms. In this chapter, we give an introduction to enzyme kinetics, which quantitatively studies how an enzyme catalyzes a reaction. Detailed discussions can be found in biochemistry and enzyme kinetics books such as [5, 18].

## 2.1 The Law of Mass Balance

The *law of mass balance* states that the input rate of mass into a system is equal to the sum of the rate of change of mass in the system in time and the rate of output from the system, as demonstrated in Fig. 2.1. Mathematically the law can be expressed as

$$\frac{dx}{dt} = \text{input rate} - \text{output rate}, \qquad (2.1)$$

where $x$ denotes the amount of mass in the system.

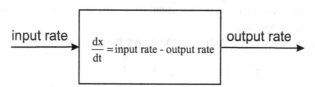

**Fig. 2.1.** The law of mass balance. The input rate of mass into a system is equal to the sum of the rate of change of mass amount $x$ in the system in time and the rate of output from the system

Liu W.: Introduction to Modeling Biological Cellular Control Systems.
DOI 10.1007/978-88-470-2490-8_2, © Springer-Verlag Italia 2012

Consider a simple system where the input rate is equal to 0 and the output rate is proportional to $x$. Then the mass balance law (2.1) gives

$$\frac{dx}{dt} = -kx, \qquad (2.2)$$

$$x(0) = x_0, \qquad (2.3)$$

where $k$ is a positive constant and $x_0$ is the initial condition. Solving the equation (2.2) gives

$$x = x_0 e^{-kt}. \qquad (2.4)$$

*Half-life* of a substance undergoing decay is the period of time that it takes for the substance to decrease by half. Let $t_{1/2}$ denote the half-life of a substance. It then follows from (2.4) that

$$\frac{1}{2}x_0 = x_0 e^{-kt_{1/2}}.$$

Solving this equation gives

$$k = \frac{\ln 2}{t_{1/2}}. \qquad (2.5)$$

This formula is useful for estimating the rate constant $k$ if the half-life of a substance is known.

## 2.2 The Law of Mass Action

Consider a chemical reaction:

$$aA + bB \underset{k_-}{\overset{k_+}{\rightleftharpoons}} cC + dD,$$

where $A, B$ are reactants, $C, D$ are products, $a, b, c, d$ are the stoichiometric coefficients of the balanced reaction, and $k_+, k_-$ are positive constants called *rate constants*. *Reaction rate* (also called *reaction velocity*) is defined to be the rate of change of concentration of reactant or product with respect to the change of time. Mathematically, it is given by the derivative $\frac{dA}{dt}, \frac{dB}{dt}$, or $\frac{dC}{dt}$.

The *law of mass action* is a mathematical model that explains and predicts the kinetics of a chemical reaction. It states that each individual forward reaction rate $v_+$ or backward reaction rate $v_-$ is proportional to the product of the concentrations of the participating molecules.

Applying the law of mass action to the above reaction, we obtain

$$v_+ = k_+[A]^a[B]^b, \quad v_- = k_-[C]^c[D]^d,$$

where the square bracket $[\cdot]$ denotes the concentration of a chemical. The exponents $a, b, c, d$ are called *orders*; $a$ is called the order with respect to $A$, $b$ is the order with respect to $B$, and so on. Orders are usually positive integers, but they may be negative integers, zero, or even fractions. The sum of all orders of the algebraic expression is the overall order of reaction. The exponents $a, b, c, d$ may be different from the stoichiometric coefficients of the balanced reaction and must be determined experimentally.

Note that a great number of chemical reactions do not follow the law of mass action. Those reactions that follow the law of mass action are called *elementary reactions*.

Using the law of mass balance, we derive a rate equation

$$\frac{d[C]}{dt} = v_+ - v_- = k_+ [A]^a [B]^b - k_- [C]^c [D]^d.$$

At equilibrium, concentrations are not changing. So $\frac{d[C]}{dt} = 0$ and then

$$k_+ [A]^a [B]^b = k_- [C]^c [D]^d.$$

The *equilibrium constant* of reaction is defined by

$$K_{eq} = \frac{k_+}{k_-} = \frac{[C]^c [D]^d}{[A]^a [B]^b}. \tag{2.6}$$

In the same way, we can derive the rate equations for other chemicals $A, B, D$ as follows:

$$\frac{d[A]}{dt} = -k_+ [A]^a [B]^b + k_- [C]^c [D]^d,$$

$$\frac{d[B]}{dt} = -k_+ [A]^a [B]^b + k_- [C]^c [D]^d,$$

$$\frac{d[D]}{dt} = k_+ [A]^a [B]^b - k_- [C]^c [D]^d.$$

Since $\frac{d[A]}{dt} + \frac{d[C]}{dt} = 0$, $[A] + [C] = A_0$ is constant, that is, the quantity $[A] + [C]$ is conserved. It is clear that other three quantities $[A] + [D], [B] + [C], [B] + [D]$ are also conserved.

For dimerization of two monomers of the same species $A$ to produce species $B$:

$$A + A \overset{k_+}{\underset{k_-}{\rightleftharpoons}} B,$$

a caution needs to be paid to the use of the law of mass balance. Since every $B$ is produced from and split into two of $A$, the rate equation for $A$ is

$$\frac{d[A]}{dt} = -2k_+ [A]^2 + 2k_- [B].$$

The rate of production of $B$ is half that of $A$:

$$\frac{d[B]}{dt} = -\frac{1}{2} \frac{d[A]}{dt} = k_+ [A]^2 - k_- [B].$$

## 2.3 The Michaelis-Menten Equation

An enzymatic reaction is schematically described in Fig. 2.2. In this reaction, a substrate binds to an enzyme to form an enzyme/substrate complex. After undergoing a reaction, the enzyme/substrate complex is converted into an enzyme/product complex and then a product is released. For the convenience of mathematical modeling, we represent this enzymatic reaction by the following chemical reaction equation

$$E + S \underset{k_{-1}}{\overset{k_1}{\rightleftharpoons}} C \overset{k_2}{\rightarrow} P + E,$$

where $E$ is an enzyme, $S$ is a substrate, $C$ is a complex formed from $E$ and $S$, $P$ is a product, and $k_1$, $k_{-1}$, $k_2$ are rate constants. The application of the law of mass action and the law of mass balance yields a system of ordinary differential equations governing the reaction

$$\frac{d[S]}{dt} = -k_1[E][S] + k_{-1}[C], \tag{2.7}$$

$$\frac{d[E]}{dt} = -k_1[E][S] + (k_{-1} + k_2)[C], \tag{2.8}$$

$$\frac{d[C]}{dt} = k_1[E][S] - (k_{-1} + k_2)[C], \tag{2.9}$$

$$\frac{d[P]}{dt} = k_2[C]. \tag{2.10}$$

Since $\frac{d[E]}{dt} + \frac{d[C]}{dt} = 0$, we have

$$[E] + [C] = [E_0], \tag{2.11}$$

where $[E_0]$ is the concentration of total available enzymes.

The mathematical model (2.7)-(2.10) for the enzyme kinetics is complex and needs to be simplified so that it is useful in modeling more sophisticated biochemical reactions. Experimental studies have shown that the concentration of the complex

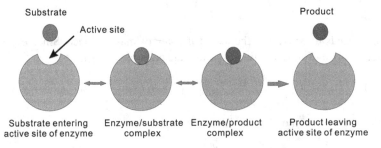

Substrate                                                          Product

Active site

Substrate entering        Enzyme/substrate    Enzyme/product        Product leaving
active site of enzyme          complex              complex         active site of enzyme

**Fig. 2.2.** A schematic description of an enzymatic reaction

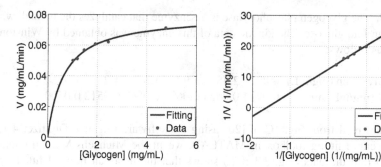

**Fig. 2.3.** The kinetics of glycogen phosphorylase follows the Michaelis-Menten equation. *Left:* the kinetic data of glycogen phosphorylase obtained by Winston *et al* [20] are fitted by the Michaelis-Menten equation (2.14). *Right:* A Lineweaver-Burk plot. The data are read from [20] using the software Engauge Digitizer 4.1

reaches an equilibrium state much faster than the substrate. At equilibrium, the concentration $[C]$ does not change in time. Thus we may assume that

$$\frac{d[C]}{dt} = 0,$$

and then

$$k_1[E][S] - (k_{-1} + k_2)[C] = 0. \tag{2.12}$$

This assumption is called the *quasi-steady state approximation.*

Solving the equations (2.11) and (2.12), we obtain

$$[C] = \frac{[E_0][S]}{[S] + \frac{k_{-1} + k_2}{k_1}}, \quad [E] = \frac{[E_0]\frac{k_{-1} + k_2}{k_1}}{[S] + \frac{k_{-1} + k_2}{k_1}}. \tag{2.13}$$

It then follows from (2.10) that

$$V = \frac{d[P]}{dt} = k_2[C] = \frac{k_2[E_0][S]}{[S] + \frac{k_{-1} + k_2}{k_1}} = \frac{V_{max}[S]}{[S] + K_m}, \tag{2.14}$$

where

$$V_{max} = k_2[E_0], \tag{2.15}$$

$$K_m = \frac{k_{-1} + k_2}{k_1}. \tag{2.16}$$

$V_{max}$ is called the maximal velocity and $K_m$ is called the *Michaelis-Menten constant.* The equation (2.14) describes the rate of reaction and is called the *Michaelis-Menten equation* [12].

*Example 1.* The glycogen phosphorylase is an enzyme that catalyzes the breakdown of glycogen into glucose. The kinetic data of this enzyme was obtained by Winston *et al* [20] as follows:

Glycogen (mg/mL): 4.82    2.48    1.98    1.5    1.2    1.0
V (mg/mL/min):        0.0714 0.0623 0.0608 0.0568 0.0512 0.0500

The data are read from Fig. 1C of [20] using the software Engauge Digitizer 4.1. Using the curve fitting toolbox in MATLAB, we fit the Michaelis-Menten equation (2.14) into the data. Fig. 2.3 (left) shows that the data can be well fitted by the Michaelis-Menten equation (2.14) with $K_m = 0.6328$ mg/mL and $V_{max} = 0.0802$ mg/mL/min.

As done in the above example, the parameters $K_m$ and $V_{max}$ can be determined by fitting the Michaelis-Menten equation (2.14) into data. Since fitting a nonlinear function into data is more difficult than fitting a linear function into data, we write the equation (2.14) as

$$\frac{1}{V} = \frac{1}{V_{max}} + \frac{K_m}{V_{max}} \frac{1}{[S]}. \tag{2.17}$$

Then $\frac{1}{V}$ is a linear function of $\frac{1}{[S]}$. The plot of this linear function is called a *Lineweaver-Burk plot*, as shown in Fig. 2.3 (right). The parameters $K_m$ and $V_{max}$ can be determined by fitting this linear function into data: $1/V_{max}$ is equal to the y-intercept of the line, $-1/K_m$ is equal to the x-intercept of the line, and $K_m/V_{max}$ is equal to the slope of the line. For instance, fitting the linear function into the kinetic data of glycogen phosphorylase from Example 1 gives

$$\frac{1}{V} = 7.75 \frac{1}{[S]} + 12.55.$$

Thus we obtain that

$$\frac{1}{V_{max}} = 12.55, \quad \frac{K_m}{V_{max}} = 7.75,$$

and then

$$V_{max} = \frac{1}{12.55} = 0.0796812749, \quad K_m = \frac{7.75}{12.55} = 0.61752988,$$

which are very close to the estimate obtained by fitting the Michaelis-Menten equation (2.14) into the data as done in Example 1.

Consider a reversible reaction:

$$E + S \underset{k_{-1}}{\overset{k_1}{\rightleftharpoons}} C \underset{k_{-2}}{\overset{k_2}{\rightleftharpoons}} P + E.$$

The system of ordinary differential equations for the dynamics of this reaction reads

$$\frac{d[S]}{dt} = -k_1[E][S] + k_{-1}[C], \tag{2.18}$$

$$\frac{d[E]}{dt} = -k_1[E][S] - k_{-2}[E][P] + (k_{-1} + k_2)[C], \tag{2.19}$$

$$\frac{d[C]}{dt} = k_1[E][S] + k_{-2}[E][P] - (k_{-1} + k_2)[C], \tag{2.20}$$

$$\frac{d[P]}{dt} = k_2[C] - k_{-2}[E][P]. \tag{2.21}$$

The quasi-steady state approximation gives

$$-k_1[E][S] - k_{-2}[E][P] + (k_{-1} + k_2)[C] = 0. \tag{2.22}$$

Solving the equations (2.11) and (2.22), we obtain

$$[C] = \frac{k_1[E_0][S] + k_{-2}[E_0][P]}{k_{-1} + k_2 + k_1[S] + k_{-2}[P]}, \quad [E] = \frac{[E_0](k_{-1} + k_2)}{k_{-1} + k_2 + k_1[S] + k_{-2}[P]}. \tag{2.23}$$

It then follows from (2.21) that

$$\begin{aligned}
V &= \frac{d[P]}{dt} \\
&= \frac{k_1 k_2 [E_0][S] + k_2 k_{-2}[E_0][P]}{k_{-1} + k_2 + k_1[S] + k_{-2}[P]} - \frac{k_{-2}[E_0](k_{-1} + k_2)[P]}{k_{-1} + k_2 + k_1[S] + k_{-2}[P]} \\
&= \frac{\frac{V_{max}^f}{K_{ms}}[S] - \frac{V_{max}^b}{K_{mp}}[P]}{1 + \frac{[S]}{K_{ms}} + \frac{[P]}{K_{mp}}},
\end{aligned} \tag{2.24}$$

where

$$V_{max}^f = k_2[E_0], \tag{2.25}$$

$$V_{max}^b = k_{-1}[E_0], \tag{2.26}$$

$$K_{ms} = \frac{k_{-1} + k_2}{k_1}, \tag{2.27}$$

$$K_{mp} = \frac{k_{-1} + k_2}{k_{-2}}. \tag{2.28}$$

In mathematical analysis, sometimes it is useful to nondimensionalize a system. We use the system (2.7)-(2.9) to demonstrate how to do so. Let $[S_0]$ denote the initial concentration of $S$. Since $[S]$ and $[S_0]$ have the same unit, the fraction $s = \frac{[S]}{[S_0]}$ is dimensionless. Therefore we divide the equation (2.7) by $[S_0]$ to obtain

$$\frac{ds}{dt} = -k_1[E]s + \frac{k_{-1}}{[S_0]}[C].$$

To nondimensionalize $[E]$, we divide this equation by $[E_0]$ to obtain

$$\frac{ds}{d([E_0]t)} = -k_1 es + \frac{k_{-1}}{[S_0]}c,$$

where $e = \frac{[E]}{[E_0]}$ and $c = \frac{[C]}{[E_0]}$. To nondimensionalize this equation, we divide it by $k_1$ to obtain

$$\frac{ds}{d\tau} = -es + \frac{k_{-1}}{k_1[S_0]}c, \qquad (2.29)$$

where $\tau = k_1[E_0]t$. Applying the above nondimensionalizing process to the equations (2.8) and (2.9), we obtain

$$\frac{[E_0]}{[S_0]}\frac{de}{d\tau} = -es + \frac{k_{-1}+k_2}{k_1[S_0]}c, \qquad (2.30)$$

$$\frac{[E_0]}{[S_0]}\frac{dc}{d\tau} = es - \frac{k_{-1}+k_2}{k_1[S_0]}c. \qquad (2.31)$$

There are other ways to nondimensionalize the system, which are left as an exercise.

Typically, the concentration of an enzyme is much smaller than that of the substrate. Hence the parameter $\varepsilon = [E_0]/[S_0]$ is very small. In mathematical analysis, it is important to study the behavior of the system as $\varepsilon \to 0$. This is a singular perturbation problem and its detailed discussions are beyond this text and are referred to [19].

## 2.4 Bi-substrate Enzymes

An enzymatic reaction in which two substrates bind to an enzyme in order is schematically described in Fig. 2.4. It can be represented by the following chemical reaction equations

$$E + A \underset{k_{-1}}{\overset{k_1}{\rightleftharpoons}} C_A,$$

$$C_A + B \underset{k_{-2}}{\overset{k_2}{\rightleftharpoons}} C_{AB} \overset{k_3}{\rightarrow} P + E.$$

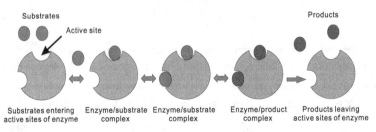

**Fig. 2.4.** A schematic description of an enzymatic reaction with two active sites

As before, we have

$$\frac{d[C_A]}{dt} = k_1[E][A] + k_{-2}[C_{AB}] - k_{-1}[C_A] - k_2[C_A][B], \qquad (2.32)$$

$$\frac{d[C_{AB}]}{dt} = -(k_{-2} + k_3)[C_{AB}] + k_2[C_A][B], \qquad (2.33)$$

$$\frac{d[P]}{dt} = k_3[C_{AB}]. \qquad (2.34)$$

Solving the system

$$k_1[E][A] + k_{-2}[C_{AB}] - k_{-1}[C_A] - k_2[C_A][B] = 0,$$
$$-(k_{-2} + k_3)[C_{AB}] + k_2[C_A][B] = 0,$$
$$[E] + [C_A] + [C_{AB}] = [E_0],$$

we obtain

$$[C_{AB}] = \frac{k_1 k_2 [E_0][A][B]}{k_1 k_2 [A][B] + k_1 k_{-2}[A] + k_1 k_3[A] + k_{-2}k_{-1} + k_3 k_{-1} + k_2 k_3[B]}, \qquad (2.35)$$

and then the velocity

$$\frac{d[P]}{dt} = k_3[C_{AB}]$$

$$= \frac{k_1 k_2 k_3 [E_0][A][B]}{k_1 k_2 [A][B] + k_1 k_{-2}[A] + k_1 k_3[A] + k_{-2}k_{-1} + k_3 k_{-1} + k_2 k_3[B]}$$

$$= \frac{K_1 V_{max}[A][B]}{K_m + K_1 K_m[A] + K_d[B] + K_1[A][B]}, \qquad (2.36)$$

where

$$K_m = \frac{k_{-2} + k_3}{k_2}, \quad K_1 = \frac{k_1}{k_{-1}}, \quad K_d = \frac{k_3}{k_{-1}}, \quad V_{max} = k_3[E_0].$$

## 2.5 Inhibitors

Biochemical reactions occurring in cells can be grouped into metabolic pathways containing sequences of chemical reactions, as shown in Fig. 2.5. Each reaction in the sequences of chemical reactions is catalyzed by a specific enzyme, and the product of one reaction is the substrate for the next one. The compounds formed at each step are the metabolic intermediates (or metabolites) that lead ultimately to the formation of an end product.

Cells are always in a homeostatic condition, and therefore the amount of product present or produced is always within certain range of concentrations. If a metabolic pathway is producing more of the end product than it needs, the end product or an inhibitor may bind to one or more of the enzymes in the pathway to inhibit the reaction, increasing the Michaelis-Menten constant $K_m$ or decreasing the maximal velocity $V_{max}$ [5]. Many naturally occurring compounds and pharmaceutical compounds are inhibitors.

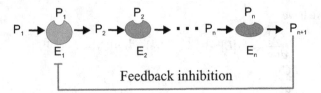

Fig. 2.5. A generic metabolic pathway. The enzyme $E_1$ converts its substrate $P_1$ into the metabolite $P_2$, the enzyme $E_2$ converts the metabolite $P_2$ into the metabolite $P_3$, and ultimately the enzyme $E_n$ converts the metabolite $P_n$ into the end product $P_{n+1}$. If the metabolic pathway is producing more of the end product than it needs, the end product or a byproduct may bind to the active sites of one or more of the regulatory enzymes of the pathway, preventing the binding of a substrate molecule, thus inhibiting the reaction

### 2.5.1 Competitive Inhibition

If an inhibitor competes with the substrate molecule for active sites, it is called a *competitive inhibitor*. For example, malonate is a competitive inhibitor of succinic dehydrogenase [5, p. 396]. An enzymatic reaction with competitive inhibition is schematically described in Fig. 2.6. It can be represented by the following chemical reaction equations:

$$E + S \underset{k_{-1}}{\overset{k_1}{\rightleftharpoons}} C_1 \overset{k_2}{\rightarrow} P + E,$$

$$E + I \underset{k_{-3}}{\overset{k_3}{\rightleftharpoons}} C_2,$$

where $I$ denotes the inhibitor. It follows from the law of mass action and the law of

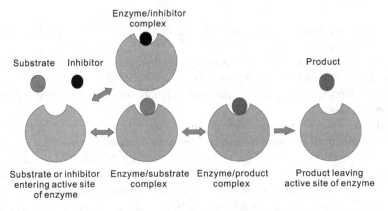

Fig. 2.6. A schematic description of an enzymatic reaction with competitive inhibition

mass balance that

$$\frac{d[C_1]}{dt} = k_1[E][S] - (k_{-1} + k_2)[C_1], \tag{2.37}$$

$$\frac{d[C_2]}{dt} = k_3[E][I] - k_{-3}[C_2], \tag{2.38}$$

$$\frac{d[P]}{dt} = k_2[C_1]. \tag{2.39}$$

As before, we have

$$[E] + [C_1] + [C_2] = [E_0]. \tag{2.40}$$

Using the quasi-steady state approximation

$$\frac{d[C_1]}{dt} = \frac{d[C_2]}{dt} = 0,$$

we obtain

$$k_1[E][S] - (k_{-1} + k_2)[C_1] = 0, \quad k_3[E][I] - k_{-3}[C_2] = 0. \tag{2.41}$$

Solving the equations (2.40) and (2.41), we obtain

$$[C_1] = \frac{[E_0][S]}{[S] + K_i K_m [I] + K_m}, \tag{2.42}$$

where

$$K_m = \frac{k_{-1} + k_2}{k_1}, \quad K_i = \frac{k_3}{k_{-3}}.$$

Defining

$$V_{max} = k_2[E_0], \tag{2.43}$$

we obtain the velocity of the reaction

$$V = \frac{d[P]}{dt} = k_2[C_1] = \frac{V_{max}[S]}{[S] + K_m(1 + K_i[I])}. \tag{2.44}$$

Note that the effect of the inhibitor is to increase the Michaelis-Menten constant from $K_m$ to $K_m(1 + K_i[I])$ and then decrease the velocity of reaction, while leaving the maximum velocity unchanged.

Consider a competitive inhibition in which two substrates bind to an enzyme in order:

$$E + A \underset{k_{-1}}{\overset{k_1}{\rightleftharpoons}} C_A,$$

$$C_A + B \underset{k_{-2}}{\overset{k_2}{\rightleftharpoons}} C_{AB} \overset{k_3}{\rightarrow} P + E,$$

$$E + I \underset{k_{-4}}{\overset{k_4}{\rightleftharpoons}} C_I.$$

As before, we have

$$k_1[E][A] + k_{-2}[C_{AB}] - k_{-1}[C_A] - k_2[C_A][B] = 0,$$
$$-(k_{-2} + k_3)[C_{AB}] + k_2[C_A][B] = 0,$$
$$k_4[E][I] - k_{-4}[C_I] = 0,$$
$$[E] + [C_A] + [C_{AB}] + [C_I] = [E_0].$$

Solving this system, we obtain

$$[C_{AB}] = \frac{K_1[B][A][E_0]}{K_m(K_4[I] + K_1[A] + 1) + K_4 K_d[I][B] + K_1[A][B] + K_d[B]}, \quad (2.45)$$

where

$$K_m = \frac{k_{-2} + k_3}{k_2}, \quad K_1 = \frac{k_1}{k_{-1}}, \quad K_4 = \frac{k_4}{k_{-4}}, \quad K_d = \frac{k_3}{k_{-1}}.$$

Then the velocity of the reaction is given by

$$V = \frac{d[P]}{dt} = k_3[C_{AB}]$$

$$= \frac{K_1 V_{max}[B][A]}{K_m(K_4[I] + K_1[A] + 1) + K_4 K_d[I][B] + K_1[A][B] + K_d[B]}, \quad (2.46)$$

where $V_{max} = k_3[E_0]$.

## 2.5.2 Uncompetitive Inhibition

If an inhibitor can bind to the enzyme-substrate complex rather than the free enzyme, it is called a *uncompetitive inhibitor* [5, p. 396]. Such a uncompetitive inhibition can be represented by the following chemical reaction equations:

$$E + S \underset{k_{-1}}{\overset{k_1}{\rightleftharpoons}} C_1 \overset{k_2}{\to} P + E,$$

$$C_1 + I \underset{k_{-3}}{\overset{k_3}{\rightleftharpoons}} C_2.$$

Similar to the situation of competitive inhibition, we can derive the equations for $[C_1]$ and $[C_2]$ as follows

$$k_1[E][S] - (k_{-1} + k_2)[C_1] - k_3[C_1][I] + k_{-3}[C_2] = 0,$$
$$k_3[C_1][I] - k_{-3}[C_2] = 0,$$
$$[E] + [C_1] + [C_2] = [E_0].$$

Solving these equations, we obtain

$$[C_1] = \frac{[E_0][S]}{K_m + [S](1 + K_i[I])},$$

where

$$K_m = \frac{k_{-1} + k_2}{k_1}, \quad K_i = \frac{k_3}{k_{-3}}.$$

Then the velocity of the reaction is

$$V = \frac{d[P]}{dt} = k_2[C_1] = \frac{V_{max}[S]}{K_m + [S](1 + K_i[I])} = \frac{V_{max}}{1 + K_i[I]} \frac{[S]}{\frac{K_m}{1 + K_i[I]} + [S]}. \tag{2.47}$$

Thus, a uncompetitive inhibitor decreases both the Michaelis-Menten constant and the maximum velocity.

### 2.5.3 Noncompetitive Inhibition

If an inhibitor can bind to both the free enzyme and the enzyme-substrate complex, it is called a *noncompetitive inhibitor* [5, p. 397]. This more complicated situation is called *noncompetitive inhibition*. A situation of noncompetitive inhibition is illustrated in Fig. 2.7 and can be represented by the following chemical reaction equations:

$$E + S \underset{k_{-1}}{\overset{k_1}{\rightleftharpoons}} C_{es} \overset{k_2}{\rightarrow} E + P,$$

$$E + I \underset{k_{-3}}{\overset{k_3}{\rightleftharpoons}} C_{ei},$$

$$C_{es} + I \underset{k_{-3}}{\overset{k_3}{\rightleftharpoons}} C_{eis},$$

$$C_{ei} + S \underset{k_{-1}}{\overset{k_1}{\rightleftharpoons}} C_{eis}.$$

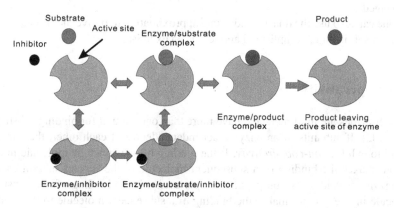

**Fig. 2.7.** A schematic description of an enzymatic reaction with noncompetitive inhibition

For simplicity, we use a simple equilibrium analysis developed in [12] by assuming that each reaction is at equilibrium. For the reaction from $E + S$ to $C_{es}$, we derive that

$$k_1[E][S] = k_{-1}[C_{es}].$$

Repeating this for all other reactions, we obtain

$$([E_0] - [C_{es}] - [C_{ei}] - [C_{eis}])[S] - K_m[C_{es}] = 0,$$
$$([E_0] - [C_{es}] - [C_{ei}] - [C_{eis}])[I] - K_i[C_{ei}] = 0,$$
$$[C_{ei}][S] - K_m[C_{eis}] = 0,$$
$$[C_{es}][I] - K_i[C_{eis}] = 0,$$

where

$$K_m = \frac{k_{-1}}{k_1}, \quad K_i = \frac{k_{-3}}{k_3}.$$

As before, we have used

$$[E] + [C_{es}] + [C_{ei}] + [C_{eis}] = [E_0].$$

Solving this system, we obtain

$$[C_{es}] = \frac{K_i[S][E_0]}{([S] + K_m)(K_i + [I])}.$$

Thus the velocity of the reaction is

$$V = \frac{d[P]}{dt} = k_2[C_{es}] = \frac{K_i V_{max}}{(K_i + [I])} \frac{[S]}{([S] + K_m)}, \tag{2.48}$$

where $V_{max} = k_2[E_0]$. In contrast to the competitive inhibition, the noncompetitive inhibition decreases the maximum velocity of the reaction, while leaving $K_m$ unchanged.

One can also use the quasi-steady state approximation to derive the rate equation, but the result is more complicated and left as an exercise.

## 2.6 Cooperativity

Many enzymes are multimers and have more than one subunit for binding substrates [5, p. 402]. If subunits of an enzyme act independently of each other, the enzyme is said to exhibit *non-cooperativity*. If the binding of one substrate molecule in one subunit makes the binding of a subsequent molecule to the second subunit easier, the enzyme is said to exhibit *positive cooperativity*. If the binding of one substrate molecule in one subunit makes the binding of a subsequent molecule to the second subunit more difficult, the enzyme is said to exhibit *negative cooperativity*.

Suppose that an enzyme can bind two substrate molecules. The reaction can be represented by

$$E + S \underset{k_{-1}}{\overset{k_1}{\rightleftharpoons}} C_1 \overset{k_2}{\longrightarrow} P + E,$$

$$C_1 + S \underset{k_{-3}}{\overset{k_3}{\rightleftharpoons}} C_2 \overset{k_4}{\longrightarrow} P + C_1.$$

It follows from the law of mass action and the law of mass balance that

$$\frac{d[C_1]}{dt} = k_1[E][S] + (k_{-3} + k_4)[C_2] - k_3[C_1][S] - (k_{-1} + k_2)[C_1],$$

$$\frac{d[C_2]}{dt} = k_3[C_1][S] - (k_{-3} + k_4)[C_2],$$

$$\frac{d[P]}{dt} = k_2[C_1] + k_4[C_2].$$

As before, we have

$$[E] + [C_1] + [C_2] = [E_0]. \tag{2.49}$$

Using the Quasi-steady state approximation $\frac{d[C_1]}{dt} = \frac{d[C_2]}{dt} = 0$ and solving for $[C_1]$ and $[C_2]$, we obtain

$$[C_1] = \frac{(k_{-3} + k_4)k_1[E_0][S]}{[S]^2 k_1 k_3 + k_1[S](k_{-3} + k_4) + (k_{-1} + k_2)(k_{-3} + k_4)}$$

$$= \frac{K_2[E_0][S]}{K_1 K_2 + K_2[S] + [S]^2},$$

$$[C_2] = \frac{k_3 k_1[E_0][S]^2}{[S]^2 k_1 k_3 + k_1[S](k_{-3} + k_4) + (k_{-1} + k_2)(k_{-3} + k_4)}$$

$$= \frac{[E_0][S]^2}{K_1 K_2 + K_2[S] + [S]^2},$$

where

$$K_1 = \frac{k_{-1} + k_2}{k_1}, \quad K_2 = \frac{k_{-3} + k_4}{k_3}.$$

Thus the velocity of the reaction is

$$V = \frac{d[P]}{dt} = k_2[C_1] + k_4[C_2] = \frac{(k_2 K_2 + k_4[S])[E_0][S]}{K_1 K_2 + K_2[S] + [S]^2}. \tag{2.50}$$

We now examine three cases. First we consider the case of *non-cooperativity* where the binding sites act independently and identically. Then $k_1 = 2k_3 = 2k_+$, $2k_{-1} = k_{-3} = 2k_-$ and $2k_2 = k_4$, where $k_+$ and $k_-$ are the forward and backward reaction rates for the individual binding sites. The factors of 2 occur because two identical binding sites are involved in the reaction, doubling the amount of the reaction. Denoting $K = \frac{k_- + k_2}{k_+}$, we obtain $K_1 = K/2, K_2 = 2K$ and then

$$V = \frac{2k_2 E_0 (K + [S])[S]}{K^2 + 2K + [S]^2} = \frac{2k_2 [E_0][S]}{K + [S]}.$$

Thus the rate of reaction with two binding sites is exactly twice that for the one binding site.

Next we consider the case of *positive cooperativity* where the binding of the first substrate molecule is slow and the first binding makes the second binding fast. This implies that $k_1 \to 0$ and $k_3 \to \infty$, but $k_1 k_3$ is kept as a constant. In this case, the velocity of reaction is

$$V = \frac{V_{max}[S]^2}{K_m^2 + [S]^2},$$

where $K_1 K_2 = K_m^2$ and $V_{max} = k_4[E_0]$. In general, if the enzyme has $n$ binding sites, then

$$V = \frac{V_{max}[S]^n}{K_m^n + [S]^n}. \tag{2.51}$$

This rate equation is called the *Hill equation*. The exponent $n$ is usually determined from experimental data and may be non-integer, not equal to the number of the active sites.

Note that the equation (2.51) can be written as

$$\ln\left(\frac{V}{V_{max} - V}\right) = n \ln S - n \ln K_m. \tag{2.52}$$

So the plot of $\ln\left(\frac{V}{V_{max} - V}\right)$ against $\ln S$ (called a Hill plot) is a straight line and the slope of the line is the Hill exponent $n$.

An enzyme can also exhibit *negative cooperativity*, in which the first binding decreases the next binding rate. This can be modeled by decreasing $k_3$ in the rate equation (2.50).

*Example 2.* The sarcoplasmic or endoplasmic reticulum $Ca^{2+}$-ATPase (SERCA) is an enzyme that resides on the membrane of intracellular sarcoplasmic or endoplasmic reticulum organelles and pumps $Ca^{2+}$ from the cytosol into the organelles. The

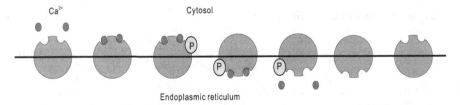

**Fig. 2.8.** A schematic description of calcium uptake by the sarcoplasmic or endoplasmic reticulum $Ca^{2+}$-ATPase (SERCA)

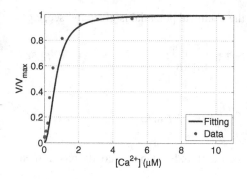

**Fig. 2.9.** Fitting of the Hill function (2.51) into the kinetic data of the sarcoplasmic or endoplasmic reticulum $Ca^{2+}$-ATPase (SERCA) with $n = 2.1$ and $K_m = 0.44$ ($\mu$M). The data are read from [14] using the software Engauge Digitizer 4.1

calcium uptake cycle by SERCA, as demonstrated in Fig. 2.8, is thought to include a binding of two $Ca^{2+}$ ions to the cytosolic portion of the $Ca^{2+}$-ATPase ($Ca_2^{2+}$-$E_1$), an ATP-dependent phosphorylation ($Ca_2^{2+}$-$E_1$-P), a translocation of $Ca^{2+}$ to the lumenal portion of the $Ca^{2+}$-ATPase ($Ca_2^{2+}$-$E_2$-P), a presumably sequential dissociation of $Ca^{2+}$ to the $Ca^{2+}$ store lumen ($E_2$-P), a dephosphorylation of the enzyme ($E_2$), and finally a regain of the original conformation ($E_1$) (see [7, 16]). Kinetic data of this enzyme were obtained by Lytton *et al* [14] as follows:

$$[Ca^{2+}](\mu M): \quad 0.01 \quad 0.0298 \ 0.0505 \ 0.101 \ 0.199 \ 0.298$$
$$V/V_{max}: \qquad\quad 0.044 \ 0.049 \quad 0.047 \quad 0.091 \ 0.153 \ 0.352$$
$$[Ca^{2+}](\mu M): 0.493 \ 1.03 \qquad 2.07 \qquad 3.09 \quad 5.07 \quad 10.4$$
$$V/V_{max}: \qquad\quad 0.584 \ 0.815 \quad 0.924 \quad 0.964 \ 0.970 \ 0.976$$

The data are read from Fig. 4 of [14] using the software Engauge Digitizer 4.1. Fig. 2.9 indicates that the data can be well fitted by the Hill function (2.51) with $n = 2.1$ and $K_m = 0.44$ ($\mu$M). Thus a positive cooperativity exists in SERCA.

## 2.7 Chemical Potential

The rate constants depend on energy because reactants in a chemical reaction have to accumulate enough energy to break down bonds before the reaction could occur. In a chemical reaction, the energy is expressed by chemical potential. Thus, the *chemical potential* of a substance, denoted by $\mu$, is defined as the change in the Gibbs free energy with respect to the amount of the substance at a constant temperature and pressure. For the precise definition of Gibbs free energy, we refer to a thermodynamics book such as [2, 9, 10]. The chemical potential is related to the concentration of the substance as follows (see, e.g., [2, p. 171])

$$\mu = \mu^\circ + RT \ln(c), \tag{2.53}$$

where $c$ is the concentration of the substance, $R$ is the gas constant, $T$ is the absolute temperature in Kelvin, and $\mu^\circ$ is the standard chemical potential of the substance.

For the chemical reaction

$$aA + bB \rightleftharpoons cC + dD,$$

it follows from (2.53) that the change in the chemical potential is given by (see, e.g., [2, 3])

$$\Delta \mu = \Delta \mu^\circ + RT \ln \frac{[C]^c [D]^d}{[A]^a [B]^b}. \qquad (2.54)$$

At equilibrium, $\Delta \mu = 0$. So

$$\Delta \mu^\circ = -RT \ln \frac{[C]^c_{eq} [D]^d_{eq}}{[A]^a_{eq} [B]^b_{eq}} = -RT \ln K_{eq}, \qquad (2.55)$$

where $K_{eq} = \frac{[C]^c_{eq} [D]^d_{eq}}{[A]^a_{eq} [B]^b_{eq}}$ is the equilibrium constant under standard conditions. It then follows that

$$\Delta \mu = -RT \ln K_{eq} + RT \ln \frac{[C]^c [D]^d}{[A]^a [B]^b} = -RT \ln \frac{K_{eq} [A]^a [B]^b}{[C]^c [D]^d}. \qquad (2.56)$$

## 2.8 The Arrhenius Formula

Reaction rates almost always increase when temperature is raised. For a reaction

$$B + C \xrightarrow{k} P,$$

the increase in the rate constant $k$ is often found to obey an equation suggested by Svante Arrhenius in 1889 (see, e.g., [1, p. 464]):

$$\ln k = \ln A - \frac{E_a}{RT}, \qquad (2.57)$$

where $A$ and $E_a$ are positive constants that depend on the reaction, $R$ is the gas constant, and $T$ is the absolute temperature in Kelvin. The constant $A$ is called the *frequency factor* and the constant $E_a$ is called the *activation energy*. In applications, it is more convenient to write Arrhenius' formula in the exponential form

$$k = A \exp \left( -\frac{E_a}{RT} \right). \qquad (2.58)$$

According to Eyring rate theory, the frequency factor $A$ is given by (see, e.g., [12])

$$A = \frac{kT}{h},$$

where $k$ is Boltzmann's constant and $h$ is Planck's constant.

## 2.9 Effects of Energy

We briefly discuss effects of energy on rate constants. For detailed discussions, we refer to [18]. Consider the enzymatic reaction:

$$E + S \underset{k_{-1}}{\overset{k_1}{\rightleftharpoons}} C \xrightarrow{k_2} P + E.$$

By (2.14), the velocity of this reaction is given by

$$V = \frac{d[P]}{dt} = \frac{k_1 k_2 [E_0][S]}{k_1[S] + k_{-1} + k_2} = \frac{K_1 k_2 [E_0][S]}{1 + \frac{K_1 k_2}{k_1} + K_1[S]},$$ (2.59)

where $K_1 = \frac{k_1}{k_{-1}}$ is the equilibrium constant. Using the free energy equation (2.55) and Arrhenius' formula (2.58), we obtain

$$k_1 = \frac{kT}{h} \exp\left(\frac{-\Delta G_1}{RT}\right),$$

$$k_2 = \frac{kT}{h} \exp\left(\frac{-\Delta G_2}{RT}\right),$$

$$K_1 = \exp\left(\frac{-\Delta G_1^\circ}{RT}\right),$$

where $\Delta G_1$ and $\Delta G_2$ are the activation energies for corresponding reactions, and $\Delta G_1^\circ$ is the standard free energy change. Substituting these equations into (2.59) gives

$$V = \frac{\frac{kT[E_0][S]}{h} \exp\left(\frac{-\Delta G_1^\circ - \Delta G_2}{RT}\right)}{1 + \exp\left(\frac{\Delta G_1 - \Delta G_1^\circ - \Delta G_2}{RT}\right) + [S] \exp\left(\frac{-\Delta G_1^\circ}{RT}\right)}.$$ (2.60)

## 2.10 Effects of pH

We briefly discuss effects of pH on rate constants. For detailed discussions, we refer to [18]. The effects of pH can be treated as the competitive inhibition:

$$E + S \underset{k_{-1}}{\overset{k_1}{\rightleftharpoons}} C_1 \overset{k_2}{\rightarrow} P + E,$$

$$E + H^+ \underset{k_{-3}}{\overset{k_3}{\rightleftharpoons}} C_2.$$

By (2.44), the velocity of the reaction is given by

$$V = \frac{V_{max}[S]}{[S] + K_m(1 + K_i[H^+])},$$ (2.61)

where

$$K_m = \frac{k_{-1} + k_2}{k_1}, \quad K_i = \frac{k_3}{k_{-3}}, \quad V_{max} = k_2[E_0].$$

Note that the effect of the proton $H^+$ is to increase the Michaelis-Menten constant from $K_m$ to $K_m(1 + K_i[H^+])$.

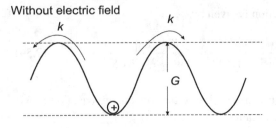

Without electric field

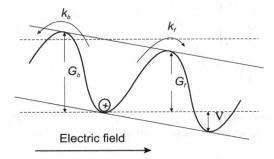

With electric field

Electric field

**Fig. 2.10.** An ion hopping model. *Above:* Energy minima are the preferred sites for ions. Ions would pause there until they acquire enough energy to hop over an energy barrier with an energy difference of $\Delta G$ to a neighboring preferred position. *Below:* An applied external electric field lowers the energy barrier on one side of the ion and raise it on the other

## 2.11 The Ion Hopping Model

Consider how an ion moves through an ordered solid like a crystal. As proposed in [11], the crystal lattice creates preferred resting positions for the mobile ion, with energetically unfavorable regions between. The structure might be represented as a periodic potential energy diagram as in Fig. 2.10A. Energy minima are the preferred sites. Ions would pause there until they acquire enough energy to hop over an energy barrier with an energy difference of $\Delta G$ to a neighboring preferred position.

Suppose that an external electric field is applied across the crystal. Then the applied field lowers the energy barrier on one side of the ion and raise it on the other as shown in Fig. 2.10B. The ion drifts down the field-electrodiffusion.

From Arrhenius' law (2.58), the rate constants in the hopping model satisfy

$$k_f = A_f \exp\left(-\frac{\Delta G_f}{RT}\right),$$

$$k_b = A_b \exp\left(-\frac{\Delta G_b}{RT}\right).$$

Assume that the electric field produces an electric potential drop of $\Delta V$ from one barrier to the next. It follows from (B.2) in Appendix B that the potential energy barriers for moving the ion from an energy minimum to the nearest maximum are changed by $\pm F z \Delta V/2$, where $F$ is Faraday's constant and $z$ is the charge of the ion.

Then the forward and backward hopping rate constants become

$$k_f = A_f \exp\left(\frac{-\Delta G + 0.5 F z \Delta V}{RT}\right), \tag{2.62}$$

$$k_b = A_b \exp\left(\frac{-\Delta G - 0.5 F z \Delta V}{RT}\right). \tag{2.63}$$

## Exercises

**2.1.** Assume that the half-life for maximal Protein Kinase B (PKB) activation is 1 minute (see, e.g., [17]). Use (2.5) to calculate the activation rate constant.

**2.2.** Consider the differential equations governing an enzymatic reaction

$$\frac{d[S]}{dt} = -k_1[E][S] + k_{-1}[C],$$

$$\frac{d[E]}{dt} = -k_1[E][S] + (k_{-1} + k_2)[C],$$

$$\frac{d[C]}{dt} = k_1[E][S] - (k_{-1} + k_2)[C].$$

Introducing dimensionless variables

$$s = \frac{[S]}{K_m}, \quad e = \frac{[E]}{K_m}, \quad c = \frac{[C]}{K_m}, \quad \tau = (k_{-1} + k_2)t,$$

where $K_m = \frac{k_{-1} + k_2}{k_1}$, derive the dimensionless equations

$$\frac{ds}{d\tau} = -es + \frac{k_{-1}}{k_{-1} + k_2} c,$$

$$\frac{de}{d\tau} = -es + c,$$

$$\frac{dc}{d\tau} = es - c.$$

**2.3.** ([12]) The process of glucose transport can be modeled as follows: We suppose that the glucose transporter (T) has two conformational states, $T_i$ and $T_e$, with its glucose binding site exposed on the cell interior (subscript $i$) or exterior (subscript $e$) of the membrane, respectively. The glucose on the interior $G_i$ can bind with $T_i$ and the glucose on the exterior can bind with $T_e$ to form the complex $C_i$ or $C_e$, respectively. Finally, a conformational change transforms $C_i$ into $C_e$ and vice versa. These statements are summarized as follows:

$$T_e + G_e \underset{k_-}{\overset{k_+}{\rightleftharpoons}} C_e,$$

$$T_i + G_i \underset{k_-}{\overset{k_+}{\rightleftharpoons}} C_i,$$

$$T_e \underset{k}{\overset{k}{\rightleftharpoons}} T_i,$$

$$C_e \underset{k}{\overset{k}{\rightleftharpoons}} C_i.$$

1. Find the differential equations for $[T_i]$, $[T_e]$, $[C_i]$, and $[C_e]$.
2. Find the steady state flux $J = k_-[C_i] - k_+[G_i][T_i]$ by setting all derivatives to zero and solving the resulting algebraic system.

**2.4.** ([12]) Consider the following trimerization reaction in which three monomers of $A$ combine to form the trimer $C$:

$$A + A \underset{k_{-1}}{\overset{k_1}{\rightleftharpoons}} B,$$

$$A + B \underset{k_{-2}}{\overset{k_2}{\rightleftharpoons}} C.$$

Find the rate of production of the trimer $C$ using the quasi-steady-state approximation.

**2.5.** ([12]) Consider an enzymatic reaction in which an enzyme can be activated or inactivated by the same chemical substance as follows:

$$E + X \underset{k_{-1}}{\overset{k_1}{\rightleftharpoons}} E_1,$$

$$E_1 + X \underset{k_{-2}}{\overset{k_2}{\rightleftharpoons}} E_2,$$

$$E_1 + S \overset{k_3}{\longrightarrow} P + Q + E.$$

Suppose further that $X$ is supplied at a constant rate and removed at a rate proportional to its concentration. Use quasi-steady-state analysis to find the dimensionless equation describing the degradation of $X$:

$$\frac{dx}{d\tau} = \gamma - x - \frac{\beta xy}{1 + x + y + \frac{\alpha}{\delta}x^2}.$$

**2.6.** ([12]) Using the quasi-steady-state approximation, show that the velocity of the reaction for an enzyme with noncompetitive inhibition in Fig. 2.7 is given by

$$V = \frac{k_1 k_2 k_{-3}[E_0][S](k_{-3} + k_1[S] + k_{-1} + k_3[I])}{(k_3[I] + k_{-3})[(k_1[S] + k_{-1})^2 + (k_1[S] + k_{-1})(k_3[I] + k_2 + k_{-3}) + k_{-3}k_2]}.$$

**2.7.** ([12]) Data for an enzymatic reaction are as follows:

$$[S](mM) \quad 0.1 \quad 0.2 \quad 0.5 \quad 1.0 \quad 2.0 \quad 3.5 \quad 5.0$$
$$V(mM/s) \quad 0.04 \quad 0.08 \quad 0.17 \quad 0.24 \quad 0.32 \quad 0.39 \quad 0.42$$

where $[S]$ denotes the concentration of substrate and $V$ denotes the corresponding rate of reaction.

1. Plot $V$ against $[S]$. Is this a Michaelis-Menten type reaction? If yes, fit Michaelis-Menten function (2.14) into the data to determine $K_m$ and $V_{max}$.
2. Plot $1/V$ against $1/[S]$. Can these data be well fitted by a straight line? If yes, use the straight line fit and (2.17) to determine $K_m$ and $V_{max}$.

**2.8.** ([12]) Data for an enzymatic reaction are as follows:

$$[S](mM) \quad 0.2 \quad 0.5 \quad 1.0 \quad 1.5 \quad 2.0 \quad 2.5 \quad 3.5 \quad 4.0 \quad 4.5 \quad 5.0$$
$$V(mM/s) \quad 0.01 \quad 0.06 \quad 0.27 \quad 0.50 \quad 0.67 \quad 0.78 \quad 0.89 \quad 0.92 \quad 0.94 \quad 0.95$$

where $[S]$ denotes the concentration of substrate and $V$ denotes the corresponding rate of reaction. The maximum velocity $V_{max}$ of the reaction is known to be 1 mM/s.

1. Plot $V$ against $[S]$. Is this a Hill type reaction? If yes, fit the Hill function (2.51) into the data to determine $n, K_m$, and $V_{max}$.
2. Plot $\ln\left(\frac{V}{V_{max}-V}\right)$ against $\ln[S]$. Can these data be well fitted by a straight line? If yes, use the straight line fit and (2.52) to determine $K_m$ and the Hill exponent $n$.

**2.9.** ([12]) In the case of noncompetitive inhibition, the inhibitor combines with the enzyme-substrate complex to give an inactive enzyme-substrate-inhibitor complex which cannot undergo further reaction, but the inhibitor does not combine directly with free enzyme or affect its reaction with substrate. Use the quasi-steady-state approximation to show that the velocity of this reaction is

$$V = \frac{V_{max}[S]}{K_m + [S] + \frac{[I]}{K_i}[S]}.$$

**2.10.** ([4, 8, 13]) Copper, zinc superoxide dismutase catalyzes the conversion of superoxide $O_2^-$ into hydrogen peroxide $H_2O_2$ according to the following enzymatic reactions

$$ECu^{2+} + O_2^- \xrightarrow{k} ECu^+ + O_2,$$

$$ECu^+ + O_2^- + 2H^+ \underset{k_-}{\overset{k}{\rightleftharpoons}} ECu^{2+} + H_2O_2,$$

where $k = 2.4 \times 10^9$ M$^{-1} \cdot$s$^{-1}$ and $k_- = 3.1$ M$^{-1} \cdot$s$^{-1}$. Derive the differential equations for the concentrations of $O_2^-$ and $H_2O_2$ using the quasi-steady-state approximation.

**2.11.** ([15]) Manganese superoxide dismutase catalyzes the conversion of superoxide $O_2^-$ into hydrogen peroxide $H_2O_2$ according to the following enzymatic reactions

$$E_A + O_2^- \xrightarrow{k_1} E_B + O_2,$$

$$E_B + O_2^- + 2H^+ \xrightarrow{k_1} E_A + H_2O_2,$$

$$E_B + O_2^- \xrightarrow{k_2} E_C,$$

$$E_C \xrightarrow{k_3} E_A,$$

where $E_A$ is the native enzyme, $E_B$, $E_C$ are reduced forms of the enzyme, $k_1 = 5.6 \times 10^8$ M$^{-1} \cdot$ s$^{-1}$, $k_2 = 4.8 \times 10^7$ M$^{-1} \cdot$ s$^{-1}$, and $k_3 = 70$ s$^{-1}$. Derive the differential equations for the concentrations of $O_2^-$ and $H_2O_2$ using the quasi-steady-state approximation.

**2.12.** ([6]) Xanthine oxidase (XO) catalyzes the conversion of hypoxanthine into xanthine and then into uric acid. A kinetic scheme of this sequential enzymatic reactions is as follows

$$E + H + O_2 \underset{k_{-1}}{\overset{k_1}{\rightleftharpoons}} EHO \xrightarrow{k_2} E + X + O_2^-,$$

$$E + X \underset{k_{-3}}{\overset{k_3}{\rightleftharpoons}} EX \xrightarrow{k_4} E + U,$$

$$E + U \underset{k_{-5}}{\overset{k_5}{\rightleftharpoons}} EU,$$

where $E$ is the xanthine oxidase, $H$ is the hypoxanthine, $X$ is the xanthine, and $U$ is the uric acid.

1. Using a simple steady-state approximation, show that the concentrations of hypoxanthine ($[H]$), xanthine ($[X]$), uric acid ($[U]$), and superoxide ($[O_2^-]$) satisfy the following differential equations

$$\frac{d[H]}{dt} = -\frac{V_{max}^H [H]}{K_m^H + [H] + \frac{K_m^H [X]}{K_m^X} + \frac{K_m^H [U]}{K_1}},$$

$$\frac{d[X]}{dt} = \frac{V_{max}^H [H] K_m^X - V_{max}^X [X] K_m^H}{K_m^H K_m^X + K_m^X [H] + K_m^H [X] + \frac{K_m^H K_m^X [U]}{K_1}},$$

$$\frac{d[U]}{dt} = \frac{V_{max}^X [X]}{K_m^X + [X] + \frac{K_m^X [H]}{K_m^H} + \frac{K_m^X [U]}{K_1}},$$

$$\frac{d[O_2^-]}{dt} = \frac{V_{max}^H [H]}{K_m^H + [H] + \frac{K_m^H [X]}{K_m^X} + \frac{K_m^H [U]}{K_1}}.$$

Because the simple steady-state approximation is used, the above equations cannot be derived in a usual way. For instance, the $\frac{d[H]}{dt}$ equation cannot be obtained from $\frac{d[H]}{dt} = k_{-1}[EHO] - k_1[E][H][O_2]$. Instead, $\frac{d[H]}{dt} = -\frac{d[O_2^-]}{dt}$ because the consumption rate of $H$ should be equal to the production rate of $O_2^-$. This remark applies to $X$: $\frac{d[X]}{dt} = k_2[EHO] - k_4[EX]$.

2. Solve the system numerically with the following values of parameters (estimated with the concentration of 0.27 mg/ml of xanthine oxidase [6]): $V_{max}^H = 1.69 \ \mu M \cdot s^{-1}$, $K_m^H = 1.86 \ \mu M$, $V_{max}^X = 2.07 \ \mu M \cdot s^{-1}$, $K_m^X = 3.38 \ \mu M$, and $K_1 = 178 \ \mu M$.

# References

1. Atkins P.W., Beran J.A.: General Chemistry, Second Edition. Scientific American Books, New York (1992).
2. Ball D.W.: Physical Chemistry. Thomson Brooke/Cole, California (2003).
3. Bolsover S.R., Hyams J.S., Shephard E.A., White H.A., Wiedemann C.G.: Cell Biology, A Short Course, Seond Edition. John Wiley & Sons, Inc., Hoboken, New Jersey (2004).
4. Bray R.C., Cockle S.A., Fielden E.M., Roberts P.B., Rotilio G., Calabrese L.: Reduction and inactivation of superoxide dismutase by hydrogen peroxide. Biochem. J. **139**, 43-48 (1974).
5. Devlin T.M.: Biochemistry. Wiley-Liss, Hoboken, NJ (2006).
6. Escribano J., Garcia-Canovas F., Garcia-Carmona F.: A kinetic study of hypoxanthine oxidation by milk xanthine oxidase. Biochem. J. **254**, 829-833 (1988).
7. Favre C.J., Schrenzel J., Jacquet J., Lew D.P., Krause K.-H.: Highly supralinear feedback inhibition of $Ca^{2+}$ uptake by the $Ca^{2+}$ load of intracellular stores. J. Biol. Chem. **271**, 14925-14930 (1996).
8. Fielden E.M., Roberts P.B., Bray R.C., Lowe D.J., Mautner G.N., Rotilio G., Calabrese L.: The mechanism of action of superoxide dismutase from pulse radiolysis and electron paramagnetic resonance. Biochem. J. **139**, 49-60 (1974).
9. Greiner W., Neise L., Stocker H.: Thermodynamics and Statistical Mechanics. Springer-Verlag, New York (1995).
10. Hammes G.G.: Thermodynamics and Kinetics for the Biological Sciences. Wiley-Interscience, New York (2000).
11. Hille B.: Ion Channels of Excitable Memebranes. Sinauer Associates, INC., Sunderland, Massachusetts (2001).
12. Keener J., Sneryd J.: Mathematical Physiology I: Cellular Physiology, II: Systems Physiology, Second Edition. Springer, New York (2009).
13. Liochev S.I., Fridovich I.: Copper, zinc superoxide dismutase and $H_2O_2$. Effects of bicarbonate on inactivation and oxidations of NADPH and URATE, and on consumption of $H_2O_2$. J. Biol. Chem. **277**, 34674-34678 (2002).
14. Lytton J., Westlin M., Burk S.E., Shull G.E., MacLennan D.H.: Functional comparisons between isoforms of the sarcoplasmic or endoplasmic reticulum family of calcium pumps. J. Biol. Chem. **267**, 14483-14489 (1992).
15. McAdam M.E., Fox R.A., Lavelle F., Fielden E.M.: A pulse-radiolysis study of the manganese-containing superoxide dismutase from bacillus stearothermophilus. Biochem. J. **165**, 71-79 (1977).

16. Raeymaekers L., Vandecaetsbeek I., Wuytack F., Vangheluwe P.: Modeling $Ca^{2+}$ dynamics of mouse cardiac cells points to a critical role of SERCA's affinity for $Ca^{2+}$. Biophys. J. **100**, 1216-1225 (2011).
17. Sedaghat A.R., Sherman A., Quon, M.J.: A mathematical model of metabolic insulin signaling pathways. Am. J. Physiol. Endocrinol. Metab. **283**, E1084-E1101 (2002).
18. Taylor K.B.: Enzyme Kinetics and Mechanisms. Kluwer Academic Publishers, Dordrecht, The Netherlands (2002).
19. White R.B.: Asymptotic Analysis of Differential Equations. Imperial College Press, London (2005).
20. Winston G.W., Reitz R.C.: Effects of chronic ethanol ingestion on liver glycogen phosphorylase in male and female rats. Am. J. Clin. Nutr. **34**, 2499-2507 (1981).

# 3

# Preliminary Systems Theory

In this chapter, we introduce preliminary systems theory. This includes controllability and observability of a system, stability of equilibrium points of a system, feedback control of a system, and parametric sensitivity of a system. These theories will be used to analyze biological cellular control systems.

In this book, $\mathbb{N}$ denotes the set of all nonnegative natural numbers, $\mathbb{C}$ denotes the set of complex numbers, $\mathbb{R}^n$ denotes the $n$-dimensional Euclidean space, and $\mathbb{R} = \mathbb{R}^1$ denotes the real line. Points in $\mathbb{R}^n$ will be denoted by $\mathbf{x} = (x_1, \cdots, x_n)$, and its norm is defined by

$$\|\mathbf{x}\| = \left( \sum_{i=1}^{n} x_i^2 \right)^{\frac{1}{2}}.$$

The inner product of $\mathbf{x}$ and $\mathbf{y}$ is defined by

$$\mathbf{x} \cdot \mathbf{y} = \sum_{i=1}^{n} x_i y_i.$$

## 3.1 Elementary Matrix Algebra

Let $m$ and $n$ be positive integers. An $m \times n$ *matrix* $\mathbf{A}$ is a rectangular array of numbers with $m$ rows and $n$ columns:

$$\mathbf{A} = \begin{bmatrix} a_{11} & a_{12} & \cdots & a_{1n} \\ a_{21} & a_{22} & \cdots & a_{2n} \\ \vdots & \vdots & \cdots & \vdots \\ a_{m1} & a_{m2n} & \cdots & a_{mn} \end{bmatrix}.$$

For example,

$$\mathbf{A} = \begin{bmatrix} 1 & 3 & 1 \\ -1 & 0 & 1 \\ -1 & -3 & -1 \end{bmatrix}.$$

Liu W.: Introduction to Modeling Biological Cellular Control Systems.
DOI 10.1007/978-88-470-2490-8_3, © Springer-Verlag Italia 2012

is a $3 \times 3$ matrix. It is convenient to write $\mathbf{A} = [a_{ij}]$. The size of a matrix tells how many rows and columns it has. We say that two matrices $\mathbf{A} = [a_{ij}]$ and $\mathbf{B} = [b_{ij}]$ are equal if they have the same size and if $a_{ij} = b_{ij}$ for all $i$ and $j$. If $m = n$, the matrix is called a *square matrix*.

### 3.1.1 Matrix Sums

If $\mathbf{A} = [a_{ij}]$ and $\mathbf{B} = [b_{ij}]$ are $m \times n$ matrices, then the *sum* $\mathbf{A} + \mathbf{B}$ is defined by

$$\mathbf{A} + \mathbf{B} = [a_{ij} + b_{ij}].$$

The sum $\mathbf{A} + \mathbf{B}$ is defined only when $\mathbf{A}$ and $\mathbf{B}$ have the same size.

*Example 3.* Let

$$\mathbf{A} = \begin{bmatrix} 4 & 0 & 5 \\ -1 & 3 & 2 \end{bmatrix}, \quad \mathbf{B} = \begin{bmatrix} 1 & 1 & 1 \\ 3 & 5 & 7 \end{bmatrix}, \quad \mathbf{C} = \begin{bmatrix} 2 & -3 \\ -1 & 3 \end{bmatrix}.$$

Then

$$\mathbf{A} + \mathbf{B} = \begin{bmatrix} 5 & 1 & 6 \\ 2 & 8 & 9 \end{bmatrix},$$

but $\mathbf{A} + \mathbf{C}$ is not defined because $\mathbf{A}$ and $\mathbf{C}$ have different sizes.

### 3.1.2 Scalar Multiple

If $r$ is a scalar and $\mathbf{A} = [a_{ij}]$ is a matrix, then the *scalar multiple* $r\mathbf{A}$ is defined by

$$r\mathbf{A} = [ra_{ij}].$$

We define $-\mathbf{A}$ as $(-1)\mathbf{A}$ and $\mathbf{A} - \mathbf{B}$ as $\mathbf{A} + (-1)\mathbf{B}$.

*Example 4.* Let $\mathbf{A}$ and $\mathbf{B}$ be the matrices in Example 3. Then

$$4\mathbf{A} = 4 \begin{bmatrix} 4 & 0 & 5 \\ -1 & 3 & 2 \end{bmatrix} = \begin{bmatrix} 16 & 0 & 20 \\ -4 & 12 & 8 \end{bmatrix}$$

and

$$\mathbf{A} - \mathbf{B} = \begin{bmatrix} 4 & 0 & 5 \\ -1 & 3 & 2 \end{bmatrix} - \begin{bmatrix} 1 & 1 & 1 \\ 3 & 5 & 7 \end{bmatrix} = \begin{bmatrix} 3 & -1 & 4 \\ -4 & -2 & -5 \end{bmatrix}.$$

**Theorem 1.** *Let* $\mathbf{A}$, $\mathbf{B}$, *and* $\mathbf{C}$ *be matrices of the same size,* $\mathbf{0}$ *denote the matrix with all zero entries, and let* $r$ *and* $s$ *be scalars. Then the following hold:*

- $\mathbf{A} + \mathbf{B} = \mathbf{B} + \mathbf{A}$;
- $(\mathbf{A} + \mathbf{B}) + \mathbf{C} = \mathbf{A} + (\mathbf{B} + \mathbf{C})$;
- $\mathbf{A} + \mathbf{0} = \mathbf{A}$;
- $r(\mathbf{A} + \mathbf{B}) = r\mathbf{A} + r\mathbf{B}$;
- $(r + s)\mathbf{A} = r\mathbf{A} + s\mathbf{A}$;
- $r(s\mathbf{A}) = (rs)\mathbf{A}$.

### 3.1.3 Matrix Multiplication

If $\mathbf{A} = [a_{ij}]$ is an $m \times n$ matrix and $\mathbf{B} = [b_{ij}]$ is an $n \times p$ matrix, then the *product*
$\mathbf{AB} = [c_{ij}]$ is the $m \times p$ matrix whose entries are given by

$$\mathbf{AB} = [c_{ij}] = [a_{i1}b_{1j} + a_{i2}b_{2j} + \cdots + a_{in}b_{nj}].$$

*Example 5.* Let

$$\mathbf{A} = \begin{bmatrix} 4 & 0 \\ -1 & 3 \end{bmatrix}, \quad \mathbf{B} = \begin{bmatrix} 1 & 1 & 1 \\ 3 & 5 & 7 \end{bmatrix}.$$

Then

$$\mathbf{AB} = \begin{bmatrix} 4\cdot1+0\cdot3 & 4\cdot1+0\cdot5 & 4\cdot1+0\cdot7 \\ -1\cdot1+3\cdot3 & -1\cdot1+3\cdot5 & -1\cdot1+3\cdot7 \end{bmatrix} = \begin{bmatrix} 4 & 4 & 4 \\ 8 & 14 & 20 \end{bmatrix}.$$

In general, $\mathbf{AB} \neq \mathbf{BA}$. For example, if

$$\mathbf{A} = \begin{bmatrix} 0 & 1 \\ 2 & 3 \end{bmatrix}, \quad \mathbf{B} = \begin{bmatrix} 2 & 3 \\ 4 & 5 \end{bmatrix},$$

then

$$\mathbf{AB} = \begin{bmatrix} 4 & 5 \\ 16 & 21 \end{bmatrix}, \quad \mathbf{BA} = \begin{bmatrix} 6 & 11 \\ 10 & 19 \end{bmatrix}.$$

So $\mathbf{AB} \neq \mathbf{BA}$.

### 3.1.4 Powers of a Matrix

If $\mathbf{A}$ is an $n \times n$ matrix and $k$ is a positive integer, then the $k$-th *power* of $\mathbf{A}$ is defined
by

$$\mathbf{A}^k = \underbrace{\mathbf{A} \cdots \mathbf{A}}_{k}.$$

### 3.1.5 Transpose of a Matrix

If $\mathbf{A} = [a_{ij}]$ is an $m \times n$ matrix, then *transpose* of $\mathbf{A}$, denoted by $\mathbf{A}^*$, is the $n \times m$ matrix
$\mathbf{A}^* = [b_{ij}]$ with $b_{ij} = a_{ji}$.

*Example 6.* Let

$$\mathbf{A} = \begin{bmatrix} 4 & 0 & 5 \\ -1 & 3 & 2 \end{bmatrix}.$$

Then

$$\mathbf{A}^* = \begin{bmatrix} 4 & -1 \\ 0 & 3 \\ 5 & 2 \end{bmatrix}.$$

The identity matrix $\mathbf{I} = [a_{ij}]$ is the $n \times n$ matrix with $a_{ii} = 1$ and $a_{ij} = 0$ for $i \neq j$.
For example

$$\mathbf{I} = \begin{bmatrix} 1 & 0 \\ 0 & 1 \end{bmatrix}.$$

**Theorem 2.** *Let* **A** *be an* $m \times n$ *matrix, and let* **B** *and* **C** *have sizes for which the indicated sums and products are defined. The the following hold:*

- $\mathbf{A}(\mathbf{BC}) = (\mathbf{AB})\mathbf{C}$ *(associative law of multiplication);*
- $\mathbf{A}(\mathbf{B}+\mathbf{C}) = \mathbf{AB}+\mathbf{AC}$ *(left distributive law);*
- $(\mathbf{B}+\mathbf{C})\mathbf{A} = \mathbf{BA}+\mathbf{CA}$ *(right distributive law);*
- $r(\mathbf{AB}) = (r\mathbf{A})\mathbf{B} = \mathbf{A}(r\mathbf{B})$ *for any scalar r;*
- $(\mathbf{AB})^* = \mathbf{B}^*\mathbf{A}^*;$
- $\mathbf{IA} = \mathbf{AI} = \mathbf{A}.$

### 3.1.6 The Determinant

Let **A** be an $n \times n$ matrix. We define the *determinant* $\det(\mathbf{A})$ of **A** by induction as follows. If $n = 1$, we define $\det(\mathbf{A}) = a_{11}$. Suppose that $\det(\mathbf{A})$ has been defined for $n = k \geq 1$. Given an element $a_{ij}$ of a matrix **A**, $\mathbf{M}_{ij}$, the minor of $a_{ij}$, is the matrix obtained from **A** by deleting the $i$th row and the $j$th column. $A_{ij}$, the cofactor of $a_{ij}$, is defined by

$$A_{ij} = (-1)^{i+j}\det(\mathbf{M}_{ij}).$$

For $n = k+1$, we define

$$\det(\mathbf{A}) = a_{11}A_{11} + a_{21}A_{21} + \cdots + a_{n1}A_{n1}.$$

If

$$\mathbf{A} = \begin{bmatrix} a_{11} & a_{12} \\ a_{21} & a_{22} \end{bmatrix},$$

then

$$\det(\mathbf{A}) = a_{11}a_{22} - a_{21}a_{12}.$$

If

$$\mathbf{A} = \begin{bmatrix} a_{11} & a_{12} & a_{13} \\ a_{21} & a_{22} & a_{23} \\ a_{31} & a_{32} & a_{33} \end{bmatrix},$$

then

$$\det(\mathbf{A}) = a_{11}\det\begin{bmatrix} a_{22} & a_{23} \\ a_{32} & a_{33} \end{bmatrix} - a_{21}\det\begin{bmatrix} a_{12} & a_{13} \\ a_{32} & a_{33} \end{bmatrix} + a_{31}\det\begin{bmatrix} a_{12} & a_{13} \\ a_{22} & a_{23} \end{bmatrix}.$$

### 3.1.7 Eigenvalues

An $n \times 1$ or $1 \times n$ matrix is called a column or row vector, respectively. Let $\mathbf{A} = [a_{ij}]$ be an $n \times n$ matrix. A number $\lambda$ is called an *eigenvalue* of **A** if there is a nonzero solution $(x_1, x_2, \cdots, x_n)$ of the following linear system

$$a_{11}x_1 + a_{12}x_2 + \cdots + a_{1n}x_n = \lambda x_1,$$
$$a_{21}x_1 + a_{22}x_2 + \cdots + a_{2n}x_n = \lambda x_2,$$
$$\vdots$$
$$a_{n1}x_1 + a_{n2}x_2 + \cdots + a_{nn}x_n = \lambda x_n.$$

The solution vector $(x_1, x_2, \cdots, x_n)$ is called an *eigenvector* corresponding to $\lambda$. Introducing the vector

$$\mathbf{x} = \begin{bmatrix} x_1 \\ x_2 \\ \vdots \\ x_n \end{bmatrix},$$

the above system can be written in the matrix form

$$\mathbf{A}\mathbf{x} = \lambda \mathbf{x}.$$

The eigenvalues of $\mathbf{A}$ are solutions of the equation

$$\det(\mathbf{A} - \lambda \mathbf{I}) = 0.$$

This equation is called the *characteristic equation* of $\mathbf{A}$.

*Example 7.* Find eigenvalues and eigenvectors of

$$\mathbf{A} = \begin{bmatrix} 2 & 2 \\ 1 & 3 \end{bmatrix}.$$

The equation for the eigenvalues is

$$\det(\mathbf{A} - \lambda \mathbf{I}) = \det \begin{bmatrix} 2 - \lambda & 2 \\ 1 & 3 - \lambda \end{bmatrix} = (2 - \lambda)(3 - \lambda) - 2 = 0.$$

Solving the equation, we obtain the eigenvalues

$$\lambda_1 = 1, \quad \lambda_2 = 4.$$

For the eigenvalue $\lambda_1 = 1$, we have the system for the corresponding eigenvector

$$x_1 + 2x_2 = 0,$$
$$x_1 + 2x_2 = 0.$$

Taking $x_2 = 1$ and solving the equation for $x_1$, we obtain an eigenvector

$$\mathbf{v}_1 = \begin{bmatrix} -2 \\ 1 \end{bmatrix}.$$

For the eigenvalue $\lambda_2 = 4$, we have the system for the corresponding eigenvector

$$-2x_1 + 2x_2 = 0,$$
$$x_1 - x_2 = 0.$$

Taking $x_2 = 1$ and solving the equation for $x_1$, we obtain an eigenvector

$$\mathbf{v}_2 = \begin{bmatrix} 1 \\ 1 \end{bmatrix}.$$

### 3.1.8 Rank of a Matrix

The set of vectors $v_1, \cdots v_k$ is *linearly dependent* if there exist $c_1, \cdots, c_k$, not all zero, such that

$$c_1 v_1 + \cdots + c_k v_k = 0.$$

Otherwise, it is said to be *linearly independent*. The vectors

$$v_1 = \begin{bmatrix} 1 \\ 2 \end{bmatrix}, \quad v_2 = \begin{bmatrix} 2 \\ 4 \end{bmatrix}$$

are linearly dependent because $-2v_1 + v_2 = 0$. The vectors

$$v_1 = \begin{bmatrix} 1 \\ 0 \\ 0 \end{bmatrix}, \quad v_2 = \begin{bmatrix} 0 \\ 1 \\ 0 \end{bmatrix}, \quad v_3 = \begin{bmatrix} 0 \\ 0 \\ 1 \end{bmatrix}$$

are linearly independent because $c_1 v_1 + c_2 v_2 + c_3 v_3 = 0$ implies that $c_1 = c_2 = c_3 = 0$.

The *rank* of $A$ is the maximum number of linearly independent row (or column) vectors of $A$. For example, the first two rows of the matrix

$$\begin{bmatrix} 1 & 3 & 1 \\ -1 & 0 & 1 \\ -1 & -3 & -1 \end{bmatrix}$$

is linearly independent, but the three row vectors are linearly dependent. So the rank of the matrix is 2.

The rank of $A$ can be calculated by reducing $A$ to a triangular matrix through basic row or column operations. The rank of $A$ is equal to the number of non-zero row vectors in the reduced triangular matrix. For instance, adding the first row of the above matrix $A$ to the second and third rows, we reduce $A$ to the following triangular matrix

$$\begin{bmatrix} 1 & 3 & 1 \\ 0 & 3 & 2 \\ 0 & 0 & 0 \end{bmatrix}.$$

So the rank of $A$ is 2.

## 3.2 Stability of Equilibrium Points

As demonstrated in Chapter 2, states of a dynamical biological system are frequently described by a system of ordinary differential equations

$$\frac{dx_1}{dt} = f_1(t, x_1, x_2, \cdots, x_n, u_1, u_2, \cdots, u_p), \tag{3.1}$$

$$\frac{dx_2}{dt} = f_2(t, x_1, x_2, \cdots, x_n, u_1, u_2, \cdots, u_p), \tag{3.2}$$

$$\vdots$$

$$\frac{dx_n}{dt} = f_n(t, x_1, x_2, \cdots, x_n, u_1, u_2, \cdots, u_p), \tag{3.3}$$

$$x_1(0) = x_1^0, \ x_2(0) = x_2^0, \ \cdots, \ x_n(0) = x_n^0, \tag{3.4}$$

where $f_i$ are given functions, $u_i$ are specified input *control variables*, and $x_1^0, x_2^0, \cdots, x_n^0$ are specified initial conditions. We call the variables $x_1, x_2, \cdots, x_n$ *state variables*. Sometimes another system

$$y_1 = g_1(t, x_1, x_2, \cdots, x_n, u_1, u_2, \cdots, u_p), \tag{3.5}$$

$$y_2 = g_2(t, x_1, x_2, \cdots, x_n, u_1, u_2, \cdots, u_p), \tag{3.6}$$

$$\vdots$$

$$y_m = g_m(t, x_1, x_2, \cdots, x_n, u_1, u_2, \cdots, u_p) \tag{3.7}$$

is associated with (3.1)-(3.3). We call the variables $y_1, y_2, \cdots, y_m$ outputs. They consist of variables of particular interest in a biological system, for example, a product of a chemical reaction.

*Example 8.* Consider the enzymatic reaction:

$$u \rightarrow E + S \underset{k_{-1}}{\overset{k_1}{\rightleftharpoons}} C \overset{k_2}{\rightarrow} P + E,$$

where $u$ is the rate of substrate input. In Section 2.3, we showed that the system of differential equations governing the states $S$, $E$, $P$ is

$$\frac{d[S]}{dt} = -k_1[E][S] + k_{-1}([E_0] - [E]) + u, \tag{3.8}$$

$$\frac{d[E]}{dt} = -k_1[E][S] + (k_{-1} + k_2)([E_0] - [E]), \tag{3.9}$$

$$\frac{d[P]}{dt} = k_2([E_0] - [E]). \tag{3.10}$$

In this reaction, the output could be the product:

$$y = [P].$$

### 3.2.1 Definition of Stability

If the functions $f_i$ in the system (3.1)-(3.3) do not depend on $t$, the system is said to be *autonomous*. Otherwise, it is said to be *nonautonomous*. Evidently, the system (3.8)-(3.10) is autonomous.

Let

$$\mathbf{x} = \begin{bmatrix} x_1 \\ x_2 \\ \vdots \\ x_n \end{bmatrix}, \quad \mathbf{f}(\mathbf{x}) = \begin{bmatrix} f_1(\mathbf{x}) \\ f_2(\mathbf{x}) \\ \vdots \\ f_n(\mathbf{x}) \end{bmatrix}.$$

The autonomous system (3.1)-(3.3) can be written as the following vector form

$$\frac{d\mathbf{x}}{dt} = \mathbf{f}(\mathbf{x}). \tag{3.11}$$

The solutions $\bar{\mathbf{x}} = (\bar{x}_1, \bar{x}_2, \cdots \bar{x}_n)^*$ of the system

$$f_1(x_1, x_2, \cdots, x_n) = 0, \tag{3.12}$$

$$f_2(x_1, x_2, \cdots, x_n) = 0, \tag{3.13}$$

$$\vdots$$

$$f_n(x_1, x_2, \cdots, x_n) = 0 \tag{3.14}$$

is called an *equilibrium point* (or steady state) of the system (3.11). If the states of the system (3.11) start at the equilibrium point $\bar{\mathbf{x}}$, they remain at $\bar{\mathbf{x}}$ for all future time since $\frac{dx_1}{dt} = 0, \frac{dx_2}{dt} = 0, \cdots, \frac{dx_n}{dt} = 0$ at the equilibrium point.

**Definition 1.** *Let $W$ be a closed set in $\mathbb{R}^n$. The set $W$ is said to be positively invariant for the system (3.11) if for any initial conditions $\mathbf{x}(0) \in W$, the solutions $\mathbf{x}(t)$ of (3.11) are in $W$ for all $t \geq 0$.*

**Definition 2.** *Let $W$ be a closed positively-invariant set for the system (3.11) in $\mathbb{R}^n$. The equilibrium point $\bar{\mathbf{x}} = (\bar{x}_1, \bar{x}_2, \cdots \bar{x}_n)^* \in W$ of the system (3.11) is:*

- *stable in $W$ if, for any $\varepsilon > 0$, there is $\delta = \delta(\varepsilon) > 0$ such that*

$$\|\mathbf{x}(t) - \bar{\mathbf{x}}\| < \varepsilon \quad \text{for } \|\mathbf{x}(0) - \bar{\mathbf{x}}\| < \delta, \mathbf{x}(0) \in W, \text{ and } t \geq 0;$$

- *unstable in $W$ if it is not stable in $W$;*
- *asymptotically stable in $W$ if there is $\delta > 0$ such that*

$$\lim_{t \to \infty} \mathbf{x}(t) = \bar{\mathbf{x}} \quad \text{for } \|\mathbf{x}(0) - \bar{\mathbf{x}}\| < \delta \text{ and } \mathbf{x}(0) \in W;$$

- *exponentially stable in $W$ if there are $C, \omega, \delta > 0$ such that*

$$\|\mathbf{x}(t) - \bar{\mathbf{x}}\| \leq Ce^{-\omega t} \quad \text{for } \|\mathbf{x}(0) - \bar{\mathbf{x}}\| < \delta, \mathbf{x}(0) \in W, \text{ and } t \geq 0.$$

*If the above statements are true for any $\mathbf{x}(0) \in W$, that is, $\delta = \infty$, then the equilibrium point is said to be globally stable, globally asymptotically stable, or globally exponentially stable.*

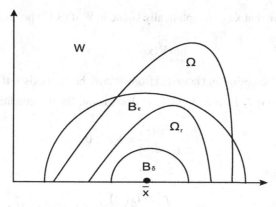

**Fig. 3.1.** Illustration of sets in the proof of Theorem 3 with $W = \{(x_1, x_2) \in \mathbb{R}^2 \mid x_1 \geq 0, x_2 \geq 0\}$

### 3.2.2 Lyapunov's Stability Theorem

We now establish stability tests.

**Theorem 3.** *Let W be a closed positively-invariant set for the system (3.11) in $\mathbb{R}^n$. Let $\bar{\mathbf{x}} = (\bar{x}_1, \bar{x}_2, \cdots \bar{x}_n)^* \in W$ be the equilibrium point of the system (3.11) and $\Omega \subset W$ be an open set in W containing $\bar{\mathbf{x}}$. Let $V : \Omega \to \mathbb{R}$ be a continuously differentiable function such that*

$$V(\bar{\mathbf{x}}) = 0, \quad V(\mathbf{x}) > 0 \text{ in } \Omega - \{\bar{\mathbf{x}}\}, \tag{3.15}$$

*and*

$$\frac{dV(\mathbf{x})}{dt} \leq 0 \text{ in } \Omega. \tag{3.16}$$

*Then $\bar{\mathbf{x}}$ is stable in W. Moreover, if*

$$\frac{dV(\mathbf{x})}{dt} < 0 \text{ in } \Omega - \{\bar{\mathbf{x}}\}, \tag{3.17}$$

*then $\bar{\mathbf{x}}$ is asymptotically stable in W.*

*Proof.* For any $\varepsilon > 0$, define (see Fig. 3.1)

$$B_\varepsilon(\bar{\mathbf{x}}) = \{\mathbf{x} \in W \mid \|\mathbf{x} - \bar{\mathbf{x}}\| \leq \varepsilon\}.$$

Then there exist $r, \delta > 0$ such that

$$\Omega_r = \{\mathbf{x} \in W \mid V(\mathbf{x}) \leq r\} \subset B_\varepsilon(\bar{\mathbf{x}}),$$

and

$$B_\delta(\bar{\mathbf{x}}) = \{\mathbf{x} \in W \mid \|\mathbf{x} - \bar{\mathbf{x}}\| \leq \delta\} \subset \Omega_r.$$

Hence, for $\mathbf{x}(0) \in B_\delta(\bar{\mathbf{x}})$, we have $V(\mathbf{x}(0)) \leq r$. It then follows from (3.16) that

$$V(\mathbf{x}(t)) \leq V(\mathbf{x}(0)) \leq r,$$

which implies that $\mathbf{x}(t) \in B_\varepsilon(\bar{\mathbf{x}})$. So $\bar{\mathbf{x}}$ is stable in W.

Next we prove that $\bar{\mathbf{x}}$ is asymptotically stable in $W$ if (3.17) holds. We first show that

$$\lim_{t \to \infty} V(\mathbf{x}(t)) = 0.$$

Since $V(\mathbf{x}(t))$ is decreasing and bounded from below, the limit exists. If $\lim_{t \to \infty} V(\mathbf{x}(t)) = L > 0$, then $V(\mathbf{x}(t)) \geq L$ since $V(\mathbf{x}(t))$ is decreasing. By the condition (3.17), we deduce that

$$\max_{V(\mathbf{x}) \geq L} \frac{dV(\mathbf{x})}{dt} = V_m < 0.$$

It then follows that

$$V(\mathbf{x}(t)) = V(\mathbf{x}(0)) + \int_0^t \frac{dV(\mathbf{x}(s))}{ds} ds \leq V(\mathbf{x}(0)) + V_m t.$$

Then $V(\mathbf{x}(t)))$ will become negative for sufficiently large $t$. This is a contradiction.

For any $\varepsilon > 0$, as argued at the beginning of the proof, there exists $r > 0$ such that $\Omega_r \subset B_\varepsilon(\bar{\mathbf{x}})$. Then there exists $T > 0$ such that $\mathbf{x}(t) \in \Omega_r \subset B_\varepsilon(\bar{\mathbf{x}})$ for $t \geq T$. This proves that $\lim_{t \to \infty} \mathbf{x}(t) = \bar{\mathbf{x}}$. □

A continuously differentiable function $V(\mathbf{x})$ satisfying (3.15) and (3.16) is called a *Lyapunov function*. The construction of a Lyapunov function for a given system is usually difficult.

We use Lyapunov's stability theorem, Theorem 3, to investigate the stability of the equilibrium point of the system (3.8)-(3.9). Solving the system

$$-k_1[\bar{E}][\bar{S}] + k_{-1}([E_0] - [\bar{E}]) + u = 0, \tag{3.18}$$
$$-k_1[\bar{E}][\bar{S}] + (k_{-1} + k_2)([E_0] - [\bar{E}]) = 0, \tag{3.19}$$

we obtain the equilibrium point of the system (3.8)-(3.9)

$$[\bar{E}] = \frac{k_2[E_0] - u}{k_2}, \tag{3.20}$$

$$[\bar{S}] = \frac{(k_{-1} + k_2)u}{k_1(k_2[E_0] - u)}. \tag{3.21}$$

Since a biological system normally returns to its equilibrium after an initial disturbance, we can expect that the solution $([S], [E])$ should converge to the equilibrium point $([\bar{S}], [\bar{E}])$ as $t \to \infty$. Indeed, the numerical solution (Fig. A.2 in Section A.11) of the system (3.8)-(3.9) obtained in Section A.11 shows that this is true. This can be further proved analytically. To this end, we define the closed set

$$K_{E_0} = \{(x_1, x_2) \in \mathbb{R}^2 \mid x_1 \geq 0 \text{ and } 0 \leq x_2 \leq [E_0]\}.$$

**Theorem 4.** *If $([S](0), [E](0)) \in K_{E_0}$ and the input $u$ is nonnegative and constant, then the solutions $([S](t), [E](t))$ of (3.8)-(3.9) are in $K_{E_0}$ for all $t \geq 0$.*

*Proof.* We need to prove that

$$[E](t) \geq 0, \qquad (3.22)$$
$$[E](t) \leq [E_0], \qquad (3.23)$$
$$[S](t) \geq 0 \qquad (3.24)$$

for all $t \geq 0$. Define

$$t_1 = \max\{T \mid [E](t) \geq 0 \text{ for all } 0 \leq t \leq T\},$$
$$t_2 = \max\{T \mid [E](t) \leq [E_0] \text{ for all } 0 \leq t \leq T\},$$
$$t_3 = \max\{T \mid [S](t) \geq 0 \text{ for all } 0 \leq t \leq T\}.$$

We prove that $t_0 = \min\{t_1, t_2, t_3\} = \infty$ by argument of contradiction in three cases.

(1). If $t_0 = t_1 < \infty$, then $[E](t) \geq 0, [E](t) \leq [E_0], [S](t) \geq 0$ for all $0 \leq t \leq t_0$ and $[E](t_0) = 0$. It then follows from (3.9) that

$$\frac{dE}{dt}\bigg|_{t_0} = -k_1[E](t_0)[S](t_0) + (k_{-1} + k_2)([E_0] - [E](t_0)) = (k_{-1} + k_2)[E_0] > 0.$$

Thus $[E](t)$ is increasing near $t_0$ and then $[E](t) < [E](t_0) = 0$ for some $t < t_0$. This is a contradiction.

(2). If $t_0 = t_2 < \infty$, then $[E](t) \geq 0, [E](t) \leq [E_0], [S](t) \geq 0$ for all $0 \leq t \leq t_0$ and $[E](t_0) = [E_0]$. It then follows from (3.9) that

$$\frac{dE}{dt}\bigg|_{t_0} = -k_1[E](t_0)[S](t_0) + (k_{-1} + k_2)([E_0] - [E](t_0)) = -k_1[E_0][S](t_0).$$

If $[S](t_0) > 0$, then $\frac{dE}{dt}\big|_{t_0} < 0$. So $[E](t)$ is decreasing near $t_0$ and then $[E](t) > [E](t_0) = [E_0]$ for some $t < t_0$. This is a contradiction. If $[S](t_0) = 0$, we consider two cases: $u \equiv 0$ and $u > 0$. If $u \equiv 0$, then the system (3.8)-(3.9) has the unique solution $[S](t) = 0$ and $[E](t) = [E_0]$ for all $t \geq t_0$. This is in contradiction with $t_0 < \infty$. If $u > 0$, it follows from (3.8) that

$$\frac{dS}{dt}\bigg|_{t_0} = -k_1[E](t_0)[S](t_0) + k_{-1}([E_0] - [E](t_0)) + u = u > 0.$$

Thus $[S](t)$ is increasing near $t_0$ and then $[S](t) < [S](t_0) = 0$ for some $t < t_0$. This is a contradiction.

(3). If $t_0 = t_3 < \infty$, then $[E](t) \geq 0, [E](t) \leq [E_0], [S](t) \geq 0$ for all $0 \leq t \leq t_0$ and $[S](t_0) = 0$. It then follows from (3.8) that

$$\frac{dS}{dt}\bigg|_{t_0} = -k_1[E](t_0)[S](t_0) + k_{-1}([E_0] - [E](t_0)) + u = k_{-1}([E_0] - [E](t_0)) + u.$$

We consider two cases: $u \equiv 0$ and $u > 0$. If $u > 0$, then $\frac{dS}{dt}\big|_{t_0} > 0$. Thus $[S](t)$ is increasing near $t_0$ and then $[S](t) < [S](t_0) = 0$ for some $t < t_0$. This is a contradiction.

If $u \equiv 0$, we consider two cases: $[E](t_0) = [E_0]$ and $[E](t_0) < [E_0]$. If $[E](t_0) = [E_0]$, then the system (3.8)-(3.9) has the unique solution $[S](t) = 0$ and $[E](t) = [E_0]$ for all $t \geq t_0$. This is in contradiction with $t_0 < \infty$. If $[E](t_0) < [E_0]$, then $\left.\frac{dS}{dt}\right|_{t_0} > 0$. Thus $[S](t)$ is increasing near $t_0$ and then $[S](t) < [S](t_0) = 0$ for some $t < t_0$. This is a contradiction.                                                                $\square$

Using Lyapunov's stability theorem, Theorem 3, we obtain the following stability theorem.

**Theorem 5.** *If* $([S](0), [E](0)) \in K_{E_0}$ *and* $u = 0$, *then the solutions* $([S](t), [E](t))$ *of* (3.8)-(3.9) *satisfy*

$$\lim_{t \to \infty} S(t) = 0, \quad \lim_{t \to \infty} E(t) = E_0.$$

*Proof.* First note that Theorem 4 ensures that $K_{E_0}$ is a closed positively-invariant set for the system (3.8)-(3.9). Define

$$V(E, S) = \frac{k(k_{-1} + k_2)}{k_{-1}} S + E_0 - E, \tag{3.25}$$

where $\frac{k_{-1}}{k_{-1} + k_2} < k < 1$. Differentiating $V$ with respect to $t$ and using the equations (3.8)-(3.9), we obtain

$$\begin{aligned}
\frac{dV}{dt} &= \frac{k(k_{-1} + k_2)}{k_{-1}} \frac{dS}{dt} - \frac{dE}{dt} \\
&= -\frac{k_1[k(k_2 + k_{-1}) - k_{-1}]}{k_{-1}} ES - (1 - k)(k_{-1} + k_2)(E_0 - E) \\
&\leq 0.
\end{aligned}$$

Hence, it is clear that the function $V$ satisfies all conditions (3.15)-(3.17) with $W = K_{E_0}$ and $\bar{x} = (0, E_0)$. It therefore follows from the Lyapunov's stability theorem, Theorem 3, that

$$\lim_{t \to \infty} S(t) = 0, \quad \lim_{t \to \infty} E(t) = E_0$$

for $(S(0), E(0)) \in K_{E_0}$.                                                    $\square$

### 3.2.3 Lyapunov's Indirect Method

To establish another stability test, we look at the stability of the linear system

$$\frac{dx}{dt} = Ax, \tag{3.26}$$

$$x(0) = x_0, \tag{3.27}$$

where $A$ is an $n \times n$ matrix. The system has the equilibrium point $\bar{x} = 0$, that is, $A0 = 0$. The stability of the equilibrium point $0$ can be characterized by the locations of the eigenvalues of $A$.

Given a polynomial $f(\lambda) = (\lambda - \lambda_1)^{q_1} \cdots (\lambda - \lambda_n)^{q_n}$, the **algebraic multiplicity** of the root $\lambda_i$ $(i = 1, \cdots, n)$ of $f(\lambda)$ is defined to be $q_i$.

**Theorem 6.** *The equilibrium point* **0** *of* (3.26) *is globally stable if and only if all eigenvalues of* **A** *satisfy* $\text{Re}\lambda_i \leq 0$ *and for every eigenvalue with* $\text{Re}\lambda_i = 0$ *and algebraic multiplicity* $q_i \geq 2$, $\text{rank}(\mathbf{A} - \lambda_i I) = n - q_i$, *where n is the dimension of* **x**. *The equilibrium point* **0** *of* (3.26) *is globally exponentially stable if and only if all eigenvalues of* **A** *satisfy* $\text{Re}\lambda_i < 0$.

The proof of the theorem is omitted and referred to [2, Theorem 4.5, p. 134].

*Example 9.* Consider the system

$$\frac{d\mathbf{x}}{dt} = \mathbf{A}\mathbf{x}, \tag{3.28}$$

where

$$\mathbf{A} = \begin{bmatrix} 0 & 1 \\ -1 & 0 \end{bmatrix}.$$

The characteristic equation of **A** is

$$\det(\lambda\mathbf{I} - \mathbf{A}) = \lambda^2 + 1 = 0.$$

Solving the equation, we obtain the eigenvalues $\lambda = \pm i$. It is clear that the algebraic multiplicity of both $i$ and $-i$ is 1 and the ranks of $i\mathbf{I} - \mathbf{A}$ and $-i\mathbf{I} - \mathbf{A}$ are equal to 1. So the conditions of Theorem 6 is satisfied and then the equilibrium **0** is stable. In fact, the solution of the system is given by

$$\begin{bmatrix} x_1 \\ x_2 \end{bmatrix} = c_1 \begin{bmatrix} \cos t \\ -\sin t \end{bmatrix} + c_2 \begin{bmatrix} \sin t \\ \cos t \end{bmatrix},$$

which shows that the equilibrium point **0** is stable.

*Example 10.* Consider the system

$$\frac{d\mathbf{x}}{dt} = \mathbf{A}\mathbf{x}, \tag{3.29}$$

where

$$\mathbf{A} = \begin{bmatrix} 0 & 1 \\ -5 & -2 \end{bmatrix}.$$

The characteristic equation of **A** is

$$\lambda^2 + 2\lambda + 5 = 0.$$

Then the eigenvalues are

$$\lambda = -1 \pm 2i.$$

Thus, by Theorem 6, the equilibrium 0 is exponentially stable. In fact, the solutions are

$$\begin{bmatrix} x_1 \\ x_2 \end{bmatrix} = c_1 \begin{bmatrix} \cos 2t \\ -\cos 2t - 2\sin 2t \end{bmatrix} e^{-t}$$

$$+ c_2 \begin{bmatrix} \sin 2t \\ -\sin 2t + 2\cos 2t \end{bmatrix} e^{-t},$$

which decay exponentially.

For higher order systems ($n \geq 3$), it may be impossible to solve characteristic polynomials explicitly. In this case, Routh-Hurwitz's stability criterion is needed for determining the signs of eigenvalues without actually solving for them. The proof of the criterion is long and is referred to [1].

**Theorem 7 (Routh-Hurwitz's Criterion).** *Suppose that all the coefficients of the polynomial* $f(s) = a_0 s^n + a_1 s^{n-1} + \cdots + a_{n-1} s + a_n$ *are positive. Construct the following table:*

$$
\begin{aligned}
s^n & : a_0 \ a_2 \ a_4 \ a_6 \ \cdots \\
s^{n-1} & : a_1 \ a_3 \ a_5 \ a_7 \ \cdots \\
s^{n-2} & : b_1 \ b_2 \ b_3 \ b_4 \ \cdots \\
s^{n-3} & : c_1 \ c_2 \ c_3 \ c_4 \ \cdots
\end{aligned}
$$

$$
\vdots \quad \vdots \quad \vdots \quad \vdots \quad \vdots \quad \vdots
$$

$$
\begin{aligned}
s^2 & : d_1 \ d_2 \\
s^1 & : e_1 \\
s^0 & : f_1
\end{aligned}
$$

*where*

$$
b_1 = -\frac{1}{a_1} \begin{vmatrix} a_0 & a_2 \\ a_1 & a_3 \end{vmatrix}, \ b_2 = -\frac{1}{a_1} \begin{vmatrix} a_0 & a_4 \\ a_1 & a_5 \end{vmatrix}, \ b_3 = -\frac{1}{a_1} \begin{vmatrix} a_0 & a_6 \\ a_1 & a_7 \end{vmatrix}, \cdots
$$

$$
c_1 = -\frac{1}{b_1} \begin{vmatrix} a_1 & a_3 \\ b_1 & b_2 \end{vmatrix}, \ c_2 = -\frac{1}{b_1} \begin{vmatrix} a_1 & a_5 \\ b_1 & b_3 \end{vmatrix}, \ c_3 = -\frac{1}{b_1} \begin{vmatrix} a_1 & a_7 \\ b_1 & b_4 \end{vmatrix}, \cdots
$$

$$
\vdots
$$

*and any undefined entries are set to zero. Then all zeros of* $f$ *have negative real parts if and only if all entries in the first column of the table are well defined and positive.*

*Example 11.* Consider the polynomial $f(s) = (s+1)(s+2)(s+3) = s^3 + 6s^2 + 11s + 6$, which has three negative zeros: -1, -2, -3. Routh's table is as follows

$$
\begin{aligned}
s^3 & : 1 \ 11 \\
s^2 & : 6 \ 6 \\
s & : 10 \ 0 \\
s^0 & : 6
\end{aligned}
$$

All entries in the first column of the table are positive.

We now return to the nonlinear system (3.11). By a linear approximation, we have

$$
f_i(\mathbf{x}) \approx f_i(\bar{\mathbf{x}}) + \sum_{j=1}^{n} \frac{\partial f_i}{\partial x_j}(\bar{\mathbf{x}})(x_j - \bar{x}_j) = \sum_{j=1}^{n} \frac{\partial f_i}{\partial x_j}(\bar{\mathbf{x}})(x_j - \bar{x}_j).
$$

Let

$$
\mathbf{A} = \begin{bmatrix} \frac{\partial f_1}{\partial x_1}(\bar{\mathbf{x}}) & \frac{\partial f_1}{\partial x_2}(\bar{\mathbf{x}}) & \cdots & \frac{\partial f_1}{\partial x_n}(\bar{\mathbf{x}}) \\ \frac{\partial f_2}{\partial x_1}(\bar{\mathbf{x}}) & \frac{\partial f_2}{\partial x_2}(\bar{\mathbf{x}}) & \cdots & \frac{\partial f_2}{\partial x_n}(\bar{\mathbf{x}}) \\ \vdots & \vdots & \cdots & \vdots \\ \frac{\partial f_n}{\partial x_1}(\bar{\mathbf{x}}) & \frac{\partial f_n}{\partial x_2}(\bar{\mathbf{x}}) & \cdots & \frac{\partial f_n}{\partial x_n}(\bar{\mathbf{x}}) \end{bmatrix}.
$$

This matrix is called *Jacobian matrix* of **f**. Then the nonlinear system (3.11) can be linearly approximated at $\bar{\mathbf{x}}$ by the following linear system

$$
\frac{d\mathbf{x}}{dt} = \mathbf{A}\mathbf{x}.
$$

Thus the stability of the equilibrium point $\bar{\mathbf{x}}$ of the nonlinear system (3.11) is determined by the stability of the linearized system.

**Theorem 8.** *Let $\bar{\mathbf{x}} = (\bar{x}_1, \bar{x}_2, \cdots \bar{x}_n)^*$ be the equilibrium point of the system (3.11). Let $\Omega \subset \mathbb{R}^n$ be an open set containing $\bar{\mathbf{x}}$ and let $\mathbf{f} : \Omega \to \mathbb{R}^n$ be continuously differentiable. Let $\mathbf{A}$ be Jacobian matrix of $\mathbf{f}$. Then:*

- *the equilibrium point $\bar{\mathbf{x}}$ is exponentially stable if $\mathrm{Re}\lambda_i < 0$ for all eigenvalues of $\mathbf{A}$;*
- *the equilibrium point $\bar{\mathbf{x}}$ is unstable if $\mathrm{Re}\lambda_i > 0$ for one or more of the eigenvalues of $\mathbf{A}$.*

This theorem is called *Lyapunov's indirect method*. Its proof is omitted and referred to [2].

We now use Lyapunov's indirect method to study the stability of the reaction system (3.8)-(3.9). For simplicity, we assume that $u = 0$. Then the system has the equilibrium point $\bar{S} = 0$ and $\bar{E} = E_0$. By a direct calculation, we can obtain Jacobian matrix of the system at the equilibrium point as follows

$$
\mathbf{A} = \begin{bmatrix} -k_1 E_0 & -k_{-1} \\ -k_1 E_0 & -k_{-1} - k_2 \end{bmatrix}.
$$

The characteristic equation of **A** is

$$
\lambda^2 + (k_{-1} + k_2 + k_1 E_0)\lambda + k_1 k_2 E_0 = 0.
$$

Then real parts of the eigenvalues are negative and so the equilibrium point is exponentially stable.

## 3.2.4 Invariance Principle

Consider the system

$$
\frac{dx}{dt} = y, \tag{3.30}
$$

$$
\frac{dy}{dt} = -\sin x - y. \tag{3.31}
$$

Define

$$V(x,y) = (1 - \cos x) + \frac{1}{2}y^2.$$

Then we have

$$\frac{dV}{dt} = \frac{dx}{dt}\sin x + y\frac{dy}{dt} = -y^2. \tag{3.32}$$

Thus $\frac{dV}{dt} = 0$ on the whole line $y = 0$ and then the condition (3.17) is not satisfied. Hence we need to modify Theorem 3 and introduce an invariance principle developed by LaSalle.

Let $\mathbf{x}(t)$ be a solution of (3.11). A point $\mathbf{p}$ is said to be a $\omega$-*limit point* of $\mathbf{x}(t)$ if there is a sequence $\{t_n\}$, with $t_n \to \infty$ as $n \to \infty$, such that $\mathbf{x}(t_n) \to \mathbf{p}$ as $n \to \infty$. The set $\omega(\mathbf{x}(0))$ of all $\omega$-limit points of $\mathbf{x}(t)$ is called the $\omega$-*limit set* of $\mathbf{x}(t)$.

Let $\mathbf{p}$ be a point and $W$ a set. The distance between $\mathbf{p}$ and $W$ is defined by

$$d(\mathbf{p},W) = \inf_{\mathbf{x}\in W} \|\mathbf{p} - \mathbf{x}\|.$$

For a solution $\mathbf{x}(t)$ of (3.11), if $d(\mathbf{x}(t),W) \to 0$ as $t \to \infty$, then we denote $\mathbf{x}(t) \to W$ as $t \to \infty$.

**Lemma 1.** *If a solution $\mathbf{x}(t)$ of (3.11) is bounded for $t \geq 0$, then its $\omega$-limit set $\omega(\mathbf{x}(0))$ is a nonempty, compact, positively invariant set. Moreover, $\mathbf{x}(t) \to \omega(\mathbf{x}(0))$ as $t \to \infty$.*

The proof is referred to [2, p. 127, Lemma 4.1]. We now present LaSalle's theorem, called LaSalle's invariance principle.

**Theorem 9.** *Let $K \subset \Omega$ be a compact set that is positively invariant under (3.11). Let $V : \Omega \to \mathbb{R}$ be a continuously differentiable function such that $\frac{dV}{dt} \leq 0$ in $K$. Let*

$$Z = \left\{ \mathbf{x} \in K \mid \frac{dV(\mathbf{x})}{dt} = 0 \right\}.$$

*If $M$ is the largest invariant set in $Z$, then $\mathbf{x}(t) \to M$ as $t \to \infty$ for every solution $\mathbf{x}(t)$ of (3.11) starting in $K$.*

*Proof.* Let $\mathbf{x}(t)$ be a solution of (3.11) starting in $K$. Since $\frac{dV(\mathbf{x}(t))}{dt} \leq 0$ in $K$, $V(\mathbf{x}(t))$ is decreasing. Because $K$ is compact and $V(\mathbf{x}(t))$ is continuous, $V(\mathbf{x}(t))$ is bounded below and then converges to some $a$. For any $\mathbf{p} \in \omega(\mathbf{x}(0))$, we have

$$V(\mathbf{p}) = \lim_{n\to\infty} V(\mathbf{x}(t_n))) = a.$$

It therefore follows that $\frac{dV(\mathbf{x})}{dt} = 0$ on $\omega(\mathbf{x}(0))$. Since $\omega(\mathbf{x}(0))$ is invariant, we derive that $\omega(\mathbf{x}(0)) \subset M \subset Z$. It then follows from Lemma 1 that $\mathbf{x}(t) \to \omega(\mathbf{x}(0)) \subset M$ as $t \to \infty$. □

We now go back to the system (3.30)–(3.31). Define

$$K = \left\{ (x,y) \in \mathbb{R}^2 \mid (1 - \cos x) + \frac{1}{2}y^2 \leq \frac{1}{2} \right\}.$$

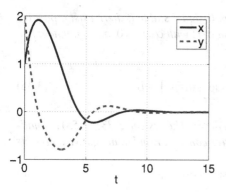

**Fig. 3.2.** A numerical solution of the system (3.30)–(3.31)

It follows from (3.32) that $K$ is positively invariant, but not compact because it is unbounded. Since $(1 - \cos x) + \frac{1}{2}y^2 \leq \frac{1}{2}$ implies that $\cos x \geq 1/2$ and then $2n\pi - \frac{\pi}{6} \leq x \leq 2n\pi + \frac{\pi}{6}$ $(n = 0, \pm 1, \pm 2, \cdots)$, $K$ is disconnected. Then the subset of $K$ defined by

$$K_n = \left\{ (x,y) \in K \mid 2n\pi - \frac{\pi}{6} \leq x \leq 2n\pi + \frac{\pi}{6} \right\}$$

is also positively invariant. Since $K_n$ is bounded and closed, it is compact. The set $Z$ is given by

$$Z = \left\{ (x,y) \in K_n \mid \frac{dV(x,y)}{dt} = 0 \right\} = \left\{ (x,y) \in K_n \mid -y^2 = 0 \right\} = \left\{ (x,0) \in K_n \right\}.$$

For any initial condition $(x,0) \in Z$ with $x \neq 0$, it follows from the equation (3.31) that $y(t) \neq 0$ for some $t > 0$. So the largest invariant set in $Z$ is $\{(2n\pi, 0)\}$. It then follows from LaSalle's invariance principle, Theorem 9, that the equilibrium point $(2n\pi, 0)$ is asymptotically stable. This is further confirmed by a numerical solution of the system in Fig. 3.2.

### 3.2.5 Input-output Stability

Consider the system

$$\frac{d\mathbf{x}}{dt} = \mathbf{f}(t, \mathbf{x}, \mathbf{u}), \ \mathbf{x}(0) = \mathbf{x}_0, \tag{3.33}$$

$$\mathbf{y} = \mathbf{h}(t, \mathbf{x}, \mathbf{u}), \tag{3.34}$$

where $\mathbf{x} \in \mathbb{R}^n, \mathbf{u} \in \mathbb{R}^m, \mathbf{y} \in \mathbb{R}^q$, $\mathbf{f} : [0, \infty) \times \Omega \times \Omega_u \to \mathbb{R}^n$ is piecewise continuous in $t$ and locally Lipschitz in $(\mathbf{x}, \mathbf{u})$, $\mathbf{h} : [0, \infty) \times \Omega \times \Omega_u \to \mathbb{R}^q$ is piecewise continuous in $t$ and continuous in $(\mathbf{x}, \mathbf{u})$, $\Omega \subset \mathbb{R}^n$ is a domain containing $\mathbf{x} = 0$, and $\Omega_u \subset \mathbb{R}^m$ is a domain containing $\mathbf{u} = 0$. Suppose that $\mathbf{x} = 0$ is an equilibrium point of the system (3.33) at the zero input:

$$\mathbf{f}(t, 0, 0) = 0$$

for all $t$.

**Definition 3.** *The system (3.33)-(3.55) is small-signal stable if there exist a positive constant r, a nonnegative increasing function $\alpha(t)$ with $\alpha(0) = 0$, and a nonnegative constant $\beta$ such that*

$$\sup_{0 \le t \le \tau} \|\mathbf{y}(t)\| \le \alpha \left( \sup_{0 \le t \le \tau} \|\mathbf{u}(t)\| \right) + \beta \qquad (3.35)$$

*for all $\tau \in [0, \infty)$ and all $\mathbf{u}$ with $\sup_{0 \le t \le \tau} \|\mathbf{u}(t)\| \le r$. The system (3.33)-(3.55) is small-signal finite-gain stable if there exist a positive constant r and nonnegative constants $\gamma$ and $\beta$ such that*

$$\sup_{0 \le t \le \tau} \|\mathbf{y}(t)\| \le \gamma \sup_{0 \le t \le \tau} \|\mathbf{u}(t)\| + \beta \qquad (3.36)$$

*for all $\tau \in [0, \infty)$ and all $\mathbf{u}$ with $\sup_{0 \le t \le \tau} \|\mathbf{u}(t)\| \le r$.*

**Theorem 10.** *Suppose that, in some neighborhood of $(\mathbf{x} = 0, \mathbf{u} = 0)$, the function $\mathbf{f}(t, \mathbf{x}, \mathbf{u})$ is continuously differentiable, Jacobian matrices at $(\mathbf{x} = 0, \mathbf{u} = 0)$*

$$\frac{\partial \mathbf{f}}{\partial \mathbf{x}} = \begin{bmatrix} \frac{\partial f_1}{\partial x_1} & \frac{\partial f_1}{\partial x_2} & \cdots & \frac{\partial f_1}{\partial x_n} \\ \frac{\partial f_2}{\partial x_1} & \frac{\partial f_2}{\partial x_2} & \cdots & \frac{\partial f_2}{\partial x_n} \\ \vdots & \vdots & \cdots & \vdots \\ \frac{\partial f_n}{\partial x_1} & \frac{\partial f_n}{\partial x_2} & \cdots & \frac{\partial f_n}{\partial x_n} \end{bmatrix}, \quad \frac{\partial \mathbf{f}}{\partial \mathbf{u}} = \begin{bmatrix} \frac{\partial f_1}{\partial u_1} & \frac{\partial f_1}{\partial u_2} & \cdots & \frac{\partial f_1}{\partial u_m} \\ \frac{\partial f_2}{\partial u_1} & \frac{\partial f_2}{\partial u_2} & \cdots & \frac{\partial f_2}{\partial u_m} \\ \vdots & \vdots & \cdots & \vdots \\ \frac{\partial f_n}{\partial u_1} & \frac{\partial f_n}{\partial u_2} & \cdots & \frac{\partial f_n}{\partial u_m} \end{bmatrix}$$

*are bounded uniformly in t, and $\mathbf{h}(t, \mathbf{x}, \mathbf{u})$ satisfies*

$$\|\mathbf{h}(t, \mathbf{x}, \mathbf{u})\| \le \eta_1 \|\mathbf{x}\| + \eta_2 \|\mathbf{u}\| \qquad (3.37)$$

*for all t and some nonnegative constants $\eta_1, \eta_2$. If $\mathbf{x} = 0$ is an exponentially stable equilibrium point of the system (3.33) at the zero input:*

$$\mathbf{f}(t, 0, 0) = 0 \quad \text{for all } t,$$

*then there exists a constant $r_0 > 0$ such that for each $\mathbf{x}_0$ with $\|\mathbf{x}_0\| < r_0$, the system (3.33)-(3.55) is small-signal finite-gain stable.*

The proof of the theorem is referred to Corollary 5.1 of [2].

*Example 12.* Consider the single-input-single-output first-order system

$$\frac{dx}{dt} = -x + u, \quad x(0) = x_0,$$
$$y = x + u.$$

Solving the system, we obtain

$$y(t) = e^{-t} x_0 + e^{-t} \int_0^t e^s u(s) ds + u(t).$$

It then follows that for $0 \leq t \leq \tau$

$$|y(t)| \leq |x_0| + e^{-t} \int_0^t e^s \sup_{0 \leq s \leq \tau} |u(s)| ds + \sup_{0 \leq t \leq \tau} |u(t)|$$

$$\leq |x_0| + \sup_{0 \leq s \leq \tau} |u(s)|(1 - e^{-t}) + \sup_{0 \leq t \leq \tau} |u(t)|$$

$$\leq 2 \sup_{0 \leq s \leq \tau} |u(s)| + |x_0|,$$

and then

$$\sup_{0 \leq t \leq \tau} |y(t)| \leq 2 \sup_{0 \leq s \leq \tau} |u(s)| + |x_0|.$$

Hence the system is small-signal finite-gain stable.

## 3.3 Controllability and Observability

Controllability and observability are structural properties of a system. Consider a control system

$$\frac{dx}{dt} = Ax + Bu, \tag{3.38}$$

$$y = Cx + Du, \tag{3.39}$$

$$x(0) = x_0, \tag{3.40}$$

where $\mathbf{x} = (x_1, x_2, \cdots, x_n)^*$ is a state vector, $\mathbf{x}_0$ is an initial state, $\mathbf{y} = (y_1, \cdots, y_l)^*$ is an output vector, $\mathbf{u} = (u_1, \cdots, u_m)^*$ is a control vector, and $\mathbf{A}, \mathbf{B}, \mathbf{C}, \mathbf{D}$ are $n \times n, n \times m$, $l \times n, l \times m$ constant matrices, respectively.

**Definition 4.** *The system* (3.38) *or the pair* $(\mathbf{A}, \mathbf{B})$ *is controllable if for any initial state* $\mathbf{x}_0$ *and final state* $\mathbf{x}_f$, *there exists a control vector* $\mathbf{u}$ *such that* $\mathbf{x}(T) = \mathbf{x}_f$ *for some* $T > 0$.

**Theorem 11.** *The pair* $(\mathbf{A}, \mathbf{B})$ *is controllable if and only if Kalman controllability matrix defined by*

$$\mathscr{C} = [\mathbf{B} \ \mathbf{AB} \ \mathbf{A}^2\mathbf{B} \cdots \mathbf{A}^{n-1}\mathbf{B}]$$

*has a rank of n.*

The proof of the theorem is omitted and referred to [3, 4].

*Example 13.* Consider the system

$$\frac{d}{dt} \begin{bmatrix} x_1 \\ x_2 \end{bmatrix} = \begin{bmatrix} 0 & 1 \\ -1 & 0 \end{bmatrix} \begin{bmatrix} x_1 \\ x_2 \end{bmatrix} + \begin{bmatrix} 0 \\ 1 \end{bmatrix} u(t).$$

Since Kalman controllability matrix

$$\mathscr{C} = [\mathbf{B} \ \mathbf{AB}] = \begin{bmatrix} 0 & 1 \\ 1 & 0 \end{bmatrix}$$

has a rank of 2, the system is controllable.

The following example shows that not every system is controllable.

*Example 14.* The system

$$\frac{d}{dt}\begin{bmatrix} x_1 \\ x_2 \end{bmatrix} = \begin{bmatrix} 1 & 0 \\ 1 & 1 \end{bmatrix}\begin{bmatrix} x_1 \\ x_2 \end{bmatrix} + \begin{bmatrix} 0 \\ 1 \end{bmatrix}u(t)$$

is not controllable since the rank of Kalman controllability matrix

$$\mathscr{C} = [\mathbf{B} \ \mathbf{AB}] = \begin{bmatrix} 0 & 0 \\ 1 & 1 \end{bmatrix}$$

is equal to 1.

Consider an observation system

$$\frac{d\mathbf{x}}{dt} = \mathbf{Ax}, \tag{3.41}$$

$$\mathbf{y} = \mathbf{Cx}, \tag{3.42}$$

$$\mathbf{x}(0) = \mathbf{x}_0, \tag{3.43}$$

where $\mathbf{y} = (y_1, y_2, \cdots, y_l)^*$ is an output vector and $\mathbf{C}$ is an $l \times n$ constant matrix.

**Definition 5.** *The system (3.41)-(3.42) or the pair $(\mathbf{A}, \mathbf{C})$ is observable if any initial state $\mathbf{x}_0$ can be uniquely determined by the observation $\mathbf{y}(t)$ over the interval $[0, T]$ for some $T > 0$.*

We define Kalman observability matrix $\mathscr{O}$ by

$$\mathscr{O} = \begin{bmatrix} \mathbf{C} \\ \mathbf{CA} \\ \mathbf{CA}^2 \\ \vdots \\ \mathbf{CA}^{n-1} \end{bmatrix}.$$

**Theorem 12.** *The pair $(\mathbf{A}, \mathbf{C})$ is observable if and only if Kalman observability matrix $\mathscr{O}$ has a rank of $n$.*

The proof of the theorem is omitted and referred to [3, 4].

*Example 15.* Consider the system

$$\frac{d}{dt}\begin{bmatrix} x_1 \\ x_2 \end{bmatrix} = \begin{bmatrix} 0 & 1 \\ -1 & 0 \end{bmatrix}\begin{bmatrix} x_1 \\ x_2 \end{bmatrix},$$

$$y = [0 \ 1]\begin{bmatrix} x_1 \\ x_2 \end{bmatrix}.$$

Since Kalman observability matrix

$$\mathscr{O} = \begin{bmatrix} \mathbf{C} \\ \mathbf{CA} \end{bmatrix} = \begin{bmatrix} 0 & 1 \\ -1 & 0 \end{bmatrix}$$

has a rank of 2, the system is observable.

*Example 16.* The system

$$\frac{d}{dt}\begin{bmatrix} x_1 \\ x_2 \end{bmatrix} = \begin{bmatrix} 1 & 1 \\ 0 & 1 \end{bmatrix}\begin{bmatrix} x_1 \\ x_2 \end{bmatrix},$$

$$y = \begin{bmatrix} 0 & 1 \end{bmatrix}\begin{bmatrix} x_1 \\ x_2 \end{bmatrix}$$

is not observable since the rank of Kalman observability matrix

$$\mathscr{O} = \begin{bmatrix} \mathbf{C} \\ \mathbf{CA} \end{bmatrix} = \begin{bmatrix} 0 & 1 \\ 0 & 1 \end{bmatrix}$$

is equal to 1.

From Theorems 11 and 12, we can derive the following duality between controllability and observability.

**Theorem 13.** *The control system*

$$\frac{d\mathbf{x}}{dt} = \mathbf{Ax} + \mathbf{B}u,$$

$$\mathbf{x}(0) = \mathbf{x}_0$$

*is controllable if and only if its dual observation system*

$$\frac{d\mathbf{x}}{dt} = \mathbf{A}^*\mathbf{x},$$

$$\mathbf{y} = \mathbf{B}^*\mathbf{x},$$

$$\mathbf{x}(0) = \mathbf{x}_0$$

*is observable.*

## 3.4 Feedback Control

Consider the control system

$$\frac{d\mathbf{x}}{dt} = \mathbf{f}(t, \mathbf{x}, \mathbf{u}, \mathbf{v}), \tag{3.44}$$

$$\mathbf{y} = \mathbf{g}(t, \mathbf{x}, \mathbf{u}, \mathbf{v}), \tag{3.45}$$

where $\mathbf{x} = (x_1, x_2, \cdots, x_n)^*$ is a state vector, $\mathbf{u} = (u_1, u_2, \cdots, u_p)$ is a control vector, $\mathbf{v} = (v_1, v_2, \cdots, v_l)$ is a disturbance vector, and $\mathbf{y} = (y_1, y_2, \cdots, y_m)^*$ is an output vector. The problem of control is to design a controller $\mathbf{u}$ so that the output $\mathbf{y}$ tracks a specified reference signal $\mathbf{r}$:

$$\lim_{t \to \infty}(\mathbf{y}(t) - \mathbf{r}(t)) = 0.$$

This control problems is referred to as *asymptotic tracking and disturbance rejection.*

There are a number of ways to achieve the control goal. If all states are available for feedback, we can design a state feedback controller

$$\mathbf{u} = \mathbf{h}(t, \mathbf{x}, \mathbf{v}).$$

Sometimes, we use a dynamical state feedback controller

$$\mathbf{u} = \mathbf{h}(t, \mathbf{x}, \mathbf{z}, \mathbf{v}),$$
$$\frac{d\mathbf{z}}{dt} = \mathbf{w}(t, \mathbf{x}, \mathbf{z}, \mathbf{v}).$$

If only the outputs are available for feedback, we need to design an output feedback controller

$$\mathbf{u} = \mathbf{h}(t, \mathbf{y}, \mathbf{v}),$$

or a dynamic output feedback controller

$$\mathbf{u} = \mathbf{h}(t, \mathbf{y}, \mathbf{z}, \mathbf{v}),$$
$$\frac{d\mathbf{z}}{dt} = \mathbf{w}(t, \mathbf{y}, \mathbf{z}, \mathbf{v}).$$

We first show how to design feedback controllers for the following linear control systems without disturbances:

$$\frac{d\mathbf{x}}{dt} = \mathbf{A}\mathbf{x} + \mathbf{B}\mathbf{u}, \tag{3.46}$$
$$\mathbf{y} = \mathbf{C}\mathbf{x} + \mathbf{D}\mathbf{u}, \tag{3.47}$$
$$\mathbf{x}(0) = \mathbf{x}_0, \tag{3.48}$$

where $\mathbf{x} = (x_1, x_2, \cdots, x_n)^*$ is a state vector, $\mathbf{x}_0$ is an initial state caused by external disturbances, $\mathbf{y} = (y_1, \cdots, y_l)^*$ is an output vector, $\mathbf{u} = (u_1, \cdots, u_m)^*$ is a control vector, and $\mathbf{A}, \mathbf{B}, \mathbf{C}, \mathbf{D}$ are $n \times n$, $n \times m$, $l \times n$, $l \times m$ constant matrices, respectively. The equation (3.46) is called a *state equation*.

We assume that all state variables are available for feedback and design a controller of the form

$$\mathbf{u} = -\mathbf{K}\mathbf{x}. \tag{3.49}$$

Such a scheme is called a *state feedback*. The $m \times n$ matrix $\mathbf{K}$ is called a *state feedback gain matrix*. Substituting the equation (3.49) into (3.46) gives

$$\frac{d\mathbf{x}}{dt} = (\mathbf{A} - \mathbf{B}\mathbf{K})\mathbf{x}, \quad \mathbf{x}(0) = \mathbf{x}_0. \tag{3.50}$$

**Definition 6.** *The pair $(\mathbf{A}, \mathbf{B})$ or the system (3.46) is stabilizable if there exists $\mathbf{K}$ such that the solution $\mathbf{x}(t)$ of (3.50) converges to zero exponentially as $t \to \infty$ for any initial state $\mathbf{x}_0$. The matrix $\mathbf{K}$ is called the feedback matrix.*

If the pair $(\mathbf{A},\mathbf{B})$ is stabilizable, then the solution $\mathbf{x}(t)$ tends to zero. Thus we have that $\mathbf{u}(t) = -\mathbf{Kx}(t) \to \mathbf{0}$ and $\mathbf{y}(t) = \mathbf{Cx}(t) + \mathbf{Du}(t) \to \mathbf{0}$. Therefore the problem of regulating the output to zero is transformed into the stabilization of the pair $(\mathbf{A},\mathbf{B})$.

*Example 17.* Consider the system

$$\frac{d}{dt}\begin{bmatrix} x_1 \\ x_2 \end{bmatrix} = \begin{bmatrix} 0 & 1 \\ -\frac{k}{m} & 0 \end{bmatrix}\begin{bmatrix} x_1 \\ x_2 \end{bmatrix} + \begin{bmatrix} 0 \\ \frac{1}{m} \end{bmatrix} u, \tag{3.51}$$

$$y = \begin{bmatrix} 1 & 0 \end{bmatrix}\begin{bmatrix} x_1 \\ x_2 \end{bmatrix}, \tag{3.52}$$

where $m$ and $k$ are positive constants. Denote

$$\mathbf{A} = \begin{bmatrix} 0 & 1 \\ -\frac{k}{m} & 0 \end{bmatrix}, \quad \mathbf{B} = \begin{bmatrix} 0 \\ \frac{1}{m} \end{bmatrix}.$$

Consider the state feedback control of the form

$$u = -\mathbf{Kx}, \tag{3.53}$$

where $\mathbf{K} = [k_1, k_2]$. Let the reference signal be 0. Then $y$ converges to 0 if all eigenvalues of the matrix $\mathbf{A} - \mathbf{BK}$ have negative real parts. Let $\mu_1, \mu_2$ be the desired eigenvalues. Then we set

$$\det(\lambda\mathbf{I} - \mathbf{A} + \mathbf{BK}) = \lambda^2 + \frac{k_2}{m}\lambda + \frac{k+k_1}{m} = (\lambda - \mu_1)(\lambda - \mu_2).$$

Equating the coefficients gives

$$k_1 = m\mu_1\mu_2 - k, \quad k_2 = -m(\mu_1 + \mu_2).$$

We then obtain the following state feedback controller

$$u = -k_1x_1 - k_2x_2.$$

To illustrate an idea of how to design an output feedback controller, we consider the control system

$$\frac{d\mathbf{x}}{dt} = \mathbf{Ax} + \mathbf{Bu}, \tag{3.54}$$

$$\mathbf{y} = \mathbf{Cx}, \tag{3.55}$$

$$\mathbf{x}(0) = \mathbf{x}_0. \tag{3.56}$$

Because we assume that only the outputs are available for feedback, we need to estimate the state variables using only input and output measurements. The estimate $\tilde{\mathbf{x}}$ of $\mathbf{x}$ can be generated by injecting the outputs into the system as follows

$$\frac{d\tilde{\mathbf{x}}}{dt} = \mathbf{A}\tilde{\mathbf{x}} + \mathbf{Bu} + \mathbf{K}_e(\mathbf{y} - \mathbf{C}\tilde{\mathbf{x}}). \tag{3.57}$$

This output injection system is called the *state observer*, known as *Luenberger observer*. The matrix $K_e$ is an output injection matrix (also called an *observer gain matrix*). We then use the estimate $\tilde{x}$ for feedback and introduce an observer-based output feedback controller

$$u = -K\tilde{x}, \tag{3.58}$$

which leads to an observer-based output feedback control system

$$\frac{dx}{dt} = Ax - BK\tilde{x}, \tag{3.59}$$

$$y = Cx, \tag{3.60}$$

$$\frac{d\tilde{x}}{dt} = A\tilde{x} - BK\tilde{x} + K_e(y - C\tilde{x}). \tag{3.61}$$

To study the stability of this control system, we introduce the error vector

$$e = x - \tilde{x}.$$

Subtracting the equation (3.61) from the equation (3.59) gives

$$\frac{de}{dt} = Ae - K_e(y - C\tilde{x}) = Ae - K_e(Cx - C\tilde{x}) = (A - K_eC)e.$$

Also the state equation (3.59) can be written as

$$\frac{dx}{dt} = (A - BK)x + BKe.$$

Combining these two equations, we obtain

$$\frac{d}{dt}\begin{bmatrix} x \\ e \end{bmatrix} = \begin{bmatrix} A - BK & BK \\ 0 & A - K_eC \end{bmatrix}\begin{bmatrix} x \\ e \end{bmatrix}. \tag{3.62}$$

Since

$$\det\left(\lambda I - \begin{bmatrix} A - BK & BK \\ 0 & A - K_eC \end{bmatrix}\right)$$
$$= \det(\lambda I - [A - BK])\det(\lambda I - [A - K_eC]),$$

it suffices to design matrices $K$ and $K_e$ such that the real parts of the eigenvalues of $A - BK$ and $A - K_eC$ are negative. Then $x(t)$ converges to zero exponentially and then so $y(t) = Cx(t)$ does.

*Example 18.* Consider the system

$$\frac{dx}{dt} = Ax + Bu,$$
$$y = Cx,$$

where

$$A = \begin{bmatrix} 0 & 1 \\ 0 & -2 \end{bmatrix}, \quad B = \begin{bmatrix} 0 \\ 4 \end{bmatrix}, \quad C = \begin{bmatrix} 1 & 0 \end{bmatrix}.$$

Design an output feedback controller such that the eigenvalues of the closed-loop system are

$$\lambda = -2 + 2\sqrt{3}i, -2 - 2\sqrt{3}i$$

and the eigenvalues of the observer system are

$$\lambda = -8, -8.$$

The feedback gain matrix $\mathbf{K} = [k_1, k_2]$ should be selected such that

$$\det(\lambda \mathbf{I} - \mathbf{A} + \mathbf{B}\mathbf{K}) = (\lambda + 2 - 2\sqrt{3}i)(\lambda + 2 + 2\sqrt{3}i),$$

and then

$$\lambda^2 + (2 + 4k_2)\lambda + 4k_1 = \lambda^2 + 4\lambda + 16.$$

Comparing the coefficients gives

$$k_1 = 4, \quad k_2 = 0.5.$$

The observer gain matrix $\mathbf{K}_e = [k_1^e, k_2^e]^*$ should be selected such that

$$\det(\lambda \mathbf{I} - \mathbf{A} + \mathbf{K}_e \mathbf{C}) = (\lambda + 8)^2,$$

and then

$$\lambda^2 + (2 + k_1^e)\lambda + 2k_1^e + k_2^e = \lambda^2 + 16\lambda + 64.$$

Comparing the coefficients gives

$$k_1^e = 14, \quad k_2^e = 36.$$

Then the output feedback controller is

$$u = -4\tilde{x}_1 - 0.5\tilde{x}_2,$$

$$\frac{d}{dt}\begin{bmatrix} \tilde{x}_1 \\ \tilde{x}_2 \end{bmatrix} = \begin{bmatrix} 0 & 1 \\ 0 & -2 \end{bmatrix}\begin{bmatrix} \tilde{x}_1 \\ \tilde{x}_2 \end{bmatrix} - \begin{bmatrix} 0 \\ 4 \end{bmatrix}[4 \ 0.5]\begin{bmatrix} \tilde{x}_1 \\ \tilde{x}_2 \end{bmatrix} + \begin{bmatrix} 14 \\ 36 \end{bmatrix}(x_1 - \tilde{x}_1)$$

$$= \begin{bmatrix} 0 & 1 \\ -16 & -4 \end{bmatrix}\begin{bmatrix} \tilde{x}_1 \\ \tilde{x}_2 \end{bmatrix} + \begin{bmatrix} 14 \\ 36 \end{bmatrix}(x_1 - \tilde{x}_1).$$

To illustrate how to design a feedback controller for nonlinear systems, we consider the system

$$\frac{d\mathbf{x}}{dt} = \mathbf{f}(\mathbf{x}, \mathbf{u}), \tag{3.63}$$

where $\mathbf{f}(\mathbf{0}, \mathbf{0}) = \mathbf{0}$. To use the design method for linear systems, we linearize (3.63) at $(\mathbf{0}, \mathbf{0})$ to obtain the linear system

$$\frac{d\mathbf{x}}{dt} = \mathbf{A}\mathbf{x} + \mathbf{B}\mathbf{u}, \tag{3.64}$$

where

$$
A = \frac{\partial \mathbf{f}}{\partial \mathbf{x}} =
\begin{bmatrix}
\frac{\partial f_1}{\partial x_1} & \frac{\partial f_1}{\partial x_2} & \cdots & \frac{\partial f_1}{\partial x_n} \\
\frac{\partial f_2}{\partial x_1} & \frac{\partial f_2}{\partial x_2} & \cdots & \frac{\partial f_2}{\partial x_n} \\
\vdots & \vdots & \cdots & \vdots \\
\frac{\partial f_n}{\partial x_1} & \frac{\partial f_n}{\partial x_2} & \cdots & \frac{\partial f_n}{\partial x_n}
\end{bmatrix}_{(0,0)}
,\quad
B = \frac{\partial \mathbf{f}}{\partial \mathbf{u}} =
\begin{bmatrix}
\frac{\partial f_1}{\partial u_1} & \frac{\partial f_1}{\partial u_2} & \cdots & \frac{\partial f_1}{\partial u_m} \\
\frac{\partial f_2}{\partial u_1} & \frac{\partial f_2}{\partial u_2} & \cdots & \frac{\partial f_2}{\partial u_m} \\
\vdots & \vdots & \cdots & \vdots \\
\frac{\partial f_n}{\partial u_1} & \frac{\partial f_n}{\partial u_2} & \cdots & \frac{\partial f_n}{\partial u_m}
\end{bmatrix}_{(0,0)}
.
$$

We then design a state feedback controller $\mathbf{u} = -\mathbf{Kx}$ to stabilize the linear system. This controller also stabilizes the original nonlinear system (3.63). This control design method via linearization can be also applied to the design of an output feedback controller.

## 3.5 Parametric Sensitivity

Consider the system

$$
\frac{d\mathbf{x}}{dt} = \mathbf{f}(t,\mathbf{x},\mathbf{p}), \quad \mathbf{x}(0) = \mathbf{x}_0, \tag{3.65}
$$

where $\mathbf{x} \in \mathbb{R}^n$ is a state vector and $\mathbf{p} \in \mathbb{R}^m$ is a parameter vector. A change in $\mathbf{p}$ results in a change in $\mathbf{x}$. The sensitivity is a measurement of the change in $\mathbf{x}$ as $\mathbf{p}$ changes. In particular, the sensitivity of the state $x_i$ with respect to the parameter $p_j$ can be measured by $\frac{\partial x_i}{\partial p_j}$. To normalize the sensitivity, we define the sensitivity index

$$
SI_{ij} = \frac{p_j}{x_i} \frac{\partial x_i}{\partial p_j}. \tag{3.66}
$$

The system of the sensitivity $\frac{\partial x_i}{\partial p_j}$ can be obtained by differentiating (3.65) with respect to $p_j$ as follows

$$
\frac{d}{dt}\left(\frac{\partial x_i}{\partial p_j}\right) = \sum_{k=1}^{n} \frac{\partial f_i(t,\mathbf{x},\mathbf{p})}{\partial x_k}\frac{\partial x_k}{\partial p_j} + \frac{\partial f_i(t,\mathbf{x},\mathbf{p})}{\partial p_j}.
$$

This system can be written in the following matrix form

$$
\frac{d}{dt}\left(\frac{\partial \mathbf{x}}{\partial \mathbf{p}}\right) = \frac{\partial \mathbf{f}}{\partial \mathbf{x}}\frac{\partial \mathbf{x}}{\partial \mathbf{p}} + \frac{\partial \mathbf{f}}{\partial \mathbf{p}}, \tag{3.67}
$$

where $\frac{\partial \mathbf{x}}{\partial \mathbf{p}}$ is the sensitivity matrix

$$\frac{\partial \mathbf{x}}{\partial \mathbf{p}} = \begin{bmatrix} \frac{\partial x_1}{\partial p_1} & \frac{\partial x_1}{\partial p_2} & \cdots & \frac{\partial x_1}{\partial p_m} \\ \frac{\partial x_2}{\partial p_1} & \frac{\partial x_2}{\partial p_2} & \cdots & \frac{\partial x_2}{\partial p_m} \\ \vdots & \vdots & \cdots & \vdots \\ \frac{\partial x_n}{\partial p_1} & \frac{\partial x_n}{\partial p_2} & \cdots & \frac{\partial x_n}{\partial p_m} \end{bmatrix}$$

and $\frac{\partial \mathbf{f}}{\partial \mathbf{x}}, \frac{\partial \mathbf{f}}{\partial \mathbf{p}}$ are Jacobian matrices

$$\frac{\partial \mathbf{f}}{\partial \mathbf{x}} = \begin{bmatrix} \frac{\partial f_1}{\partial x_1} & \frac{\partial f_1}{\partial x_2} & \cdots & \frac{\partial f_1}{\partial x_n} \\ \frac{\partial f_2}{\partial x_1} & \frac{\partial f_2}{\partial x_2} & \cdots & \frac{\partial f_2}{\partial x_n} \\ \vdots & \vdots & \cdots & \vdots \\ \frac{\partial f_n}{\partial x_1} & \frac{\partial f_n}{\partial x_2} & \cdots & \frac{\partial f_n}{\partial x_n} \end{bmatrix}, \quad \frac{\partial \mathbf{f}}{\partial \mathbf{p}} = \begin{bmatrix} \frac{\partial f_1}{\partial p_1} & \frac{\partial f_1}{\partial p_2} & \cdots & \frac{\partial f_1}{\partial p_m} \\ \frac{\partial f_2}{\partial p_1} & \frac{\partial f_2}{\partial p_2} & \cdots & \frac{\partial f_2}{\partial p_m} \\ \vdots & \vdots & \cdots & \vdots \\ \frac{\partial f_n}{\partial p_1} & \frac{\partial f_n}{\partial p_2} & \cdots & \frac{\partial f_n}{\partial p_m} \end{bmatrix}.$$

Since $\mathbf{x}$ does not depend on $\mathbf{p}$ initially, we have $\frac{\partial \mathbf{x}}{\partial \mathbf{p}}(0) = \mathbf{0}$. Therefore, the sensitivity indices can be obtained by solving the system (3.65) and (3.67).

*Example 19.* Consider the enzymatic reaction system

$$\frac{dS}{dt} = -k_1 ES + k_{-1}(E_0 - E), \tag{3.68}$$

$$\frac{dE}{dt} = -k_1 ES + (k_{-1} + k_2)(E_0 - E). \tag{3.69}$$

Assume that the parameters have the values $k_1 = 0.5, k_2 = 10, k_{-1} = 0.1$. The Jacobian matrices are given by ($\mathbf{x} = (S, E)$ and $\mathbf{p} = (k_1, k_2, k_{-1})$)

$$\frac{\partial \mathbf{f}}{\partial \mathbf{x}} = \begin{bmatrix} -k_1 E & -k_1 S - k_{-1} \\ -k_1 E & -k_1 S - k_{-1} - k_2 \end{bmatrix}, \quad \frac{\partial \mathbf{f}}{\partial \mathbf{p}} = \begin{bmatrix} -ES & 0 & E_0 - E \\ -ES & E_0 - E & E_0 - E \end{bmatrix}.$$

Evaluating $\frac{\partial \mathbf{f}}{\partial \mathbf{x}}$ at the parameter $k_1 = 0.5, k_2 = 10, k_{-1} = 0.1$ gives

$$\frac{\partial \mathbf{f}}{\partial \mathbf{x}} = \begin{bmatrix} -0.5E & -0.5S - 0.1 \\ -0.5E & -0.5S - 10.1 \end{bmatrix}.$$

Let

$$x_1 = \frac{\partial S}{\partial k_1}, \ x_2 = \frac{\partial S}{\partial k_2}, \ x_3 = \frac{\partial S}{\partial k_{-1}}, \ x_4 = \frac{\partial E}{\partial k_1}, \ x_5 = \frac{\partial E}{\partial k_2}, \ x_6 = \frac{\partial E}{\partial k_{-1}}.$$

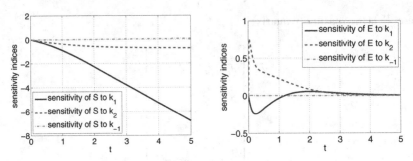

**Fig. 3.3.** Sensitivity indices of the system (3.68)-(3.69)

We then obtain the sensitivity system

$$\frac{dS}{dt} = -0.5ES + 0.1(E_0 - E),$$

$$\frac{dE}{dt} = -0.5ES + 10.1(E_0 - E),$$

$$\frac{dx_1}{dt} = -0.5Ex_1 - (0.5S + 0.1)x_4 - ES,$$

$$\frac{dx_2}{dt} = -0.5Ex_2 - (0.5S + 0.1)x_5,$$

$$\frac{dx_3}{dt} = -0.5Ex_3 - (0.5S + 0.1)x_6 + E_0 - E,$$

$$\frac{dx_4}{dt} = -0.5Ex_1 - (0.5S + 10.1)x_4 - ES,$$

$$\frac{dx_5}{dt} = -0.5Ex_2 - (0.5S + 10.1)x_5 + E_0 - E,$$

$$\frac{dx_6}{dt} = -0.5Ex_3 - (0.5S + 10.1)x_6 + E_0 - E,$$

$$S(0) = 10, \ E(0) = 0.1, \ x_1(0) = x_2(0) = x_4(0) = x_4(0) = x_5(0) = x_6(0) = 0.$$

The system is solved numerically using MATLAB. The sensitivity indices plotted in Fig. 3.3 show that the substrate $S$ is more sensitive to $k_1$, the rate constant of binding of substrate to enzyme, than other parameters and that the enzyme $E$ is more sensitive to $k_2$, the enzyme turnover rate, than other parameters. This agrees with what we could expect from biology and then provides a theoretical support for the biological observation.

# Exercises

**3.1.** Find $A + B, AB$, and $BA$ when

$$A = \begin{bmatrix} 1 & 2 \\ 7 & 10 \end{bmatrix}, \quad B = \begin{bmatrix} 5 & 0 \\ -3 & 2 \end{bmatrix}.$$

**3.2.** ([5]) Establish the distributive laws for matrix multiplication:

$$(A + B)C = AC + BC, \quad C(A + B) = CA + CB.$$

**3.3.** Prove that $(AB)^* = B^*A^*$.

**3.4.** Find the solution of the system

$$Ax = \begin{bmatrix} 1 \\ 0 \\ 1 \end{bmatrix}$$

when

$$A = \begin{bmatrix} 1 & 0 & 1 \\ 0 & 2 & 0 \\ 1 & 3 & 0 \end{bmatrix}.$$

**3.5.** Find $A'(t)$ and $\int_0^t A(s)ds$ if

$$A(t) = \begin{bmatrix} 5 & 3t^2 \\ \sin t & \cos 2t \end{bmatrix}.$$

**3.6.** Calculate the determinant $\det(A)$ of each matrix

$$A = \begin{bmatrix} 1 & 4 & 1 \\ 5 & 1 & 0 \\ 1 & 1 & 7 \end{bmatrix}, \quad A = \begin{bmatrix} 2 & -2 \\ 1 & 8 \end{bmatrix}.$$

**3.7.** Find eigenvalues and their corresponding eigenvectors of each matrix

$$A = \begin{bmatrix} 1 & 4 & 3 \\ 4 & 1 & 1 \\ 3 & 1 & 7 \end{bmatrix}, \quad A = \begin{bmatrix} 2 & -2 \\ 5 & 3 \end{bmatrix}.$$

**3.8.** Find the rank of each matrix

$$A = \begin{bmatrix} 1 & 4 & 1 \\ 5 & 1 & 0 \\ 1 & 1 & 7 \\ 1 & 8 & 10 \end{bmatrix}, \quad A = \begin{bmatrix} 2 & -2 & 0 & 1 & 7 & 9 \\ 1 & 8 & 6 & 2 & 9 & 1 \\ 3 & 6 & 6 & 3 & 16 & 10 \end{bmatrix}.$$

**3.9.** Calmodulin contains four copies of a $Ca^{2+}$-binding EF-hand, each of which binds one $Ca^{2+}$ ion. Calmodulin senses the rise of the cytosolic calcium concentration and transmits the calcium signal to calcineurin. The biochemical reactions in this $Ca^{2+}$ sensing and signal transduction process can be described as follows:

$$4Ca^{2+} + \text{calmodulin} \underset{k_{-1}}{\overset{k_1}{\rightleftharpoons}} \quad CaM \overset{k}{\rightarrow} \text{calmodulin} + [Ca^{2+}]_{out},$$

$$CaM + \text{calcineurin} \underset{k_{-2}}{\overset{k_2}{\rightleftharpoons}} \quad CaN,$$

where $CaM$ denotes the $Ca^{2+}$-bound calmodulin, $CaN$ denotes the $CaM$-bound calcineurin, $[Ca^{2+}]_{out}$ denotes $Ca^{2+}$ leaving the cell, and $k$'s are reaction rate constants:

1. Derive differential equations for $Ca^{2+}$, CaM, and CaN.
2. Prove that, for nonnegative initial conditions with $0 \le [CaM](0) + [CaN](0) \le [CaM_0], 0 \le [CaN](0) \le [CaN_0]$, the solutions of the derived system satisfy

$$[Ca^{2+}](t), [CaM](t), [CaN](t) \ge 0,$$
$$[CaM](t) + [CaN](t) \le [CaM_0], \quad [CaN](t) \le [CaN_0],$$
$$\lim_{t\to\infty}[Ca^{2+}](t) = \lim_{t\to\infty}[CaM](t) = \lim_{t\to\infty}[CaN](t) = 0,$$

where $[CaM_0]$ is the total concentration of $Ca^{2+}$-free, $Ca^{2+}$-bound, and $Ca^{2+}$-calcineurin-bound calmodulin; $[CaN_0]$ is the total concentration of $CaM$-free and $CaM$-bound calcineurin.

**3.10.** ([2]) For each of the following systems, use a quadratic Lyapunov function candidate to show that the origin is asymptotically stable:

1. $\frac{dx_1}{dt} = -x_1 + x_1 x_2, \quad \frac{dx_2}{dt} = -x_2.$

2. $\frac{dx_1}{dt} = -x_2 - x_1(1 - x_1^2 - x_2^2), \quad \frac{dx_2}{dt} = x_1 - x_2(1 - x_1^2 - x_2^2).$

3. $\frac{dx_1}{dt} = x_2(1 - x_1^2), \quad \frac{dx_2}{dt} = -(x_1 + x_2)(1 - x_1^2).$

**3.11.** ([2]) Consider the second-order system

$$\frac{dx_1}{dt} = \frac{-6x_1}{(1+x_1^2)^2} + 2x_2, \quad \frac{dx_2}{dt} = \frac{-2(x_1 + x_2)}{(1+x_1^2)^2}.$$

Let $V(x) = \frac{x_1^2}{1+x_1^2} + x_2^2$. Show that $V(x) > 0$ and $\frac{dV}{dt} < 0$ for all $x \in \mathbb{R}^2 - \{0\}$.

**3.12.** Consider the pendulum system

$$\frac{dx_1}{dt} = x_2,$$
$$\frac{dx_2}{dt} = -\frac{g}{L}\sin x_1 - \frac{k}{mL}x_2.$$

Use Lyapunov's indirect method to investigate the stability of the equilibrium points $(0, 0)$ and $(\pi, 0)$.

**3.13.** Consider the system

$$\frac{dx}{dt} = y,$$

$$\frac{dy}{dt} = -x - y^3.$$

Define

$$V(x,y) = \frac{1}{2}(x^2 + y^2).$$

Use LaSalle's invariance principle to prove that the equilibrium point at the origin is asymptotically stable.

**3.14.** Consider the system

$$\frac{d}{dt}\begin{bmatrix} x_1 \\ x_2 \\ x_3 \end{bmatrix} = \begin{bmatrix} -1 & -2 & -2 \\ 0 & -1 & 1 \\ 1 & 0 & -1 \end{bmatrix}\begin{bmatrix} x_1 \\ x_2 \\ x_3 \end{bmatrix} + \begin{bmatrix} 2 \\ 0 \\ 1 \end{bmatrix}u,$$

$$y = \begin{bmatrix} 1 & 1 & 0 \end{bmatrix}\begin{bmatrix} x_1 \\ x_2 \\ x_3 \end{bmatrix}.$$

Is the system controllable and observable?

**3.15.** Consider the system

$$\frac{d}{dt}\begin{bmatrix} x_1 \\ x_2 \\ x_3 \end{bmatrix} = \begin{bmatrix} 2 & 0 & 0 \\ 0 & 2 & 0 \\ 0 & 3 & 1 \end{bmatrix}\begin{bmatrix} x_1 \\ x_2 \\ x_3 \end{bmatrix},$$

$$y = \begin{bmatrix} 1 & 1 & 1 \end{bmatrix}\begin{bmatrix} x_1 \\ x_2 \\ x_3 \end{bmatrix}.$$

1. Show that the system is not observable.
2. Show that the system is observable if the output is given by

$$\begin{bmatrix} y_1 \\ y_2 \end{bmatrix} = \begin{bmatrix} 1 & 1 & 1 \\ 1 & 2 & 3 \end{bmatrix}\begin{bmatrix} x_1 \\ x_2 \\ x_3 \end{bmatrix}.$$

**3.16.** Consider the system

$$\frac{dx}{dt} = Ax + Bu,$$

where

$$A = \begin{bmatrix} 0 & 1 & 0 \\ 0 & 0 & 1 \\ -1 & -5 & -6 \end{bmatrix}, \quad B = \begin{bmatrix} 0 \\ 1 \\ 1 \end{bmatrix}.$$

1. Show that the pair $(\mathbf{A}, \mathbf{B})$ is controllable.
2. Find a state feedback gain matrix $\mathbf{K}$ such that the closed-loop system with the feedback control $u = -\mathbf{K}\mathbf{x}$ has the closed-loop poles (eigenvalues) $\lambda = -2 \pm 4i$ and $\lambda = -10$.

**3.17.** Consider the system

$$\dot{\mathbf{x}} = \mathbf{A}\mathbf{x} + \mathbf{B}u,$$
$$y = \mathbf{C}\mathbf{x},$$

where

$$\mathbf{A} = \begin{bmatrix} 0 & 1 & 0 \\ 0 & 0 & 1 \\ -6 & -11 & -6 \end{bmatrix}, \quad \mathbf{B} = \begin{bmatrix} 0 \\ 0 \\ 1 \end{bmatrix}, \quad \mathbf{C} = \begin{bmatrix} 1 & 0 & 0 \end{bmatrix}.$$

Design an output feedback controller by observer approach such that the desired closed-loop poles (the eigenvalues of $\mathbf{A} - \mathbf{B}\mathbf{K}$) are located at

$$\lambda = -1 + i, -1 - i, -5$$

and the desired observer poles (the eigenvalues of $\mathbf{A} - \mathbf{K}_e\mathbf{C}$) are located at

$$\lambda = -6, -6, -6.$$

**3.18.** Consider the system

$$\frac{dx_1}{dt} = x_2,$$
$$\frac{dx_2}{dt} = -k_1 \sin x_1 - (k_2 + k_3 \cos x_1)x_2.$$

Assume that the parameters have the values $k_1 = 1$, $k_2 = 1$, $k_3 = 0$. Calculate the sensitivity indices of $x_1$ and $x_2$ with respect to $k_1, k_2, k_3$.

# References

1. Coppel W.A.: Stability and Asymptotic Behavior of Differential Equations. D. C. Heath and Co., Boston, Mass. (1966).
2. Khalil H.K.: Nonlinear Systems. Prentice Hall, New Jersey (2002).
3. Liu W.: Elementary Feedback Stabilization of the Linear Reaction-convection-diffusion Equations and the Wave Equations. Mathématiques et Applications **66**, Springer, New York (2010).
4. Morris K.A.: Introduction to Feedback Control. Academic Press, San Diego (2001).
5. Waltman P.: A Second Course in Elementary Differential Equations. Dover Publications, Inc., Mineola, New York (2004).

# Control of Blood Glucose

Some molecular control mechanisms of blood glucose are described schematically in Fig. 4.1. Glucose comes from food and liver, and is utilized by brain and nerve cells (insulin-independent) via the glucose transporter 3 (GLUT3) or by tissue cells such as muscle, kidney, and fat cells (insulin-dependent) via the glucose transporter 4 (GLUT4). Glucose is transported into and out of liver cells by the concentration-driven glucose transporter 2 (GLUT2), which is insulin-independent. In response to a low blood glucose level ($< 80$ mg/dl or 4.4 mmol/L), $\alpha$ cells of the pancreas produce the hormone glucagon. The glucagon initiates a series of activations of kinases, and finally leads to the activation of the glycogen phosphorylase, which catalyzes the breakdown of glycogen into glucose. In addition, the series of activations of kinases also result in the inhibition of glycogen synthase and then stop the conversion of glucose to glycogen. In response to a high blood glucose level ($> 120$ mg/dl or 6.7 mmol/L), $\beta$ cells of the pancreas secrete insulin. Insulin triggers a series of reactions to activate the glycogen synthase, which catalyzes the conversion of glucose into glycogen. Insulin also initiates a series of activations of kinases in tissue cells to lead to the redistribution of GLUT4 from intracellular storage sites to the plasma membrane. Once at the cell surface, GLUT4 transports glucose into the muscle or fat cells.

Mathematical models play an important role in addressing the problem of control of blood glucose. They can simulate oscillations of glucose and insulin and provide theoretical insights into the glucose control mechanisms. They are also required for integrating a glucose monitoring system into insulin pump technology to form a closed-loop insulin delivery system on the feedback of blood glucose levels, the so-called "artificial pancreas" [16, 33, 38]. Therefore, many minimal mathematical models describing interaction mechanisms between glucose and insulin have been constructed since the pioneering work of Albisser et al [3, 4] and Clemens et al [11]. These models include the linear model of Ackerman et al [2] and nonlinear compartmental models proposed by researchers, including Bergman et al [6, 7, 8] , Bertoldo et al [9], Li et al [25], Man et al [28, 29, 30], Pedersen et al [34], Sturis et al [39], and Toffolo et al [40, 41]. In this chapter, we introduce a model developed in [26, 27].

Liu W.: Introduction to Modeling Biological Cellular Control Systems.
DOI 10.1007/978-88-470-2490-8_4, © Springer-Verlag Italia 2012

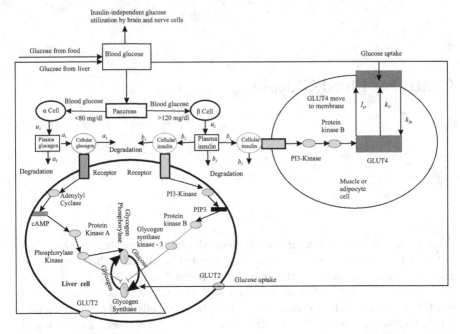

**Fig. 4.1.** A schematic description of a simple blood glucose control system. Molecular control mechanisms of blood glucose sketched in this figure are described in the text. Reproduced with permission from [26]

## 4.1 A Control System of Blood Glucose

We classify glucose into glucose in liver and glucose in plasma. Since there is an inter-conversion between glycogen and glucose in the liver, glycogen is taken into account. Let $g^y, g^l, g^p$ denote the concentrations of glycogen in the liver, glucose in the liver, and glucose in plasma, respectively. The differential equations governing $g^y$, $g^l$, and $g^p$ can be derived from the law of mass balance (2.1).

In the liver, the glycogen phosphorylase catalyzes conversion of glycogen into glucose and the glycogen synthase catalyzes conversion of glucose into glycogen. Experimental observations indicated that the glycogen phosphorylase [44, 45] and the glycogen synthase [31] follow the Michaelis-Menten equation. So the rate of conversion of glycogen into glucose is equal to $\frac{p_1 V_{max}^{gp} g^y}{K_m^{gp} + g^y}$ and the rate of conversion of glucose into glycogen is equal to $\frac{p_2 V_{max}^{gs} g^l}{K_m^{gs} + g^l}$, where $V_{max}^{gp}$, $V_{max}^{gs}$ are the maximum velocities of glycogen phosphorylase and glycogen synthase, respectively, $K_m^{gp}$, $K_m^{gs}$ are the Michaelis-Menten constants of glycogen phosphorylase and glycogen synthase, respectively, and $p_1$, $p_2$ are the percentages of activation of glycogen phosphorylase and glycogen synthase, respectively. From the view point of control theory, $p_1$ and $p_2$ serve as a feedback controller. It then follows from the law of mass balance (2.1) that

$$\frac{dg^y}{dt} = -\frac{p_1 V_{max}^{gp} g^y}{K_m^{gp} + g^y} + \frac{p_2 V_{max}^{gs} g^l}{K_m^{gs} + g^l}. \tag{4.1}$$

The glucose transporter GLUT2 transports glucose into or out of the live and follows the Michaelis-Menten equation [12]. Thus, the rate of transport of glucose into the liver can be modeled by $\frac{V_{max}^{g2} g^p}{K_m^{g2} + g^p}$ and the rate of transport of glucose from the liver into blood can be modeled by $\frac{V_{max}^{g2} g^l}{K_m^{g2} + g^l}$, where $V_{max}^{g2}$ is the maximum velocity of GLUT2 and $K_m^{g2}$ is the Michaelis-Menten constant. It then follows from the law of mass balance (2.1) that

$$\frac{dg^l}{dt} = -\frac{p_2 V_{max}^{gs} g^l}{K_m^{gs} + g^l} - \frac{V_{max}^{g2} g^l}{K_m^{g2} + g^l} + \frac{p_1 V_{max}^{gp} g^y}{K_m^{gp} + g^y} + \frac{V_{max}^{g2} g^p}{K_m^{g2} + g^p}. \tag{4.2}$$

The glucose transporter GLUT4 transports glucose into tissue cells and follows the Michaelis-Menten equation [32]. Thus, the rate of glucose transport from blood into the tissue cells can be modeled by $\frac{p_3 V_{max}^{g4} g^p}{K_m^{g4} + g^p}$, where $V_{max}^{g4}$ is the maximum velocity of GLUT4, $K_m^{g4}$ is the Michaelis-Menten constant, and $p_3$ is the percentage of GLUT4 on the cell surface. We lump all other insulin-independent glucose uptake into the glucose transporter GLUT3. Since GLUT3 follows the Michaelis-Menten equation [12], its glucose transport rate can be modeled by $\frac{V_{max}^{g3} g^p}{K_m^{g3} + g^p}$, where $V_{max}^{g3}$ is the maximum velocity of GLUT3 and $K_m^{g3}$ is the Michaelis-Menten constant. We denote by $g_{in}^p$ the exogenous glucose uptake from food. It then follows from the law of mass balance (2.1) that

$$\frac{dg^p}{dt} = -\frac{p_3 V_{max}^{g4} g^p}{K_m^{g4} + g^p} - \frac{V_{max}^{g2} g^p}{K_m^{g2} + g^p} - \frac{V_{max}^{g3} g^p}{K_m^{g3} + g^p} + \frac{V_{max}^{g2} g^l}{K_m^{g2} + g^l} + g_{in}^p. \tag{4.3}$$

The equations (4.1), (4.2), and (4.3) constitute a control system of blood glucose with the feedback controllers $p_1, p_2$, and $p_3$ to be designed.

## 4.2 Design of Feedback Controllers

Following the molecular control mechanisms of blood glucose, we design the feedback controllers $p_1, p_2$, and $p_3$ in this section.

### 4.2.1 Insulin and Glucagon Transition

Plasma insulin does not act directly on glucose metabolism, but through remote cellular insulin [8]. We assume that this is also the case for glucagon. The transition from plasma space to cellular space is described in Fig. 4.1. In this figure, $u_1$ stands for the glucagon infusion rate (GIR); $u_2$ stands for the insulin infusion rate (IIR); positive

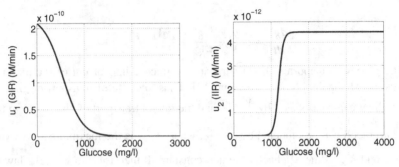

**Fig. 4.2.** *Left:* Glucagon infusion rate (4.6). *Right:* Insulin infusion rate (4.7). Reproduced with permission from [27]

constants $a$'s and $b$'s denote transition or degradation rate constants. Let $h_1^p$ and $h_2^p$ ($h$ for hormone and $p$ for plasma) denote the concentrations of the plasma glucagon and insulin, respectively. Then the transitional delay process can be modeled as follows:

$$\frac{dh_1^p}{dt} = -(a_1 + a_2)h_1^p + u_1, \tag{4.4}$$

$$\frac{dh_2^p}{dt} = -(b_1 + b_2)h_2^p + u_2. \tag{4.5}$$

Unlike the model proposed by Sturis *et al* [39], we assume that the intracellular insulin does not come back to the plasma space. This is similar to the model proposed by Bergman *et al* [8].

We use the following feedback glucagon and insulin infusion rates

$$u_1 = \frac{G_m}{1 + q_1 \exp(\alpha_1 (g^p - 1000))}, \tag{4.6}$$

$$u_2 = \frac{R_m}{1 + q_2 \exp(\alpha_2 (C_1 - g^p))}, \tag{4.7}$$

where $C_1, G_m, q_1, q_2, \alpha_1, \alpha_2, R_m$ are positive constants. $u_2$ is adopted from [39]. The constants $q_1, q_2, \alpha_1, \alpha_2$ are selected such that the glucagon secretion increases rapidly when the blood glucose level drops down to around 800 (mg/l) and the insulin secretion increases rapidly when the blood glucose level rises up to around 1500 (mg/l), as shown in Fig. 4.2.

### 4.2.2 Glucagon Signaling Pathway

For the glucagon signaling pathway, we treat it as a black box and propose a simple model as follows. We assume that the glucagon receptor recycling is a closed subsystem, that is, its synthesis is equal to its degradation. Let $H_1, R_1^i$, and $R_1$ denote the cellular glucagon, the inactive and active glucagon receptors, respectively. Then the biochemical process of activation of the receptors by the glucagon can be described by the diagram

$$H_1 + R_1^i \xrightarrow{a_4} R_1 \xrightarrow{a_5} R_1^i,$$

where $a$'s are rate constants. It then follows from the law of mass balance and the law of mass action that

$$\frac{dh_1}{dt} = -a_4 h_1 (R_1^0 - r_1) - a_3 h_1 + \frac{a_1 V_p}{V} h_1^p, \qquad (4.8)$$

$$\frac{dr_1}{dt} = a_4 h_1 (R_1^0 - r_1) - a_5 r_1, \qquad (4.9)$$

where lower case letters denote the concentrations of the corresponding molecules. In the above equations, $R_1^0$ is the total concentration of glucagon receptors, $V_p$ is the plasma glucagon volume, and $V$ is the cellular glucagon volume. We have used the receptor conserved equation

$$r_1 + r_1^i = R_1^0.$$

The term $a_1 h_1^p V_p$ is the total amount of the plasma glucagon per unit time transferred to the intracellular space and then $a_1 h_1^p V_p / V$ is the concentration of the cellular glucagon per unit time.

### 4.2.3 Insulin Signaling Pathway

A molecular mathematical model for the insulin signaling pathway in muscle cells was developed by Sedaghat *et al* [35]. We assume that the insulin signaling pathway in the liver cells is analogous to the one in the muscle cells. Thus we adopt this model for both muscle and liver cells. Since muscle and liver are two different organs, our assumption needs to be justified biologically.

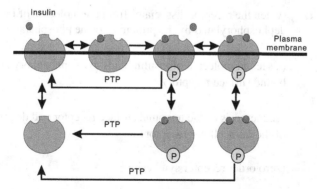

**Fig. 4.3.** Schematic description of insulin receptor binding. The binding and phosphorylation precesses are explained in the text

### 4.2.3.1 Insulin Receptor Recycling Subsystem

The binding process of insulin to its receptor is schematically descried in Fig. 4.3. Once a molecule of insulin binds to an insulin receptor, it is rapidly autophosphory-lated [1]. The phosphorylated receptor may either bind another molecule of insulin or dissociate from the first molecule of insulin. Binding of a second molecule of insulin does not affect the phosphorylation state of the receptor. However, the receptor is dephosphorylated by protein tyrosine phosphatases (PTP) when it dissociates from the first molecule of insulin. Free and phosphorylated surface receptors can be internalized. When they dissociate from the molecules of insulin, they are dephosphorylated by the protein tyrosine phosphatases.

To build a model for the subsystem, we introduce the following state variables:

$x_1^j$ = cellular insulin,

$x_2^j$ = concentration of unbound surface insulin receptors,

$x_3^j$ = concentration of unphosphorylated once-bound surface receptors,

$x_4^j$ = concentration of phosphorylated twice-bound surface receptors,

$x_5^j$ = concentration of phosphorylated once-bound surface receptors,

$x_6^j$ = concentration of unbound unphosphorylated intracellular receptors,

$x_7^j$ = concentration of phosphorylated twice-bound intracellular receptors,

$x_8^j$ = concentration of phosphorylated once-bound intracellular receptors.

The superscript $j$ is used to distinguish the insulin signaling pathways in liver and muscle: $j = l$ (l for liver) refers to the pathway in the liver while $j = m$ (m for muscle) refers to the pathway in the muscle.

The reactions between different receptor states are described as follows:

$$x_1^j + x_2^j \underset{k_{15}}{\overset{k_1}{\rightleftharpoons}} x_3^j \qquad \text{(one molecule of insulin binds to the unphosphorylated surface receptor)}$$

$$x_3^j \xrightarrow{k_3} x_5^j \qquad \text{(the once-bound surface receptor is phosphorylated)}$$

$$x_2^j \xleftarrow[k_{17}[PTP]]{} x_5^j \qquad \text{(when the receptor dissociates from the molecule of insulin, it is dephosphorylated by the protein tyrosine phosphatases (PTP))}$$

$$x_1^j + x_5^j \underset{k_{16}}{\overset{k_2}{\rightleftharpoons}} x_4^j \qquad \text{(a second molecule of insulin binds to the phosphorylated once-bound surface receptor)}$$

$$x_2^j \underset{k_{18}}{\overset{k_4}{\rightleftharpoons}} x_6^j \xrightarrow{k_{21}} \qquad \text{(endocytosis of the unbound surface receptor and degradation of the intracellular receptor)}$$

$$\xrightarrow{k_5} x_6^j \qquad \text{(zero order receptor synthesis)}$$

$$x_4^j \underset{k_{20}}{\overset{k_{19}}{\rightleftharpoons}} x_7^j \qquad \text{(endocytosis of the phosphorylated twice-bound surface receptor)}$$

$$x_5^j \underset{k_{20}}{\overset{k_{19}}{\rightleftharpoons}} x_8^j \qquad \text{(endocytosis of the phosphorylated once-bound surface receptor)}$$

$$x_6^j \xleftarrow[k_6[PTP]]{} x_7^j \qquad \text{(when insulin diffuses off the twice-bound intracellular receptor,}$$
the protein tyrosine phosphatases (PTP) dephosphorylate it)

$$x_6^j \xleftarrow[k_6[PTP]]{} x_8^j \qquad \text{(when insulin diffuses off the once-bound intracellular receptor,}$$
the protein tyrosine phosphatases (PTP) dephosphorylate it).

The protein tyrosine phosphatases that dephosphorylate the insulin receptor are explicitly represented as a multiplicative factor ([PTP]) that modulates receptor dephosphorylation rate. Using the mass action law and the mass balance law, the governing equations for the state variables can be derived from the above reactions as follows (see [35]):

$$\frac{dx_1^j}{dt} = -k_1 x_1^j x_2^j - b_3 x_1^j + b_1 V_p h_2^p / V \tag{4.10}$$

$$\frac{dx_2^j}{dt} = k_{15} x_3^j + k_{17}[PTP] x_5^j - k_1 x_1^j x_2^j + k_{18} x_6^j - k_4 x_2^j \tag{4.11}$$

$$\frac{dx_3^j}{dt} = -k_{15} x_3^j + k_1 x_1^j x_2^j - k_3 x_3^j \tag{4.12}$$

$$\frac{dx_4^j}{dt} = k_2 x_1^j x_5^j - k_{16} x_4^j + k_{20} x_7^j - k_{19} x_4^j \tag{4.13}$$

$$\frac{dx_5^j}{dt} = k_3 x_3^j + k_{16} x_4^j - k_2 x_1^j x_5^j - k_{17}[PTP] x_5^j + k_{20} x_8^j - k_{19} x_5^j \tag{4.14}$$

$$\frac{dx_6^j}{dt} = k_5 - k_{21} x_6^j + k_6[PTP](x_7^j + x_8^j) + k_4 x_2^j - k_{18} x_6^j \tag{4.15}$$

$$\frac{dx_7^j}{dt} = k_{19} x_4^j - k_{20} x_7^j - k_6[PTP] x_7^j \tag{4.16}$$

$$\frac{dx_8^j}{dt} = k_{19} x_5^j - k_{20} x_8^j - k_6[PTP] x_8^j. \tag{4.17}$$

### 4.2.3.2 Postreceptor Signaling Subsystem

It is assumed that the postreceptor signaling subsystem is a closed subsystem. Activated insulin receptors phosphorylate the insulin receptor substrate-1 (IRS-1), which then binds and activates PI 3-kinase (Fig. 4.4). Define the following state variables:

$x_9^j =$ concentration of unphosphorylated IRS-1,
$x_{10}^j =$ concentration of tyrosine-phosphorylated IRS-1,
$x_{11}^j =$ concentration of inactivated PI 3-kinase,
$x_{12}^j =$ concentration of tyrosine-phosphorylated IRS-1/activated PI 3-kinase complex.

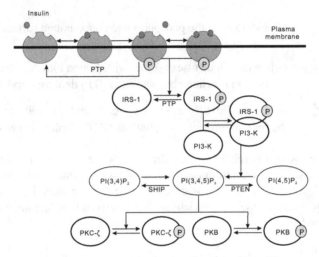

**Fig. 4.4.** Schematic description of postreceptor signaling subsystem. The postreceptor signaling precess is explained in the text

The reactions in this system are described as follows:

$$x_9^j \xrightleftharpoons[k_{22}[PTP]]{k_7(x_4^j+x_5^j)/[IR_p]} x_{10}^j \qquad \text{(phosphorylation of IRS-1 by activated insulin receptors and dephosphorylation of IRS-1 by the protein tyrosine phosphatases (PTP))}$$

$$x_{10}^j + x_{11}^j \xrightleftharpoons[k_{23}]{k_8} x_{12}^j \qquad \text{(phosphorylated IRS-1 binds and activates PI 3-kinase).}$$

The dependence of IRS-1 phosphorylation on phosphorylated surface receptors is assumed to be proportional to the fraction of phosphorylated surface receptors $(x_4^j+x_5^j)/[IR_p]$, where $[IR_p]$ is the concentration of phosphorylated surface receptors achieved after maximal insulin stimulation. Thus the rate constant for IRS-1 phosphorylation is given by $k_7(x_4^j+x_5^j)/[IR_p]$. It then follows that differential equations governing phosphorylation of IRS-1 and subsequent formation of phosphorylated IRS-1/activated PI 3-kinase complex are (see [35]):

$$\frac{dx_9^j}{dt} = k_{22}[\text{PTP}]x_{10}^j - k_7 x_9^j(x_4^j+x_5^j)/[IR_p], \qquad (4.18)$$

$$\frac{dx_{10}^j}{dt} = k_7 x_9^j(x_4^j+x_5^j)/[IR_p] + k_{23}x_{12}^j - (k_{22}[\text{PTP}] + k_8 x_{11}^j)x_{10}^j, \qquad (4.19)$$

$$\frac{dx_{11}^j}{dt} = k_{23}x_{12}^j - k_8 x_{11}^j x_{10}^j, \qquad (4.20)$$

$$\frac{dx_{12}^j}{dt} = -k_{23}x_{12}^j + k_8 x_{11}^j x_{10}^j. \qquad (4.21)$$

Activated PI3-kinase converts the substrate phosphatidylinositol 4,5-bisphosphate (PI(4,5)P$_2$) to the product phosphatidylinositol 3,4,5-trisphosphate (PI(3,4,5)P$_3$) (Fig. 4.4).  Phosphoinositol phosphatases such as SHIP2 (SH2-containing 5'-inositol phosphatase) convert PI(3,4,5)P$_3$ to phosphatidylinositol 3,4-bisphosphate (PI(3,4)P$_2$), whereas phosphoinositol phosphatases such as PTEN (phosphatase homologous to tensin ) convert PI(3,4,5)P$_3$ to PI(4,5)P$_2$. Define:

$$x_{13}^j = \text{percentage of PI(3,4,5)P3 out of the total lipid population,}$$

$$x_{14}^j = \text{percentage of PI(4,5)P2 out of the total lipid population,}$$

$$x_{15}^j = \text{percentage of PI(3,4)P2 out of the total lipid population.}$$

These conversion reactions are described as follows:

$$x_{14}^j \underset{k_{24}[PTEN]}{\overset{k_9}{\rightleftharpoons}} x_{13}^j \quad \text{(conversion between PI(4,5)P}_2 \text{ and PI(3,4,5)P}_3$$
$$\text{by PI3-kinase and PTEN)}$$

$$x_{15}^j \underset{k_{25}[SHIP]}{\overset{k_{10}}{\rightleftharpoons}} x_{13}^j \quad \text{(conversion between PI(3,4)P}_2 \text{ and PI(3,4,5)P}_3 \text{ by SHIP2).}$$

It is assumed that the rate constant $k_9$ for generation of PI(3,4,5)P$_3$ depends on $x_{12}$ linearly as follows

$$k_9 = (k_{9stimulated} - k_{9basal})x_{12}/[PI3K] + k_{9basal}, \qquad (4.22)$$

where $[PI3K]$ is the equilibrium concentration of activated PI3-kinase obtained after maximal insulin stimulation. As with $[PTP]$, the lipid phosphatase factors $[SHIP]$ and $[PTEN]$ correspond to the relative phosphatase activity in the cell and are assigned a value of 1 under normal physiological conditions. It then follows that differential equations governing these phosphatidylinositides are (see [35]):

$$\frac{dx_{13}^j}{dt} = k_9 x_{14}^j + k_{10} x_{15}^j - (k_{24}[\text{PTEN}] + k_{25}[\text{SHIP}])x_{13}^j, \qquad (4.23)$$

$$\frac{dx_{14}^j}{dt} = k_{24}[\text{PTEN}]x_{13}^j - k_9 x_{14}^j, \qquad (4.24)$$

$$\frac{dx_{15}^j}{dt} = k_{25}[\text{SHIP}]x_{13}^j - k_{10} x_{15}^j. \qquad (4.25)$$

PI(3,4,5)P$_3$ mediates activation of downstream kinases PKB (protein kinase B) and PKC-$\zeta$ (protein kinase C).  Define:

$$x_{16}^j = \text{percentage of inactivated PKB,}$$

$$x_{17}^j = \text{percentage of activated PKB,}$$

$$x_{18}^j = \text{percentage of inactivated PKC-}\zeta,$$

$$x_{19}^j = \text{percentage of activated PKC-}\zeta.$$

The activation of PKB and PKC-$\zeta$ can be described as follows:

$$x_{16}^j \underset{k_{26}}{\overset{k_{11}}{\rightleftharpoons}} x_{17}^j \quad \text{(activation of PKB mediated by PI(3,4,5)P}_3)$$

$$x_{18}^j \underset{k_{27}}{\overset{k_{12}}{\rightleftharpoons}} x_{19}^j \quad \text{(activation of PKC-}\zeta \text{ mediated by PI(3,4,5)P}_3).$$

The differential equations describing the activation are (see [35]):

$$\frac{dx_{16}^j}{dt} = k_{26}x_{17}^j - k_{11}x_{16}^j, \tag{4.26}$$

$$\frac{dx_{17}^j}{dt} = -k_{26}x_{17}^j + k_{11}x_{16}^j, \tag{4.27}$$

$$\frac{dx_{18}^j}{dt} = k_{27}x_{19}^j - k_{12}x_{18}^j, \tag{4.28}$$

$$\frac{dx_{19}^j}{dt} = -k_{27}x_{19}^j + k_{12}x_{18}^j. \tag{4.29}$$

We assume that, with maximal insulin stimulation, both PKB and PKC-$\zeta$ exist in a 10:1 unactivated-to-activated distribution at equilibrium. It therefore follows from the equations (4.26) and (4.28) that

$$k_{11} = 0.1k_{26}, \quad k_{12} = 0.1k_{27}.$$

To model the PI(3,4,5)P$_3$ mediation in activation of PKB and PKC-$\zeta$, we assume that the rate constants for activation of both PKB and PKC-$\zeta$ increase from zero to their maximal values as a linear function of the increase in PI(3,4,5)P$_3$ levels:

$$k_{11} = 0.1 \times k_{26} \times (x_{13} - 0.31)/(3.1 - 0.31), \tag{4.30}$$
$$k_{12} = 0.1 \times k_{27} \times (x_{13} - 0.31)/(3.1 - 0.31), \tag{4.31}$$

where 0.31 is the basal value of PI(3,4,5)P$_3$ and 3.10 is the value of PI(3,4,5)P$_3$ in the cell after maximal insulin stimulation.

With 80% of the metabolic insulin signaling effect attributed to PKC-$\zeta$ and 20% of the effect attributed to PKB, the insulin effect can be represented by (see [35])

$$\text{effect} = (0.2x_{17}^j + 0.8x_{19}^j)/AP_{equil}, \tag{4.32}$$

where $AP_{equil} = 100/11$ is the steady-state level of combined activity for PKB and PKC-$\zeta$ after maximal insulin stimulation.

### 4.2.3.3 GLUT4 Transport Subsystem

Under basal conditions, GLUT4 recycles between an intracellular compartment and the cell surface. Insulin stimulation leads to additional GLUT4 trafficking from the intracellular compartment to the cell surface. Define

$$x_{20}^m = \text{percentage of intracellular GLUT4,}$$
$$x_{21}^m = \text{percentage of cell surface GLUT4.}$$

The GLUT4 recycling can be described as follows:

$$x_{20}^m \underset{k_{28}}{\overset{k_{13}+I_{g4}}{\rightleftharpoons}} x_{21}^m \quad \text{(GLUT4 recycling)}$$

$$\xrightarrow{k_{14}} x_{20}^m \xrightarrow{k_{29}} \qquad \text{(zero order GLUT4 synthesis and first order GLUT4 degradation),}$$

where $I_{g4}$ denotes the insulin effect on GLUT4. Thus the GLUT4 recycling is modeled by (see [35]):

$$\frac{dx_{20}^m}{dt} = k_{28}x_{21}^m - (k_{13}+I_{g4})x_{20}^m - k_{29}x_{20}^m + k_{14}, \tag{4.33}$$

$$\frac{dx_{21}^m}{dt} = -k_{28}x_{21}^m + (k_{13}+I_{g4})x_{20}^m. \tag{4.34}$$

We assume that the insulin effect on GLUT4, $I_{g4}$, depends on its effect linearly:

$$I_{g4} = I_{max}[\text{effect}],$$

where $I_{max}$ is the effect after maximal insulin stimulation. By assuming that the basal equilibrium distribution of 4% cell surface GLUT4 and 96% GLUT4 in the intracellular pool transitions on maximal insulin stimulation to a new steady state of 40% cell surface GLUT4 and 60% intracellular GLUT4, we obtain

$$\frac{k_{13}}{k_{28}} = \frac{4}{96}, \quad \frac{k_{13}+I_{max}}{k_{28}} = \frac{40}{60}.$$

It then follows that $I_{max} = \left(\frac{40}{60} - \frac{4}{96}\right)k_{28}$ and then

$$I_{g4} = \left(\frac{40}{60} - \frac{4}{96}\right)k_{28}[\text{effect}]$$

$$= \left(\frac{40}{60} - \frac{4}{96}\right)k_{28}(0.2x_{17}^m + 0.8x_{19}^m)/AP_{equil}. \tag{4.35}$$

## 4.2.4 Dynamical Feedback Controllers

We assume that the activation level of glycogen phosphorylase is determined by the percentage of activated glucagon receptors. This leads to the feedback controller

$$p_1 = \frac{r_1}{R_1^0}. \tag{4.36}$$

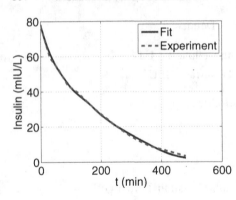

**Fig. 4.5.** Fitting the solution of (4.5) into the insulin data of Korach-André *et al* [21] (courtesy of François Péronnet). In this fitting, the glucose data of Fig. 4.7 (left) are used for $g^p$. Reproduced with permission from [26]

The activation effect on the glycogen synthase resulted from insulin is determined by (see [35])

$$I_{gs} = 0.2x_{17}^l + 0.8x_{19}^l, \tag{4.37}$$

where $x_{17}^l$ is the percentage of activated PKB and $x_{19}^l$ is the percentage of activated PKC-$\zeta$. The inactivation of glycogen synthase by glucagon is assumed to be determined by the reciprocal of the concentration of activated glucagon receptors $1/(1 + a_6 r_1)$, where $a_6$ is a positive inhibition constant. Then the control of activity of the glycogen synthase can be expressed as

$$p_2 = \frac{I_{gs}}{1 + a_6 r_1} = \frac{0.2x_{17}^l + 0.8x_{19}^l}{1 + a_6 r_1}. \tag{4.38}$$

Since only the cell surface GLUT4 transports calcium, we have

$$p_3 = x_{21}^m. \tag{4.39}$$

## 4.3 Estimation of Parameters

As shown in the above mathematical model, a mathematical model of a biological system usually contains numerous parameters. Estimation of these parameters is challenging because no universal methods can be used and in some cases no complete sets of data are available for fitting a model into data. Hence we usually have to appeal to different means, such as adoption of existing parameters in the literature, numerical simulations, use of optimization methods like the simulated annealing optimization [10, 20] and nonlinear least squares, and use of built-in parameter estimation functions like "SBparameterestimation" from the SBToolbox [47] and "sbioparamestim" from the SimBiology toolbox of MATLAB.

We now demonstrate how to fit a model into data to estimate parameters. Time-dependent insulin data was obtained by Korach-Andre *et al* [21]. We use these data to estimate the parameters $b_1, b_2, C_1, \alpha_2, q_2$, and $R_m$ by fitting the solution of (4.5) into these data. To this end, we solve (4.5) to obtain that

$$h_2^p = e^{-(b_1+b_2)t} h_2^p(0) + e^{-(b_1+b_2)t} \int_0^t \frac{R_m e^{-(b_1+b_2)s}}{1 + q_2 \exp(\alpha_2(C_1 - g^p(s)))} ds.$$

Using the method of least squares, we then fit this solution into the insulin data of Fig. 4.7 (right), as shown in Fig. 4.5. In this fitting, the glucose data of Fig. 4.7 (left) are used for $g^p$. For the parameters $b_1$ and $b_2$, we can obtain only the sum $b_1 + b_2 = 0.0171$ min$^{-1}$ and the determination of each of $b_1$ and $b_2$ becomes a problem because we could not find the experimental rate $b_2$ of plasma insulin degradation in the literature. Part of the reason is that the degradation site is not on plasma membrane. There is an intracellular pool of insulin in beta cells, which is delivered to the plasma membrane of a cell upon certain stress. After the action of insulin in target cells, the cells will uptake it via endocytosis and degrade it by endosome and lysosomes. Therefore, the overall degradation rate could be approximately equal to $b_3$ and then $b_2$ is small. Hence we tried different combinations of $b_1$ and $b_2$ and found that the combination of $b_1 = 0.0151$ min$^{-1}$ and $b_2 = 0.002$ min$^{-1}$ (much less than $b_3 = 0.01$ min$^{-1}$) gives the best fit of the simulated glucose and insulin time courses into the data as shown in Fig. 4.7. Since the unit of the insulin data is mIU/L, with the insulin conversion 1 mIU/L $=6.945 \times 10^{-12}$ M [46], the obtained value 0.64 (mIU/L/min) of $R_m$ is converted to

$$R_m = 0.64 \times 6.945 \times 10^{-12} \text{ M/min} = 4.44 \times 10^{-12} \text{ M/min}.$$

The glucagon degradation rate $a_3$ is assumed to be equal to the insulin degradation rate $b_3$. The deactivation effect constant $a_6$ of glucagon receptors on the glycogen synthase and the maximum velocities $V_{max}^{g2}, V_{max}^{g3}, V_{max}^{g4}, V_{max}^{gs}, V_{max}^{gp}$ for various enzymes are determined by hand in the way such that the simulated glucose and insulin profiles fit best into the data of Korach-Andre et al [21] (Fig. 4.7). Since the maximum velocities depend on the total concentration of an enzyme, they could vary from experiment to experiment. Thus we did not use the reported values in the literature. In fact, the reported values resulted in a worse simulation by our model (simulation not shown). The original unit mM of various Michaelis-Menten constants $K_m^{g2}, K_m^{g3}, K_m^{g4}, K_m^{gs}, K_m^{gp}$ is converted to mg/L. All other parameters are adopted from the literature as indicated in Tables 4.1 and 4.2. As always, the reliability of these parameter values is questionable biologically since they were determined by numerical simulations and data fittings, not by experiments.

## 4.4 Simulation of Glucose and Insulin Dynamics

Using the feedback control system consisting of equations (4.1)-(4.39), we reproduce qualitatively the experimental results of Korach-André et al [21] (courtesy of François Péronnet). The parameter values used in this reproduction are listed in Tables 4.1 and 4.2 and the initial data are listed in Table 4.3. The exogenous glucose input $(g_{in}^p)$ is the experimental data of Korach-André et al [21], which is reproduced

**Table 4.1.** Model parameters adapted from [35]

| Parameter | Value |
| --- | --- |
| PTP | 1.0 |
| PTEN | 1.0 |
| SHIP | 1.0 |
| $IR_p$ | $8.97 \times 10^{-13}$ (M) |
| $AP_{equil}$ | $100/11$ |
| PI3K | $5 \times 10^{-15}$ (M) |
| $k_{9stimulated}$ | $1.39$ (min$^{-1}$) |
| $k_{9basal}$ | $(0.31/99.4) \times 94/3.1 \times k_{9stimulated}$ |
| $k_1$ | $6 \times 10^7$ (M$^{-1}$min$^{-1}$) |
| $k_2$ | $6 \times 10^7$ (M$^{-1}$min$^{-1}$) |
| $k_3$ | $2500$ (min$^{-1}$) |
| $k_4$ | $0.003/9$ (min$^{-1}$) |
| $k_5$ | $10 \times k_{21}$ (M min$^{-1}$) if $x_6 + x_7 + x_8 > 10^{-13}$ |
| $k_5$ | $60 \times k_{21}$ (M min$^{-1}$) if $x_6 + x_7 + x_8 \leq 10^{-13}$ |
| $k_6$ | $0.461$ (min$^{-1}$) |
| $k_7$ | $4.16$ ( min$^{-1}$) |
| $k_8$ | $10 \times 5/70.775 \times 10^{12}$ ( min$^{-1}$) |
| $k_9$ | $(k_{9stimulated} - k_{9basal}) \times x_{12}/PI3K + k_{9basal}$ |
| $k_{10}$ | $3.1/2.9 \times 2.77$ ( min$^{-1}$) |
| $k_{11}$ | $0.1 \times k_{26} \times (x_{13} - 0.31)/(3.1 - 0.31)$ |
| $k_{12}$ | $0.1 \times k_{27} \times (x_{13} - 0.31)/(3.1 - 0.31)$ |
| $k_{13}$ | $4/96 \times 0.167$ (min$^{-1}$) |
| $k_{14}$ | $96 \times 0.001155$ (min$^{-1}$) |
| $k_{15}$ | $0.2$ (min$^{-1}$) |
| $k_{16}$ | $20$ (min$^{-1}$) |
| $k_{17}$ | $0.2$ (min$^{-1}$) |
| $k_{18}$ | $0.003$ (min$^{-1}$) |
| $k_{19}$ | $2.1 \times 10^{-3}$ (min$^{-1}$) |
| $k_{20}$ | $2.1 \times 10^{-4}$ (min$^{-1}$) |
| $k_{21}$ | $1.67 \times 10^{-18}$ (min$^{-1}$) |
| $k_{22}$ | $2.5/7.45 \times 4.16$ (min$^{-1}$) |
| $k_{23}$ | $10$ (min$^{-1}$) |
| $k_{24}$ | $94/3.1 \times k_{9stimulated}$ |
| $k_{25}$ | $2.77$ (min$^{-1}$) |
| $k_{26}$ | $10 \times \ln(2)$ (min$^{-1}$) |
| $k_{27}$ | $10 \times \ln(2)$ (min$^{-1}$) |
| $k_{28}$ | $0.167$ (min$^{-1}$) |
| $k_{29}$ | $0.001155$ (min$^{-1}$) |
| effect | $(0.2 \times x_{17} + 0.8 \times x_{19})/AP_{equil}$ |
| $I_{g4}$ | $(4/6 - 4/96) \times k_{28} \times$ effect |

**Table 4.2.** Model parameters

| Parameter | Value | Description |
|---|---|---|
| $K_m^{g2}$ | 15300 (mg/L) | Michaelis-Menten constant of GLUT2 [12] |
| $K_m^{g3}$ | 1530 (mg/L) | Michaelis-Menten constant of GLUT3 [12] |
| $K_m^{g4}$ | 774 (mg/L) | Michaelis-Menten constant of GLUT4 [32] |
| $K_m^{gs}$ | 67 (mg/L) | Michaelis-Menten constant of glycogen synthase [31] |
| $K_m^{gp}$ | 600 (mg/L) | Michaelis-Menten constant of Glycogen phosphorylase [45] |
| $V_{max}^{g2}$ | 50 (mg/L/min) | Maximum velocity of GLUT2 |
| $V_{max}^{g3}$ | 10 (mg/L/min) | Maximum velocity of GLUT3 |
| $V_{max}^{g4}$ | 4 (mg/L/min) | Maximum velocity of GLUT4 |
| $V_{max}^{gs}$ | 28.4 (mg/L/min) | Maximum velocity of glycogen synthase |
| $V_{max}^{gp}$ | 258 (mg/L/min) | Maximum velocity of Glycogen phosphorylase |
| $R_1^0$ | $9 \times 10^{-13}$ (M) | Total glucagon receptors [27] |
| $a_1$ | 0.12 (min$^{-1}$) | Plasma glucagon transitional rate [27] |
| $a_2$ | 0.3 (min$^{-1}$) | Plasma glucagon degradation rate [5, 13, 24] |
| $a_3$ | 0.01 (min$^{-1}$) | Glucagon degradation rate |
| $a_4$ | $6 \times 10^7$ (M$^{-1}$min$^{-1}$) | Glucagon association rate to its receptors [18] |
| $a_5$ | 0.2 ( min$^{-1}$) | Glucagon receptor deactivation rate [27] |
| $a_6$ | $10^3 / R_1^0$ (M$^{-1}$) | Deactivation effect of glucagon receptor on glycogen synthase |
| $b_1$ | 0.0151 (min$^{-1}$) | Plasma insulin transitional rate |
| $b_2$ | 0.002 (min$^{-1}$) | Plasma insulin degradation rate |
| $b_3$ | 0.01 (min$^{-1}$) | Insulin degradation rate [39] |
| $V$ | 11 (l) | Volume of cellular insulin space [42] |
| $V_p$ | 3 (l) | Volume of plasma insulin space [42] |
| $G_m$ | $2.23 \times 10^{-10}$ (M/min) | Maximum glucagon infusion rate [27] |
| $R_m$ | $4.44 \times 10^{-12}$ (M/min) | Maximum insulin infusion rate |
| $C_1$ | 1162 (mg/L) | |
| $\alpha_1$ | 0.005 ((mg/L)$^{-1}$) | [27] |
| $\alpha_2$ | 0.01594 ((mg/L)$^{-1}$) | |
| $q_1$ | 10 | [27] |
| $q_2$ | 1.85 | |

in Fig. 4.6. The data is converted into the glucose input rate (mg/L/min) by dividing it by 0.23 (L/kg), the effective volume of glucose distribution [21]. The model is solved numerically by using MATLAB.

Fig. 4.7 shows that the glucose and insulin dynamics simulated by the model agrees qualitatively with the experimental data, although they do not match perfectly. The difference between simulation and data could be resulted from a number of factors. First, definitely the model is just a phenomenological one and its accuracy needs to be improved. Second, due to the complexity of the model, the values of numerous parameters cannot be estimated accurately by fitting the model into the data. Third, the plasma glucose and insulin dynamics depends critically on the exogenous glucose

**Table 4.3.** Initial conditions

| Variable | Reference | Variable | Reference |
|---|---|---|---|
| $h_1^p(0) = 1.389 \times 10^{-11}$ M | | $h_2^p(0) = 5.38 \times 10^{-10}$ M | |
| $h_1(0) = 0$ | | $r_1(0) = 0$ | |
| $x_1^m(0) = 0$ | | $x_2^m(0) = 9 \times 10^{-13}$ M | [35] |
| $x_3^m(0) = 0$ | | $x_4^m(0) = 0$ | |
| $x_5^m(0) = 0$ | | $x_6^m(0) = 1 \times 10^{-13}$ M | [35] |
| $x_7^m(0) = 0$ | | $x_8^m(0) = 0$ | |
| $x_9^m(0) = 1 \times 10^{-12}$ M | [35] | $x_{10}^m(0) = 0$ | |
| $x_{11}^m(0) = 1 \times 10^{-13}$ M | [35] | $x_{12}^m(0) = 0$ | |
| $x_{13}^m(0) = 0.31$ | [35] | $x_{14}^m(0) = 99.4$ | [35] |
| $x_{15}^m(0) = 0.29$ | [35] | $x_{16}^m(0) = 100$ | [35] |
| $x_{17}^m(0) = 0$ | | $x_{18}^m(0) = 100$ | [35] |
| $x_{19}^m(0) = 0$ | [35] | $x_{20}^m(0) = 90$ | [35] |
| $x_{21}^m(0) = 10$ | [35] | $x_1^l(0) = 0$ | |
| $x_2^l(0) = 9 \times 10^{-13}$ M | | $x_3^l(0) = 0$ | |
| $x_4^l(0) = 0$ | | $x_5^l(0) = 0$ | |
| $x_6^l(0) = 1 \times 10^{-13}$ M | | $x_7^l(0) = 0$ | |
| $x_8^l(0) = 0$ | | $x_9^l(0) = 1 \times 10^{-12}$ M | |
| $x_{10}^l(0) = 0$ | | $x_{11}^l(0) = 1 \times 10^{-13}$ M | |
| $x_{12}^l(0) = 0$ | | $x_{13}^l(0) = 0.31$ | |
| $x_{14}^l(0) = 99.4$ | | $x_{15}^l(0) = 0.29$ | |
| $x_{16}^l(0) = 100$ | | $x_{17}^l(0) = 0$ | |
| $x_{18}^l(0) = 100$ | | $x_{19}^l(0) = 0$ | |
| $g^y(0) = 200$ mg/L | | $g^l(0) = 0$ | |
| $g^p(0) = 1935$ mg/L | | $g_{in}^p(0) = 25.9$ mg/L/min | [21] |

input ($g_{in}^p$). However, there might be errors in measuring the exogenous glucose input in the experiment and then the actual exogenous glucose that resulted in the plasma glucose and insulin data might be different.

## 4.5 Model Prediction of Parametrical Sensitivity

We now analyze the sensitivity of glucose and insulin to parameters. The sensitivity index of a state variable $y$ with respect to a parameter $p$ is defined by $\frac{p}{y}\frac{\partial y}{\partial p}$ [19, 43]. The sensitivity system is solved by using MATLAB over the time period from 0 to 480 and the averaged sensitivity index is calculated according to the formula $\frac{1}{480}\int_0^{480}\left|\frac{p}{y}\frac{\partial y}{\partial p}\right|dt$. The simulated indices are listed in Tables 4.4 and 4.5 in a descend-

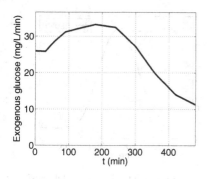

**Fig. 4.6.** Exogenous glucose input rate from the experimental data of Korach-André *et al* [21] (courtesy of François Péronnet)

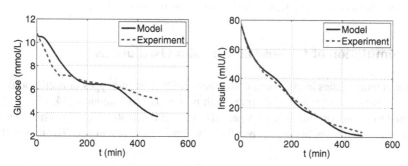

**Fig. 4.7.** Comparison of blood glucose (*left*) and insulin (*right*) dynamics simulated by the model consisting of equations (4.1)-(4.39) with the experimental data of Korach-André *et al* [21] (courtesy of François Péronnet). Reproduced with permisson from [26]

ing order and they show that both glucose and insulin are most sensitive to the parameters PTEN (the multiplicative factor modulating $k_{24}$ [35]), $k_{24}$ (the rate constant for conversion of PI(3,4,5)P$_3$ to PI(4,5)P$_2$ [35]), PTP (the multiplicative factor that modulates insulin receptor dephosphorylation rate [35]), and $V_{max}^{g4}$ (the maximum velocity of GLUT4). This numerical evidence predicts that PTEN that converts PI(3,4,5)P$_3$ to PI(4,5)P$_2$, PTP that dephosphorylates the insulin receptors, and GLUT4 that transports glucose into muscle cells play the most important role in the insulin signaling pathway.

Fig. 4.8 shows that the insulin sensitivity index with respect to the maximum velocity of GLUT4 is negative initially. This implies that an increase of the velocity will result in a decrease in the needs of plasma insulin. The figure also shows that the insulin sensitivity index with respect to the rate constant $k_{28}$ is positive. This implies that if the GLUT4 internalization rate increases, then more insulin is needed to stimulate the trafficking of GLUT4 from intracellular stores to the plasma membrane. Other simulated index figures with respect to other parameters (not shown) make a sense biologically. This might indicate that the model (4.1)-(4.39) could phenomenologically simulate the molecular mechanisms of interaction between glucose and insulin.

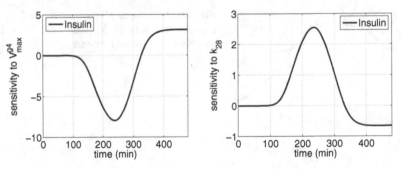

**Fig. 4.8.** Sensitivity indices of insulin to different parameters simulated by the model consisting of equations (4.1)-(4.39). Reproduced with permission from [26]

## 4.6  Simulation of Glucose and Insulin Oscillations

Experimental studies have revealed that there are at least two types of oscillations of glucose and insulin: rapid oscillations with periods of 8-15 minutes [14, 15, 17, 22, 23] and ultradian oscillations with periods of 50-200 minutes [36, 37]. The mechanisms that generate both types of oscillations remain to be elucidated. To analyze the

**Table 4.4.** Averaged glucose sensitivity indices to parameters

| Parameter | Index | Parameter | Index | Parameter | Index |
|-----------|-------|-----------|-------|-----------|-------|
| PTEN | 0.9039 | $k_{24}$ | 0.9038 | PTP | 0.8411 |
| $V_{max}^{g4}$ | 0.8293 | $k_{9stimulated}$ | 0.5315 | $AP_{equil}$ | 0.4859 |
| PI3K | 0.4812 | $k_8$ | 0.4781 | $k_{23}$ | 0.4778 |
| $IR_p$ | 0.4377 | $k_7$ | 0.4377 | $k_{22}$ | 0.4358 |
| $k_1$ | 0.4222 | $k_{9basal}$ | 0.4166 | $k_{17}$ | 0.4054 |
| $K_m^{g4}$ | 0.2805 | $k_{13}$ | 0.2570 | $k_{28}$ | 0.2535 |
| $b_1$ | 0.2182 | $R_m$ | 0.1903 | $b_3$ | 0.1837 |
| $V_{max}^{g2}$ | 0.1733 | $K_m^{g2}$ | 0.1575 | $V_{max}^{g3}$ | 0.1469 |
| $k_{14}$ | 0.0776 | $k_{29}$ | 0.0734 | $K_m^{g3}$ | 0.0729 |
| $q_2$ | 0.0369 | $b_2$ | 0.0311 | $\alpha_2$ | 0.0307 |
| $a_4$ | 0.0201 | $a_5$ | 0.0196 | $q_1$ | 0.0188 |
| $V_{max}^{gp}$ | 0.0163 | $\alpha_1$ | 0.015 | $a_1$ | 0.0145 |
| $a_2$ | 0.0142 | $k_{18}$ | 0.0132 | $a_3$ | 0.0115 |
| $k_4$ | 0.0127 | $V_{max}^{gs}$ | 0.0076 | $K_m^{gp}$ | 0.0073 |
| $a_6$ | 0.0063 | $k_{19}$ | 0.006 | SHIP | 0.0057 |
| $k_{10}$ | 0.0057 | $k_{25}$ | 0.0057 | $K_m^{gs}$ | 0.0046 |
| $k_{27}$ | 0.0005 | $k_5$ | 0.0002 | $k_2$ | 0.0001 |
| $k_{16}$ | 0.0001 | $k_{26}$ | 0.0001 | $k_3$ | 0.0 |
| $k_6$ | 0.0 | $k_{15}$ | 0.0 | $k_{20}$ | 0.0 |

**Table 4.5.** Averaged insulin sensitivity indices to parameters

| Parameter | Index | Parameter | Index | Parameter | Index |
|---|---|---|---|---|---|
| PTEN | 3.0638 | $k_{24}$ | 3.0638 | PTP | 2.6993 |
| $V_{max}^{g4}$ | 2.8045 | $k_{9stimulated}$ | 1.7166 | $AP_{equil}$ | 1.5881 |
| PI3K | 1.5542 | $k_8$ | 1.5442 | $k_{23}$ | 1.5428 |
| $IR_p$ | 1.4155 | $k_7$ | 1.4155 | $k_{22}$ | 1.4062 |
| $b_3$ | 1.3853 | $k_1$ | 1.3707 | $k_{9basal}$ | 1.3495 |
| $k_{17}$ | 1.2938 | $b_1$ | 1.0094 | $K_m^{g4}$ | 0.9355 |
| $k_{13}$ | 0.8585 | $k_{28}$ | 0.8184 | $k_{14}$ | 0.6608 |
| $V_{max}^{g2}$ | 0.6483 | $k_{29}$ | 0.6233 | $K_m^{g2}$ | 0.5900 |
| $R_m$ | 0.5326 | $V_{max}^{g3}$ | 0.5053 | $\alpha_2$ | 0.3421 |
| $K_m^{g3}$ | 0.2421 | $\alpha_1$ | 0.1445 | $q_1$ | 0.1266 |
| $a_4$ | 0.1298 | $a_5$ | 0.124 | $b_2$ | 0.126 |
| $a_1$ | 0.0938 | $a_2$ | 0.0908 | $k_{18}$ | 0.082 |
| $q_2$ | 0.0763 | $k_4$ | 0.074 | $V_{max}^{gp}$ | 0.0689 |
| $V_{max}^{gs}$ | 0.0633 | $a_6$ | 0.0628 | $a_3$ | 0.0563 |
| $K_m^{gp}$ | 0.0276 | $k_{27}$ | 0.0025 | $k_{19}$ | 0.0214 |
| SHIP | 0.019 | $k_{10}$ | 0.019 | $k_{25}$ | 3.019 |
| $K_m^{gs}$ | 0.0166 | $k_5$ | 0.0023 | $k_{26}$ | 0.0006 |
| $k_2$ | 0.0004 | $k_{16}$ | 0.0004 | $k_3$ | 0.0001 |
| $k_{15}$ | 0.0001 | $k_6$ | 0.0 | $k_{20}$ | 0.0 |

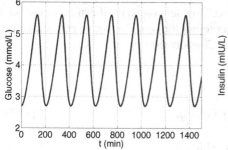 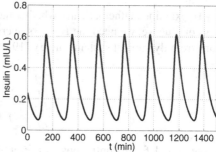

**Fig. 4.9.** Oscillations of glucose and insulin simulated by the model consisting of equations (4.1)-(4.39) with the constant exogenous glucose input $g_{in}^p = 25.7$ mg/L/min. In this simulation, the initial condition $x_3^m(0)$ (the concentration of unphosphorylated once-bound surface insulin receptors) is changed from 0 to $9 \times 10^{-10}$ M. We ran our program for 2000 minutes such that the solutions reach their homeostasis, and then we set that moment as the initial time 0. Reproduced with permission from [26]

oscillation causes theoretically, we simulate glucose and insulin oscillations with a constant exogenous glucose input. Although the constant glucose intravenous infusion of 4.5 mg/kg/min was reported in [36], we use the average rate $g_{in}^p = 25.7$

mg/L/min of the exogenous glucose input of Korach-André *et al* [21] because our model is tested by this input. From the dynamic system theory, it is well known that the period of oscillating solutions of a system depends on the initial conditions. Thus, to produce the glucose oscillation with the period of 208 minutes reported by Shapiro *et al* [36], the initial condition $x_3^m(0)$ (the concentration of unphosphorylated once-bound surface insulin receptors) is changed from 0 to $9 \times 10^{-10}$ M. In this simulation, we ran our program for 2000 minutes such that the solutions reach their homeostasis, and then we set that moment as the initial time 0. Fig. 4.9 shows that both glucose and insulin oscillate with a same period of about 208 minutes. If $x_3^m(0) = 0$, then the period is about 360 minutes (simulation not shown).

There is a discrepancy between the simulation and experimental observations reported by Shapiro *et al* [36]: the reported oscillation periods for glucose and insulin were different, 208 minutes for glucose and 106 minutes for insulin, while the simulated oscillation periods are the same for both. This discrepancy strongly suggests that this simple model is quite far away from modeling this sophisticated system accurately. On the other hand, to my knowledge in dynamic system theory, it seems that it is extremely difficult or impossible to construct a system of ordinary differential equations such that its different states converge to different periodic functions with different periods. Therefore, partial differential equations defined in different spatial domains may be needed for modeling the system more accurately. As discussed in [27], the simple static output feedback controller $u_2$ defined by (4.7) may not be able to model the complex process of insulin secretion from the pancreatic $\beta$ cells in response to the rise of blood glucose and a dynamic distributed parameter output feedback controller may be necessary. Hence, we are facing a great challenge in this modeling.

To examine mathematically whether the liver subsystem itself can cause oscillations of glucose, we need the Bendixson criterion for nonexistence of periodic solutions from dynamical systems theory [19]. Consider the system

$$\frac{dx_1}{dt} = f_1(x_1, x_2), \tag{4.40}$$

$$\frac{dx_2}{dt} = f_2(x_1, x_2), \tag{4.41}$$

where $f_1$ and $f_2$ are continuously differentiable.

**Theorem 14 (Bendixson criterion).** *If $\frac{\partial f_1}{\partial x_1} + \frac{\partial f_2}{\partial x_2}$ is not identically zero and does not change sign on a simply connected region $\Omega$ of $\mathbb{R}^2$, then the system (4.40)-(4.41) has no periodic solutions lying entirely in $\Omega$.*

*Proof.* We argue by contradiction. If the system (4.40)-(4.41) has a periodic solution lying entirely in $\Omega$, then the graph of the solution in the phase plane is a closed curve $C$. On the curve $C$, it follows from (4.40) and (4.41) that $\frac{dx_2}{dx_1} = \frac{f_2}{f_1}$. Therefore, we have

$$\int_C [f_2(x_1, x_2) dx_1 - f_1(x_1, x_2) dx_2] = 0.$$

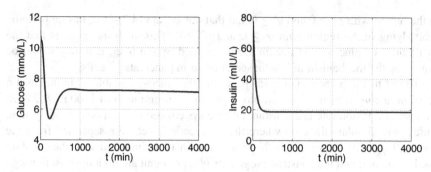

**Fig. 4.10.** Steady states of glucose and insulin simulated by the model consisting of equations (4.1)-(4.39) with constant exogenous glucose input $g_{in}^p = 25.7$ mg/L/min and the fixed glucose input $g^p = 1200$ mg/L in the insulin infusion rate $u_2$. Reproduced with permission from [26]

It then follows from Green's theorem that

$$\iint_D \left( \frac{\partial f_1}{\partial x_1} + \frac{\partial f_2}{\partial x_2} \right) dx_1 dx_2 = 0,$$

where $D$ is the interior of $C$. This is impossible because $\frac{\partial f_1}{\partial x_1} + \frac{\partial f_2}{\partial x_2}$ is not identically zero and does not change sign on $D$. Hence the system (4.40)-(4.41) has no periodic solutions lying entirely in $\Omega$. $\square$

We now consider the equations (4.1), (4.2), and (4.3) and define

$$f_1(g^y, g^l) = -\frac{V_{max}^{gp} r_1 g^y}{R_1^0(K_m^{gp} + g^y)} + \frac{V_{max}^{gs} I_{gs} g^l}{(1 + a_6 r_1)(K_m^{gs} + g^l)},$$

$$f_2(g^y, g^l) = -\frac{V_{max}^{gs} I_{gs} g^l}{(1 + a_6 r_1)(K_m^{gs} + g^l)} - \frac{V_{max}^{g2} g^l}{K_m^{g2} + g^l} + \frac{V_{max}^{gp} r_1 g^y}{R_1^0(K_m^{gp} + g^y)} + \frac{V_{max}^{g2} g^p}{K_m^{g2} + g^p},$$

$$f_3(g^l, g^p) = -\frac{x_{21}^m V_{max}^{g4} g^p}{K_m^{g4} + g^p} - \frac{V_{max}^{g2} g^p}{K_m^{g2} + g^p} - \frac{V_{max}^{g3} g^p}{K_m^{g3} + g^p} + \frac{V_{max}^{g2} g^l}{K_m^{g2} + g^l}.$$

Computing derivatives of these functions gives

$$\frac{\partial f_1(g^y, g^l)}{\partial g^y} = -\frac{V_{max}^{gp} r_1 K_m^{gp}}{R_1^0(K_m^{gp} + g^y)^2},$$

$$\frac{\partial f_2(g^y, g^l)}{\partial g^l} = -\frac{V_{max}^{gs} I_{gs} K_m^{gs}}{(1 + a_6 r_1)(K_m^{gs} + g^l)^2} - \frac{V_{max}^{g2} K_m^{g2}}{(K_m^{g2} + g^l)^2},$$

$$\frac{\partial f_3(g^l, g^p)}{\partial g^p} = -\frac{x_{21}^m V_{max}^{g4} K_m^{g4}}{(K_m^{g4} + g^p)^2} - \frac{V_{max}^{g2} K_m^{g2}}{(K_m^{g2} + g^p)^2} - \frac{V_{max}^{g3} K_m^{g3}}{(K_m^{g3} + g^p)^2}.$$

Then $\frac{\partial f_1(g^y, g^l)}{\partial g^y} + \frac{\partial f_2(g^y, g^l)}{\partial g^l}$ and $\frac{\partial f_2(g^y, g^l)}{\partial g^l} + \frac{\partial f_3(g^l, g^p)}{\partial g^p}$ are negative. Hence it follows from the Bendixson criterion that the system (4.1)-(4.2) has no periodic orbits lying in the region with $g^y \geq 0$ and $g^l \geq 0$ and that the system (4.2)-(4.3) has no periodic orbits lying in the region with $g^l \geq 0$ and $g^p \geq 0$. This analysis suggests that the oscillations of glucose and insulin are not caused by the liver, and they would be caused by the mechanism of insulin secretion from pancreatic $\beta$ cells.

To test this hypothesis numerically, we fix the glucose variable $g^p$ in the insulin infusion rate $u_2$ to be the constant blood glucose concentration of 1200 mg/L. Then the insulin infusion rate is a constant and the system becomes an open-loop system, which may simulate the case where the pancreatic $\beta$ cells are separated from the system. Fig. 4.10 shows that both glucose and insulin simulated with the constant insulin infusion and the constant exogenous glucose input are no longer oscillating.

## Exercises

**4.1.** Consider the system (4.1)-(4.3).

1. For a constant exogenous glucose input $g_{in}^p$, find the equilibrium of the system.
2. Analyze the linear stability of the equilibrium (Analyze the eigenvalues of the Jacobian matrix of the system at the equilibrium).
3. Linearize the system at the equilibrium and analyze controllability of the linearized control system.
4. Let the plasma glucose $g^p$ be the output of the system. Analyze observability of the linearized control system.

**4.2.** Use the model consisting of equations (4.1)-(4.39) to simulate glucose and insulin oscillations with different constant exogenous glucose inputs $g_{in}^p$ and different initial conditions.

**4.3.** Construct a nonlinear model (a nonlinear function of $x_{13}$) for rate constants $k_{11}$ and $k_{12}$ given in (4.30) and (4.31) and then solve the control system consisting of equations (4.1)-(4.39) with (4.30) and (4.31) replaced by your model.

**4.4.** Construct a nonlinear model (a nonlinear function of effect) for the insulin effect on GLUT4, $I_{g4}$, given in (4.35) and then solve the control system consisting of equations (4.1)-(4.39) with (4.35) replaced by your model.

**4.5.** Derive the sensitivity system of the system consisting of equations (4.1)-(4.39) and then solve it numerically.

**4.6.** Fit the following solution of (4.5)

$$h_2^p = e^{-(b_1+b_2)t} h_2^p(0) + e^{-(b_1+b_2)t} \int_0^t \frac{R_m e^{-(b_1+b_2)s}}{1 + q_2 \exp(\alpha_2(C_1 - g^p(s)))} ds$$

into the following data obtained by Korach-André *et al* [21]:

| $t(\text{min})$: | 0 | 30 | 60 | 90 | 120 | 180 | 240 |
|---|---|---|---|---|---|---|---|
| $h_2^p(\text{mIU/L})$: | 68.4 | 61.25 | 49.1 | 46.1875 | 33.2375 | 26.55 | 15.625 |
| $g^p(\text{mmol/L})$: | 10.750 | 9.450 | 7.963 | 7.150 | 7.213 | 6.725 | 6.525 |
| $t(\text{min})$: | 300 | 360 | 420 | 480 | | | |
| $h_2^p(\text{mIU/L})$: | 13.3 | 11.0625 | 6.775 | 5.77143 | | | |
| $g^p(\text{mmol/L})$: | 6.325 | 5.988 | 5.475 | 5.213 | | | |

**4.7.** Following the model of insulin signaling pathway, construct a model for the glucagon signaling pathway. Then incorporate your model into the control system consisting of equations (4.1)-(4.39).

# References

1. Ablooglu A.J., Kohanski R.A.: Activation of the insulin receptor's kinase domain changes the rate-determining step of substrate phosphorylation. Biochemistry **40**, 504-513 (2001).
2. Ackerman E., Gatewood L.C., Rosevear J.W., Molnar G.D.: Model studies of blood glucose regulation. Bull. Math. Biophys. **27**, 21-27 (1965).
3. Albisser A.M., Leibel B.S., Ewart T.G., Davidovac Z., Botz C.K., Zingg W.: An artificial endocrine pancreas. Diabetes **23**, 389-396 (1974).
4. Albisser A.M., Leibel B.S., Ewart T.G., Davidovac Z., Botz C.K., Zingg W., Schipper H., Gander R.: Clinical control of diabetes by the artificial pancreas. Diabetes **23**, 397-404 (1974).
5. Bastl C., Finkelstein F., Sherwin R., Hendler R., Felig P., Hayslett J.: Renal extraction of glucagon in rats with normal and reduced renal function. Am. J. Physiol. **233**, 67-71 (1977).
6. Bergman R.N., Ider Y.Z., Bowden C.R., Cobelli C.: Quantitative estimation of insulin sensitivity. Am. J. Physiol. Endocrinol. Metab. **236**, E667-E677 (1979).
7. Bergman R.N., Phillips L.S., Cobelli C.: Measurement of insulin sensitivity and $\beta$-cell glucose sensitivity from the response to intraveous glucose. J. Clinical Investigation. **68**, 1456-1467 (1981).
8. Bergman R.N., Finegood D.T., Ader M.: Assessment of insulin sensitivity in vivo. Endocrine Reviews. **6**, 45-86 (1985).
9. Bertoldo A., Pencek R.R., Azuma K., Price J.C., Kelley C., Cobelli C., Kelley D.E.: Interactions between delivery, transport, and phosphorylation of glucose in governing uptake into human skeletal muscle. Diabetes. **55**, 3028-3037 (2006).
10. Cerny V.: A thermodynamical approach to the travelling salesman problem: an efficient simulation algorithm. J. Optim. Theory and Appl. **45**, 41-51 (1985).
11. Clemens A.H., Chang P.H., Myers R.W.: The development of biostator, a glucose controlled insulin infusion system (GCIIS). Horm Metab Res. Suppl **7**, 23-33 (1977).
12. Colville C.A., Seatter M.J., Jess T.J., Gould G.W., Thomas H.M.: Kinetic analysis of the liver-type (GLUT2) and brain-type (GLUT3) glucose transporters in Xenopus oocytes: substrate specificities and effects of transport inhibitors. Biochem. J. **290**, 701-706 (1993).
13. Emmanouel, D., Jaspan J., Rubenstein A., Huen A., Fink E., Katz A.: Glucagon metabolism in the rat: contribution of the kidney to the metabolic clearance rate of the hormone. J. Clin. Invest. **62**, 6-13 (1978).

14. Goodner C.J., Walike B.C., Koerker D.J., Ensinck J.W., Brown A.C., Chideckel E.W., Palmer J., Kalnasy L.: Insulin, glucagon, and glucose exhibit synchronous, sustained oscillations in fasting monkeys. Science **195**, 177-179 (1977).

15. Hansen B.C., Jen K.C., Pek S.B., Wolfe R.A.: Rapid oscillations in plasma insulin, glucagon, and glucose in obese and normal weight humans. Journal of Clinical Endocrinology & Metabolism. **54**, 785-792 (1982).

16. Hovorka R.: Continuous glucose monitoring and closed-loop systems. Diabetic Medicine **23**, 1-12 (2006).

17. Jaspan J.B., Lever E., Polonsky K.S., Cauter E.V.: In vivo pulsatility of pancreatic islet peptides. Am. J. Physiol. Endocrinol. Metab. **251**, E215-E226 (1986).

18. Kahn C.R.: Membrane receptors for hormones and neurotransmitters. J. Cell Biology. **70**, 261-286 (1976).

19. Khalil H.K.: Nonlinear Systems. Prentice Hall, New Jersey (2002).

20. Kirkpatrick S., Gelatt C.D., Vecchi, M. P.: Optimization by Simulated Annealing. Science, New Series **220**, 671-680 (1983).

21. Korach-André M., Roth H., Barnoud D., Péan M. Péronnet F., Leverve X.: Glucose appearance in the peripheral circulation and liver glucose output in men after a large $^{13}$C starch meal. Am. J. Clin. Nutr. **80**, 881-886 (2004).

22. Lang D.A., Matthews D.R., Peto J., Turner R.C.: Cyclic oscillations of basal plasma glucose and insulin concentrations in human beings. New England Journal of Medicine. **301** 1023-1027 (1979).

23. Lang D.A., Matthews D.R., Burnett M., Turner R.C.: Brief, irregular oscillations of basal plasma insulin and glucose concentrations in diabetic man. Diabetes **30**, 435-439 (1981).

24. Lefebvre P., Luykx A., Nizet A.: Renal handling of endogenous glucagon in the dog: comparison with insulin. Metabolism **23** 753-760 (1974).

25. Li J., Kuang Y., Mason C.C.: Modeling the glucose-insulin regulatory system and ultradian insulin secretory oscillations with two explicit time delays. J. Theo. Biol. **242**, 722-735 (2006).

26. Liu W., Hsin C., Tang F.: A molecular mathematical model of glucose mobilization and uptake. Math. Biosci. **221**, 121-129 (2009).

27. Liu W., Tang F.: Modeling a simplified regulatory system of blood glucose at molecular levels. J. Theor. Biol. **252**, 608-620 (2008).

28. Man C.D., Caumo A., Basu R., Rizza R.A., Toffolo G., Cobelli C.: Minimal model estimation of glucose absorption and insulin sensitivity from oral test: validation with a tracer method. Am. J. Physiol. Endocrinol Metab. **287**, E637-E643 (2004).

29. Man C.D., Campioni M., Polonsky K.S., Basu R., Rizza R.A., Toffolo G., Cobelli C.: Two-hour seven-sample oral glucose tolerance test and meal protocol: minimal model assessment of $\beta$-cell responsivity and insulin sensitivity in nondiabetic individuals. Diabetes. **54**, 3265-3273 (2005).

30. Man C.D., Rizza R.A., Cobelli C.: Meal simulation model of the glucose-insulin system. IEEE Tran. Biomed. Eng. **54**, 1740-1749 (2007).

31. Nakai C. Thomas J.A.: Effects of magnesium on the kinetic properties of bovine heart glycogen synthase D. J. Biol. Chem. **250**, 4081-4086 (1975).

32. Nishimura H., Pallardo F., Seidner G.A., Vannucci S., Simpson I.A., Birnbaum M.J.: Kinetics of GLUT1 and GLUT4 GlucosTe ransporters Expressed in Xenopus Oocytes. J. Biol. Chem. **268**, 8514-8520 (1993).

33. Panteleon A.E., Loutseiko M., Steil G.M., Rebrin K.: Evaluation of the effect of gain on the meal response of an automated closed-loop insulin delivery system. Diabetes **55**, 1995-2000 (2006).

34. Pedersen M.G., Toffolo G.M., Cobelli C.: Cellular modeling: insight into oral minimal models of insulin secretion. Am. J. Physiol. Endocrinol. Metab. **298**, E597-E601 (2010).

35. Sedaghat A.R., Sherman A., Quon M.J.: A mathematical model of metabolic insulin signaling pathways. Am. J. Physiol. Endocrinol. Metab. **283**, E1084-E1101 (2002).

36. Shapiro E.T., Tillil H., Polonsky K.S., Fang V.S., Rubenstein A.H., Cauter E.V.: Oscillations in insulin secretion during constant glucose infusion in normal man: relationship to changes in plasma glucose. J. Clinical Endocrinol. & Metab. **67**, 307-314 (1988).

37. Simon C., Brandenberger G., Follenius M.: Ultradian oscillations of plasma glucose, insulin, and C-peptide in man during continuous enteral nutrition. J. Clinical Endocrinol. & Metab. **64**, 669-674 (1987).

38. Steil G.M., Rebrin K., Darwin C., Hariri F., Saad M.F.: Feasibility of automating insulin delivery for the treatment of type 1 diabetes. Diabetes **55**, 3344-3350 (2006).

39. Sturis J., Polonsky K.S., Mosekilde E., Cauter E.V.: Computer model for mechanisms underlying ultradian oscillations of insulin and glucose. Am. J. Physiol. Endocrinol. Metab. **260**, E801-E809 (1991).

40. Toffolo G., Cobelli C.: The hot IVGTT two-compartment minimal model: an improved version. Am. J. Physiol. Endocrinol. Metab. **284**, E317-E321 (2003).

41. Toffolo G., Campioni M., Basu R., Rizza R.A., Cobelli C.: A minimal model of insulin secretion and kinetics to assess hepatic insulin extraction. Am. J. Physiol. Endocrinol. Metab. **290**, E169-E176 (2006).

42. Tolic I.M., Mosekilde E., Sturis J.: Modeling the insulin-glucose feedback system: the significance of pulsatile insulin secretion. J. Theoretical Biology **207**, 361-375 (2000).

43. Varma A., Morbidelli M., Wu H.: Parametric Sensitivity in Chemical Systems. Cambridge University Press, UK (1999).

44. Walcott S., Lehman S.L.: Enzyme kinetics of muscle glycogen phosphorylase b. Biochemistry **46**, 11957-11968 (2007).

45. Winston G.W., Reitz R.C.: Effects of chronic ethanol ingestion on liver glycogen phosphorylase in male and female rats. Am. J. Clin. Nutr. **34**, 2499-2507 (1981).

46. The Clinician's Ultimate Reference. http://www.globalrph.com/ (2011). Accessed 20 July 2011.

47. Systems Biology Toolbox for MATLAB. http://www.sbtoolbox.org/ (2011). Accessed 18 June 2011.

# 5

# Control of Calcium in Yeast Cells

Yeast cells uptake calcium from their environment via Mid1p, Cch1p, and other unidentified transporters [7], and maintain a normal cytosolic $Ca^{2+}$ level of 50 - 200 nM by means of a feedback control system [1, 16, 23]. The rise of cytosolic calcium activates calmodulin which in turn activates the serine/threonine phosphatase calcineurin (Fig. 5.1). The activated calcineurin de-phosphorylates Crz1p and suppresses the activity of Vcx1p. Activated Crz1p enters the nucleus and up-regulates the expression of PMR1 and PMC1 (for review, see [13]). Pmr1p pumps calcium ions into the organelle Golgi and possibly endoplasmic reticulum (ER). When the calcium concentrations in Golgi and ER exceed their resting levels of 300 $\mu$M [30] and 10 $\mu$M [1, 39], respectively, the calcium in ER and Golgi will be secreted along with the canonical secretory pathways. Pmc1p pumps calcium ions into vacuole, an organelle that stores excess ions and nutrients. While most calcium ions inside vacuoles form polyphosphate salts and are not re-usable, a small fraction of calcium ions can be channeled to the cytosol by Yvc1p. The total and free vacuolar calcium concentrations are 2 mM and 30 $\mu$M, respectively [16]. When needed, Yvc1p channels calcium to the cytosol and contributes to the rise of cytosolic calcium concentration [14].

Mathematical models for the calcium dynamics were established by Cui *et al* [9, 10] and Tang *et al* [41]. In this chapter, we present an age-dependent model developed by Tang and Liu [41]. We divide the whole system (Fig. 5.1) into several subsystems, model each subsystem individually, and then integrate them into a complete system.

## 5.1 A Model of Aging Process

As cells age, functions of proteins, such as calcium pumps, calmodulin, and calcineurin, in the cells decline and they will die at a certain generation. The *survival rate* of cells at a generation is defined to be the ratio of the number of cells that still live at the generation to the number of the whole cells in an experiment. The graph of the survival rate against generation is called a cell *survival curve*.

Liu W.: Introduction to Modeling Biological Cellular Control Systems.
DOI 10.1007/978-88-470-2490-8_5, © Springer-Verlag Italia 2012

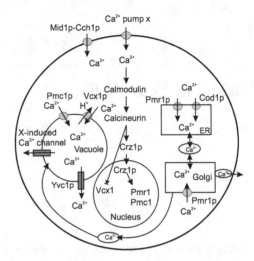

**Fig. 5.1.** A schematic description of intracellular calcium control system in a budding yeast cell. The environmental $Ca^{2+}$ ions enter the cell via an unknown $Ca^{2+}$ pump X in the normal conditions and Mid1p-Cch1p under some abnormal conditions such as depletion of secretory $Ca^{2+}$. A rise of cytosolic $Ca^{2+}$ level triggers a cascade of activations of calmodulin, calcineurin, and Crz1p, leading to the transcription of PMC1 and PMR1, and suppression of VCX1. Then Pmc1p pumps $Ca^{2+}$ into the vacuole. Pmr1p pumps $Ca^{2+}$ into the Golgi apparatus and likely ER. $Ca^{2+}$ ions in ER and Golgi can be secreted out of the cell by the conventional secretory pathway and transported to the vacuole via the multivesicular body (MVB) pathway. The organelle vacuole is an intracellular storage of calcium ions. The fission and fusion of vacuole in response to environmental osmotic pressure variation or cell cycle progression changes the vacuolar membrane and activates the channel Yvc1, which also increases the cytosolic calcium ions. Indirect evidences suggest that vacuoles release calcium ions to assist the progression of the cell cycle (see text for detail). We thus hypothesized that there is a cell cycle-dependent factor X-induced calcium channel on the vacuole membrane. Reproduced with permission from [41]

Since a cell survival curve could be the overall result of such decline, we used the experimental survival curve $S(t)$ of wild type yeast cells (BY4742) to describe the aging process of proteins (Fig. 5.2). For instance, the decay of the function of the calcium pump Pmr1p will be modeled by $S(t)R_{pmr}$, where $R_{pmr}$ is the Pmr1p maximum velocity. The survival curve $S(t)$ in Fig. 5.2 is obtained by linearly interpolating the experimental data. The unit of generation was converted into minute by the conversion: a generation = 120 minutes.

## 5.2 Calcium Uptake from Environment

Experimental observations by Kellermayer *et al* [19] and Locke *et al* [21] suggested that wild-type yeast cells uptake $Ca^{2+}$ from the growth medium using primarily a low-affinity $Ca^{2+}$ uptake system and a second high-affinity $Mg^{2+}$-resistant $Ca^{2+}$

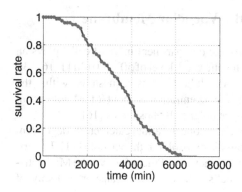

**Fig. 5.2.** Experimental survival curve of wild type yeast cells (BY4742) obtained by Tang [41]. Reproduced with permission

uptake system operating at a much lower level. The second uptake system is coordinately regulated by two distinct mechanisms. The first mechanism is similar to the store-operated calcium entry (also known as the capacitative calcium entry) in mammalian cells [32, 33, 34, 44]. Depletion of $Ca^{2+}$ in the endoplasmic reticulum (ER)/Golgi apparatus activates $Ca^{2+}$ entry across the plasma membrane. The second mechanism couples cellular $Ca^{2+}$ uptake to the level of $Ca^{2+}$ in the extracellular environment. Locke *et al* [21] showed that Cch1p and Mid1p are both required for a high-affinity $Ca^{2+}$ influx system that can be stimulated up to 25-fold in situations causing depletion of secretory $Ca^{2+}$ pools. Since the second uptake system works only in the situation of the $Ca^{2+}$ store depletion and the low extracellular $Ca^{2+}$ environment, we consider only the first primary uptake system. Experimental observations [21] indicated that the calcium uptake system follows the Michaelis-Menten equation. Hence the uptake velocity of the system can be modeled by

$$r_{ex} = \frac{u_1 R_{ex}[Ca^{2+}]_{ex}}{K_{ex} + [Ca^{2+}]_{ex}}, \qquad (5.1)$$

where $R_{ex}$ is the maximum velocity, $K_{ex}$ is the Michaelis-Menten constant, $[Ca^{2+}]_{ex}$ denotes the concentration of extracellular calcium, and $u_1$ is a feedback controller to be designed. In living cells, the metabolic byproducts such as reactive oxygen species (ROS) attack the intracellular membranes and membrane-bound channels and pumps. For example, the vacuolar membrane begins to deteriorate at generations 6 to 7 for a strain whose average life span is 15 to 17 generations [42]. Thus, we assume that intracellular pumps such as Pmr1p and Pmc1p undergo an age-dependent decline. While the plasma membrane also ages, its aging rate is different from that of intracellular membranes. On the extracellular leaflet of the plasma membrane, sterol molecules and sphingolipids form lipid rafts, a patch of ordered microdomain [2]. However, such lipid raft in intracellular membranes is much smaller and more transient than that of the plasma membrane [20]. Sterol molecules in intracellular membranes are easier than those in plasma membrane to be attacked by intracellular ROS. Consequently, the aging rate of the plasma membrane channels and pumps is different from that of intracellular membrane proteins. Therefore, to allow us to focus on the intracellular events, we assume that the plasma membrane pump X does not age in our model.

## 5.3 Calcium Movement across the Vacuolar Membrane

The vacuole, which is the major $Ca^{2+}$ storage compartment in yeast cells, plays an important role in maintaining a normal cytosolic $Ca^{2+}$ level of 50 - 200 nM [1, 16, 23]. It was estimated that over 95% of the total cellular $Ca^{2+}$ is sequestered within the vacuole [1, 23, 43]. The total vacuolar $Ca^{2+}$ concentration is about 2 mM with about 30 $\mu$M free $Ca^{2+}$ due to $Ca^{2+}$ binding to vacuolar polyphosphates [16].

In response to an increase of cytosolic $Ca^{2+}$ concentration, calcineurin-dependent $Ca^{2+}$ ATPase Pmc1p transports $Ca^{2+}$ from the cytosol into the vacuole [11]. Experimental results [40] showed that the transport kinetics of Pmc1p follows the Michaelis-Menten equation for log phase (young) cells. Therefore the transport velocity of Pmc1p can be modeled by

$$r_{pmc} = \frac{u_2 S(t) R_{pmc} [Ca^{2+}]_i}{K_{pmc} + [Ca^{2+}]_i}, \qquad (5.2)$$

where $[Ca^{2+}]_i$ denotes the cytosolic calcium concentration, $R_{pmc}$ is the maximum velocity, $K_{pmc}$ is the Michaelis-Menten's constant, and $u_2$ is a feedback controller to be designed. Hereafter, the aging function $S(t)$ is included to describe the function decline of a pump or channel.

Another known protein mediating vacuolar $Ca^{2+}$ sequestration is the $Ca^{2+}/H^+$ exchanger Vcx1p/Hum1p [12, 23, 27, 31, 43]. Experimental results [27] showed that the transport kinetics of Vcx1p also follows the Michaelis-Menten equation:

$$r_{vcx} = \frac{u_3 S(t) R_{vcx} [Ca^{2+}]_i}{K_{vcx} + [Ca^{2+}]_i}, \qquad (5.3)$$

where $R_{vcx}$ is the maximum velocity, $K_{vcx}$ is the Michaelis-Menten's constant, $u_3$ is a feedback controller to be designed.

The sequestered $Ca^{2+}$ in the vacuole is compounded with vacuolar polyphosphates in a relatively stable form [23]:

$$Ca^{2+} + polyphosphate \xrightarrow{k_f} complex.$$

Since the polyphosphate pool is huge in the vacuole and always available for calcium to bind, we can assume that the polyphosphate concentration is a constant. Then the compounding reaction rate is given by

$$r_{com} = k_f [polyphosphate][Ca^{2+}]_v = k_6 [Ca^{2+}]_v, \qquad (5.4)$$

where $k_6 = k_f [polyphosphate]$ is a positive constant. Thus this compounding reaction rate is proportional to the concentration $[Ca^{2+}]_v$ of free calcium in the vacuole.

Until recently, this large store of vacuolar $Ca^{2+}$ was thought to be relatively inert due to its association with polyphosphates [1]. However, a study found that vacuolar $Ca^{2+}$ can be released at a very slow rate of 0.016 nmol/min [16] in a regulated manner through the action of Yvc1p [29]. We assume that the transport velocity of Yvc1p can be modeled by

$$r_{yvc} = u_4 S(t) R_{yvc} \left( [Ca^{2+}]_v - [Ca^{2+}]_i \right), \tag{5.5}$$

where $R_{yvc}$ is a positive constant and $u_4$ is a feedback controller to be designed. The difference $\left( [Ca^{2+}]_v - [Ca^{2+}]_i \right)$ denotes the concentration gradient that powers the passive channel.

Experimental observations suggest that the calcium release from vacuoles may be induced by the inositol (1,4,5)-trisphosphate (IP$_3$). In sea urchin embryonic cells, the increase of cytosolic calcium during the mitotic phase is the IP$_3$-stimulated calcium release from calcium storages such as ER [6]. Vacuoles also release calcium upon IP$_3$ treatment [4]. Because there are no identified IP$_3$ receptors in yeast, we assume that there is a putative factor X on the vacuolar membrane that is responsible for IP$_3$-induced calcium release. Similar to Yvc1p, the release kinetics of $Ca^{2+}$ through the unidentified channel X is assumed to be modeled by

$$r_x = u_5 S(t) R_x \left( [Ca^{2+}]_v - [Ca^{2+}]_i \right), \tag{5.6}$$

where $R_x$ is positive constant and $u_5$ is a feedback controller to be designed.

Therefore the $Ca^{2+}$ dynamics in the vacuole can be modeled by

$$
\begin{aligned}
\frac{d[Ca^{2+}]_v}{dt} &= r_{pmc} + r_{vcx} - r_{com} - r_{yvc} - r_x + k_7 r_{gol} \\
&= \frac{u_2 S(t) R_{pmc} [Ca^{2+}]_i}{K_{pmc} + [Ca^{2+}]_i} + \frac{u_3 S(t) R_{vcx} [Ca^{2+}]_i}{K_{vcx} + [Ca^{2+}]_i} \\
&\quad - k_6 [Ca^{2+}]_v - u_4 S(t) R_{yvc} \left( [Ca^{2+}]_v - [Ca^{2+}]_i \right) \\
&\quad - u_5 S(t) R_x \left( [Ca^{2+}]_v - [Ca^{2+}]_i \right) \\
&\quad + k_7 k_9 u_6 [Ca^{2+}]_g,
\end{aligned}
\tag{5.7}
$$

where $[Ca^{2+}]_g$ is the Golgi calcium concentration and $u_6$ is a feedback controller to be designed. The term $r_{gol}$ defined in (5.9) below describes how $Ca^{2+}$ ions are transported out of the Golgi by vesicles in response to a high $Ca^{2+}$ concentration in the Golgi.

## 5.4 Calcium Movement across the Golgi Membrane

The Golgi apparatus has also been shown to play an important role in maintaining normal cytosolic $Ca^{2+}$ levels in yeast cells through the action of the Golgi-localized $Ca^{2+}$-ATPase Pmr1p [1, 24, 30, 37]. Like Pmc1p, in response to an increase of cytosolic $Ca^{2+}$ concentration, Pmr1p transports $Ca^{2+}$ from cytosol into Golgi and

experimental results [45] showed that the transport kinetics of Pmrlp also follows the Michaelis-Menten equation:

$$r_{pmr} = \frac{u_2 S(t) R_{pmr}[Ca^{2+}]_i}{K_{pmr} + [Ca^{2+}]_i},\tag{5.8}$$

where $R_{pmr}$ is the maximum velocity, $K_{pmr}$ is the Michaelis-Menten's constant, and $u_2$ is a feedback controller to be designed.

Ca$^{2+}$ within the Golgi lumen controls essential processes, such as protein processing and sorting. In resting living HeLa cells, the concentration of Ca$^{2+}$ in Golgi is about 0.3 mM [30]. Since it is not clear how the calcium homeostasis in Golgi is maintained, we assume that the velocity of transport of Ca$^{2+}$ out of Golgi by vesicles is proportional to the concentration $[Ca^{2+}]_g$ of calcium in Golgi:

$$r_{gol} = k_9 u_6 [Ca^{2+}]_g,\tag{5.9}$$

where $u_6$ is a feedback controller to be designed and $k_9$ is a positive constant.

In summary, the Ca$^{2+}$ dynamics in the Golgi apparatus can be modeled by

$$\begin{aligned}\frac{d[Ca^{2+}]_g}{dt} &= r_{pmr} - r_{gol} + r_{er}\\ &= \frac{u_2 S(t) R_{pmr}[Ca^{2+}]_i}{K_{pmr} + [Ca^{2+}]_i} - k_9 u_6 [Ca^{2+}]_g + k_8 u_7 [Ca^{2+}]_{er},\end{aligned}\tag{5.10}$$

where the term $r_{er}$ defined in (5.14) below describes Ca$^{2+}$ transport from ER.

## 5.5 Calcium Movement across the Endoplasmic Reticulum Membrane

The yeast endoplasmic reticulum (ER) appears to play a lesser role in cellular Ca$^{2+}$ storage because the free Ca$^{2+}$ concentration of this compartment has been reported to be only about 10 $\mu$M [1, 39]. It was recently shown that the COD1/SPF1 gene product is an ER-localized Ca$^{2+}$-ATPase that sequesters cytosolic Ca$^{2+}$ into the ER and follows the Michaelis-Menten equation [8]:

$$r_{cod} = \frac{S(t) R_{cod}[Ca^{2+}]_i}{K_{cod} + [Ca^{2+}]_i},\tag{5.11}$$

where $R_{cod}$ is the maximum velocity and $K_{cod}$ is the Michaelis-Menten's constant.

A number of studies have also suggested that Pmr1p [39] and the vacuolar Ca$^{2+}$-ATPase Pmc1p [5] also play a role in maintaining ER Ca$^{2+}$ stores under certain conditions. Like Pmr1p in Golgi, the transport velocity of Pmr1p is modeled by

$$r_{erpmr} = \frac{u_2 S(t) R_{erpmr}[Ca^{2+}]_i}{K_{erpmr} + [Ca^{2+}]_i},\tag{5.12}$$

and the transport velocity of Pmc1p is modeled by

$$r_{erpmc} = \frac{u_2 S(t) R_{erpmc} [Ca^{2+}]_i}{K_{erpmc} + [Ca^{2+}]_i}. \tag{5.13}$$

The ER is the site of essential cellular processes that require stringent regulation of lumenal ion levels. The ER also serves as a source of releasable $Ca^{2+}$ that can be mobilized for various cellular demands [8]. Consequently, the ion levels of the ER are under continuous, dynamic control in order to serve these essential functions. Since the control mechanism of maintaining the calcium homeostasis in ER is not clear, we assume that the velocity of transport of $Ca^{2+}$ out of ER by vesicles is proportional to the concentration $[Ca^{2+}]_{er}$ of $Ca^{2+}$ in ER:

$$r_{er} = k_8 u_7 [Ca^{2+}]_{er}, \tag{5.14}$$

where $u_7$ is a feedback controller to be designed and $k_8$ is a positive constant.

In summary, the $Ca^{2+}$ dynamics in ER can be modeled by

$$\frac{d[Ca^{2+}]_{er}}{dt} = r_{erpmr} + r_{erpmc} + r_{cod} - r_{er} + k_{10} r_{gol}$$

$$= \frac{u_2 S(t) R_{erpmr} [Ca^{2+}]_i}{K_{erpmr} + [Ca^{2+}]_i}$$

$$+ \frac{u_2 S(t) R_{erpmc} [Ca^{2+}]_i}{K_{erpmc} + [Ca^{2+}]_i} + \frac{S(t) R_{cod} [Ca^{2+}]_i}{K_{cod} + [Ca^{2+}]_i}$$

$$- k_8 u_7 [Ca^{2+}]_{er} + k_9 k_{10} u_6 [Ca^{2+}]_g. \tag{5.15}$$

Since part of calcium transported from Golgi by vesicles flows out of the cell, we have $k_7 + k_{10} < 1$.

## 5.6 A Calcium Control System

From Fig. 5.1, we derive the equation of calcium in the cytosol as follows

$$\frac{d[Ca^{2+}]_i}{dt} = r_{ex} - r_{pmc} - r_{vcx} + r_{yvc} - r_{pmr} - r_{cod} - r_{erpmr} - r_{erpmc} + r_x$$

$$= \frac{u_1 R_{ex} [Ca^{2+}]_{ex}}{K_{ex} + [Ca^{2+}]_{ex}} - \frac{u_2 S(t) R_{pmc} [Ca^{2+}]_i}{K_{pmc} + [Ca^{2+}]_i} - \frac{u_3 S(t) R_{vcx} [Ca^{2+}]_i}{K_{vcx} + [Ca^{2+}]_i}$$

$$+ u_4 S(t) R_{yvc} \left( [Ca^{2+}]_v - [Ca^{2+}]_i \right) - \frac{u_2 S(t) R_{pmr} [Ca^{2+}]_i}{K_{pmr} + [Ca^{2+}]_i}$$

$$- \frac{S(t) R_{cod} [Ca^{2+}]_i}{K_{cod} + [Ca^{2+}]_i} - \frac{u_2 S(t) R_{erpmr} [Ca^{2+}]_i}{K_{erpmr} + [Ca^{2+}]_i} - \frac{u_2 S(t) R_{erpmc} [Ca^{2+}]_i}{K_{erpmc} + [Ca^{2+}]_i}$$

$$+ u_5 S(t) R_x \left( [Ca^{2+}]_v - [Ca^{2+}]_c \right). \tag{5.16}$$

Then the equations (5.7), (5.10), (5.15), and (5.16) constitute an intracellular calcium control system in yeast cells.

## 5.7 Design of Feedback Controllers

Following molecular mechanisms, we design the feedback controllers proposed in the above sections.

### 5.7.1 Control of Calcium Uptake from Environment

Calmodulin contains four copies of a $Ca^{2+}$-binding EF-hand, each of which binds one $Ca^{2+}$ ion. In yeast, the most C-terminal EF-hand in yeast calmodulin (site IV) is defective for $Ca^{2+}$ binding [13, 22, 38]. The other three EF-hands bind $Ca^{2+}$ with high affinity. Calmodulin senses the rise of the cytosolic calcium concentration and transmits the calcium signal to calcineurin and Crz1p, which up-regulates the expression of calcium-utilizing and dissipating pumps (Fig. 5.1). The biochemical reactions in this $Ca^{2+}$ sensing and signal transduction process can be described as follows [9]:

$$3Ca^{2+} + calmodulin \underset{k_{-1}}{\overset{k_1}{\rightleftharpoons}} CaM,$$

$$CaM + calcineurin \underset{k_{-2}}{\overset{k_2}{\rightleftharpoons}} CaN,$$

where $CaM$ denotes the $Ca^{2+}$-bound calmodulin, $CaN$ denotes the $CaM$-bound calcineurin, and $k$'s are reaction rate constants. Using the law of mass balance, the kinetics of these reactions can be modeled by

$$\frac{d[calm]}{dt} = -k_1[Ca^{2+}]_i^3[calm] + k_{-1}[CaM], \tag{5.17}$$

$$\frac{d[CaM]}{dt} = k_1[Ca^{2+}]_i^3[calm] - k_{-1}[CaM] - k_2[CaM][calc] + k_{-2}[CaN], \tag{5.18}$$

$$\frac{d[calc]}{dt} = -k_2[CaM][calc] + k_{-2}[CaN], \tag{5.19}$$

$$\frac{d[CaN]}{dt} = k_2[CaM][calc] - k_{-2}[CaN]. \tag{5.20}$$

Adding the equations (5.17), (5.18), and (5.20) together gives

$$\frac{d}{dt}([calm] + [CaM] + [CaN]) = 0,$$

which implies

$$[calm](t) + [CaM](t) + [CaN](t) = [CaM_0], \tag{5.21}$$

where $[CaM_0]$ denotes the total concentration of $Ca^{2+}$-free and $Ca^{2+}$-bound calmodulin. Adding the equations (5.19) and (5.20) together gives

$$\frac{d}{dt}([calc] + [CaN]) = 0,$$

which implies

$$[calc](t) + [CaN](t) = [CaN_0],\qquad(5.22)$$

where $[CaN_0]$ denotes the total concentration of $CaM$-free and $CaM$-bound calcineurin. It then follows from (5.17)-(5.22) that

$$
\begin{aligned}
\frac{d[CaM]}{dt} &= k_1 S(t)[Ca^{2+}]_i^3([CaM_0] - [CaM] - [CaN]) - k_{-1} S(t)[CaM] \\
&\quad - k_2 S(t)[CaM]([CaN_0] - [CaN]) + k_{-2} S(t)[CaN],\qquad(5.23)
\end{aligned}
$$

$$\frac{d[CaN]}{dt} = k_2 S(t)[CaM]([CaN_0] - [CaN]) - k_{-2} S(t)[CaN].\qquad(5.24)$$

In the above equations, the aging function $S(t)$ is included to describe the decline of functions of calmodulin and calcineurin caused by reactive oxygen species (ROS).

It has been reported that calmodulin has the ability of directly inhibiting $Ca^{2+}$ influx pathways [46]. Hence Cui et al [10] proposed the following feedback controller

$$u_1 = \frac{1}{1 + I_{ex}[CaM]},\qquad(5.25)$$

where $I_{ex} > 0$ is an inhibition constant.

### 5.7.2 Control of $Ca^{2+}/H^+$ Exchanger Vcx1p

The mechanism of regulating Vcx1p has not yet been completely understood. The only knowledge is that calcineurin inhibits Vcx1 function possibly by posttranslational mechanisms [11, 12]. Thus the feedback controller $u_3$ for Vcx1p was designed by Cui et al [9] as follows

$$u_3 = \frac{1}{1 + k_5[CaN]},\qquad(5.26)$$

where $k_5$ is a positive constant.

### 5.7.3 Control of Calcium Pumps Pmc1p and Pmr1p

Since the translocation of Crz1p is similar to that of NFAT (nuclear factor of activated T-cells) in mammalian cells, we introduce the model of NFAT developed by Salazar et al [36].

Nuclear factors of activated T-cells (NFAT) are highly phosphorylated proteins. They are transcription factors and are regulated by the calcium-dependent phosphatase calcineurin. NFAT1, one of NFAT family members, is phosphorylated on fourteen conserved phosphoserine residues in its regulatory domain, thirteen of which are dephosphorylated upon stimulation. Dephosphorylation of all thirteen residues is required to mask a nuclear export signal, cause full exposure of a nuclear localization signal, and promote transcriptional activity [28].

NFAT1 can exist in two global conformations. An active conformational NFAT1, in which the nuclear localization signal is exposed and the nuclear export signal is

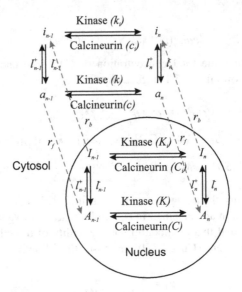

**Fig. 5.3.** Conformational switch model for activation of NFAT1. Cytoplasmic NFAT1 in the active conformation and inactive conformation with $n$ phosphorylated residues are denoted by $a_n$ and $i_n$, respectively, $n = 0, 1, 2, \cdots, N$. The corresponding nuclear NFAT1 in the active conformation and inactive conformation with $n$ phosphorylated residues are denoted by $A_n$ and $I_n$, respectively. Cytoplasmic NFAT1 in the active conformation is imported into the nucleus with a rate constant $r_f$. Conversely, nuclear NFAT1 in the inactive conformation are exported with a rate constant $r_b$. Conformational transitions, phosphorylation, and dephosphorylation can take place both in the cytoplasm and in the nucleus

masked, is imported into the nucleus, as demonstrated in Fig. 5.3. An inactive conformational NFAT1, in which the nuclear localization signal is masked and the nuclear export signal is exposed, is exported from the nucleus. The switching between the two conformations is regulated by the phosphorylation state in such a way that the probability of the active conformation is high in the dephosphorylated state and low in the phosphorylated state [36].

Calcineurin cleaves the phosphate group in the NFAT1 regulatory domain in a sequential order, proceeding from the serine-rich region 1 (SRR-1) to the more C-terminal SP motifs 2 and 3. The main kinases, glycogen synthase kinase-3 (GSK-3) and casein kinase I (CK-I), sequentially phosphorylate NFAT1 from SP motifs to the SRR-1. Thus we assume that dephosphorylation and phosphorylation are sequential and proceed in the reverse order. For example, the changes of phosphorylation state can be written as

$$0000\text{PPPPPPPPP} \underset{\text{kinases}}{\overset{\text{calcineurin}}{\rightleftharpoons}} 00000\text{PPPPPPPP}$$

where 0s and Ps denote unphosphorylated and phosphorylated sites, respectively.

In this sequential model, $a_n$ and $i_n$ ($A_n$ and $I_n$) will be used to denote the active and inactive conformations in cytosol (nucleus), respectively, with $n$ phosphorylated sites. Then the percentage of the active conformation in cytosol is given by

$$\phi_a = \frac{\sum_{n=0}^{N} a_n}{\sum_{n=0}^{N}(a_n + i_n)} \tag{5.27}$$

and the percentage of the active conformation in nucleus is given by

$$\phi_A = \frac{\sum_{n=0}^{N} A_n}{\sum_{n=0}^{N}(A_n + I_n)}. \tag{5.28}$$

We assume that the total number of NFAT1 is equal to 1. Let $Z$ denote the nuclear fraction of NFAT1. Then the inactive conformational NFAT1 in nucleus is $(1-\phi_A)Z$, the cytoplasmic fraction is $1-Z$, and the active conformational NFAT1 in cytosol is $\phi_a(1-Z)$. Since the active conformational NFAT1 in cytosol is imported into the nucleus and the inactive conformational NFAT1 in nucleus is exported from the nucleus, $Z$ is governed by the differential equation [36]

$$\frac{dZ}{dt} = r_f \phi_a(1-Z) - r_b(1-\phi_A)Z, \tag{5.29}$$

where $r_f$ and $r_b$ are the rate constants.

We now use the simple equilibrium analysis to express $\phi_a$ and $\phi_A$ in terms of rate constants. To this end, we assume the first-order kinetics

$$a_{n+1} \underset{k}{\overset{c}{\rightleftharpoons}} a_n, \quad A_{n+1} \underset{K}{\overset{C}{\rightleftharpoons}} A_n, \quad i_{n+1} \underset{k_i}{\overset{c_i}{\rightleftharpoons}} i_n, \quad I_{n+1} \underset{K_i}{\overset{C_i}{\rightleftharpoons}} I_n$$

for dephosphorylation and phosphorylation, respectively, as shown in Fig. 5.3. The rate constant $c, C, c_i$, or $C_i$ is the activity of calcineurin with respect to cleaving the phosphate group $n+1$. The rate constant $k, K, k_i$, or $K_i$ is the kinase activity with respect to adding the same phosphate group. When the dephosphorylation and phosphorylation rates are equal, we obtain that

$$\frac{a_{n+1}}{a_n} = \frac{k}{c}, \quad \frac{A_{n+1}}{A_n} = \frac{K}{C}, \quad \frac{i_{n+1}}{i_n} = \frac{k_i}{c_i}, \quad \frac{I_{n+1}}{I_n} = \frac{K_i}{C_i}. \tag{5.30}$$

At equilibrium of the conformational transition

$$a_n \underset{l_n^-}{\overset{l_n^+}{\rightleftharpoons}} i_n, \quad A_n \underset{l_n^-}{\overset{l_n^+}{\rightleftharpoons}} I_n,$$

we derive that

$$L_n = \frac{l_n^+}{l_n^-} = \frac{i_n}{a_n} = \frac{I_n}{A_n}. \tag{5.31}$$

By (5.30) and (5.31), we obtain

$$a_n = \left(\frac{k}{c}\right)^n a_0, \quad i_n = \left(\frac{k_i}{c_i}\right)^n i_0 = \left(\frac{k_i}{c_i}\right)^n a_0 L_0 = \left(\frac{k\lambda}{c}\right)^n a_0 L_0,$$

where
$$\lambda = \frac{k_i c}{k c_i}.$$

It then follows from (5.27) that

$$
\begin{aligned}
\phi_a &= \frac{\sum_{n=0}^{N} \left(\frac{k}{c}\right)^n a_0}{\sum_{n=0}^{N} \left(\left(\frac{k}{c}\right)^n a_0 + \left(\frac{k\lambda}{c}\right)^n a_0 L_0\right)} \\
&= \frac{\sum_{n=0}^{N} \left(\frac{k}{c}\right)^n}{\sum_{n=0}^{N} \left(\left(\frac{k}{c}\right)^n + \left(\frac{k\lambda}{c}\right)^n L_0\right)}.
\end{aligned}
\tag{5.32}
$$

Assuming that $\frac{k_i c}{k c_i} = \frac{K_i C}{K C_i}$, in the same way, we obtain from (5.28) that

$$
\phi_A = \frac{\sum_{n=0}^{N} \left(\frac{K}{C}\right)^n}{\sum_{n=0}^{N} \left(\left(\frac{K}{C}\right)^n + \left(\frac{K\lambda}{C}\right)^n L_0\right)}.
\tag{5.33}
$$

We now use the NFAT kinetics model (5.29) to model the nuclear import and export of Crz1p. For simplicity, we assume that $k/c = K/C$ in (5.29) and that the kinase level $k$ is a constant. Furthermore we may express $c/k$ as the concentration $[CaN]$ of activated calcineurin in a relative unit. Let $h$ denote the total nuclear fraction of Crz1p. Then it follows from (5.29) that

$$
\frac{dh}{dt} = k_3 \phi\left(\frac{1}{[CaN]}\right)(1-h) - k_4 \left[1 - \phi\left(\frac{1}{[CaN]}\right)\right] h,
\tag{5.34}
$$

where $k_3$ denotes the rate constant of import of Crz1p into the nucleus and $k_4$ denotes the rate constant of export of Crz1p from the nucleus. The function $\phi$ is given by

$$
\phi(x) = \frac{1}{1 + L_0 \frac{\left((\lambda x)^{N+1}-1\right)(x-1)}{(\lambda x - 1)\left(x^{N+1}-1\right)}}.
\tag{5.35}
$$

With notations for NFAT, we can derive that

$$
h(t) = \sum_{n=0}^{N}(A_n + I_n) = A_0 \left(\frac{x^{N+1}-1}{x-1} + L_0 \frac{(\lambda x)^{N+1}-1}{\lambda x - 1}\right)
\tag{5.36}
$$

and then

$$
A_0 = \frac{h(t)}{\frac{x^{N+1}-1}{x-1} + L_0 \frac{(\lambda x)^{N+1}-1}{\lambda x - 1}} = h(t)\theta(x),
$$

where $x = K/C$ and

$$
\theta(x) = \frac{1}{\frac{x^{N+1}-1}{x-1} + L_0 \frac{(\lambda x)^{N+1}-1}{\lambda x - 1}}.
\tag{5.37}
$$

Since the synthesis of Pmc1p and Pmr1p depends on the transcriptionally active Crz1p fraction in the nucleus and the fully dephosphorylated nuclear fraction of Crz1p is required to promote the transcriptional activity, the feedback controller $u_2$ for Pmc1p and Pmr1p can be modeled by

$$u_2 = h(t)\theta\left(\frac{1}{[CaN]}\right).$$

(5.38)

### 5.7.4 Control of Channel Yvc1p

Zhou et al [47] found that $Ca^{2+}$ presented from the cytoplasmic side activates Yvc1p. There is very little Yvc1p activity at or $< 10^{-6}$ M cytoplasmic $Ca^{2+}$, but the activities are clearly present in $10^{-4}$ M [47]. Hence, we propose a feedback controller for Yvc1p as follows

$$u_4 = \frac{[Ca^{2+}]_i}{K_{yvc} + [Ca^{2+}]_i},$$

(5.39)

where $K_{yvc}$ is a positive constant.

### 5.7.5 Control of X-induced Calcium Channel on the Vacuolar Membrane

We have assumed that the unidentified channel X on the vacuolar membrane may work in such a way as $IP_3$ receptors work. Thus, the unidentified protein X may be similar to $IP_3$. Since the synthesis of $IP_3$ depends on cell cycle, we assume that the synthesis of the protein X also depends on cell cycle. Therefore, we introduce a simple cell cycle model developed by Norel et al [26].

It has been suggested that the basic mechanism of activation and inactivation of the M (mitosis) phase promoting factor (MPF) underlies the cell cycle progression in all eukaryotic organisms [26]. MPF is activated by cyclin and inactivated by MPF-inactivase. During the interphase, cyclin accumulates until the rate of MPF activation by cyclin exceeds the rate of inactivation of MPF by MPF-inactivase. As a result, active MPF accumulates, which leads to a series of modifications of other mitotic substrates.

For formally describing MPF activation, we assume that (i) in the early embryos, cyclin synthesis suffices for the activation of MPF and for the induction of mitosis with each cyclin molecule activating more than one pre-MPF molecule, and that (ii) MPF activity is autocatalytic [26]. Let $C^*$ and $M^*$ denote the concentrations of cyclin and active MPF at any time. Then the equations that describe the dynamics of MPF and cyclin are [26]

$$\frac{dM^*}{dt^*} = e^*C^* + f^*C^*(M^*)^2 - \frac{g^*M^*}{M^* + K_m^*},$$

(5.40)

$$\frac{dC^*}{dt^*} = i^* - j^*M^*.$$

(5.41)

In equation (5.40), the first term describes the activation of MPF by cyclin, the second term implies that MPF activity is autocatalytic, and the third term describes the inactivation of MPF by a putative inactivase, which follows the Michaelis-Menten equation. In equation (5.41), the first term is the constant rate of accumulation of cyclin and the second term assumes that the degradation rate of cyclin is proportional to the concentration of active MPF.

It is convenient to introduce dimensionless variables. Thus we define

$$[M] = \frac{M^*}{K_m^*}, \quad [C] = \frac{C^*}{K_m^*}, \quad t = \frac{j^* t^*}{p},$$

where $p$ is a dimensionless scalar. Substituting these dimensionless variables into (5.40)-(5.41), we obtain

$$\frac{d[C]}{dt} = p(i - [M]), \tag{5.42}$$

$$\frac{d[M]}{dt} = p\left(k_{11}[C] + k_{12}[C][M]^2 - \frac{k_{13}[M]}{[M]+1}\right), \tag{5.43}$$

where

$$k_{11} = \frac{e^*}{j^*}, \quad k_{12} = \frac{f^*(K_m^*)^2}{j^*}, \quad k_{13} = \frac{g^*}{j^* K_m^*}, \quad i = \frac{i^*}{j^* K_m^*}.$$

The models for IP$_3$ synthesis have been developed by many researcher, for examples, Baran [3] and Keizer et al [18] to mention a few. Since the unidentified protein X may be similar to IP$_3$, we adopt the IP$_3$ model from [3] and [18] with a modification as the model of the protein X as follows

$$\frac{d[X]}{dt} = \frac{k_{14}[M]}{k_{15}+[M]} \frac{[Ca^{2+}]_i}{k_{16}+[Ca^{2+}]_i} - k_{17}[X], \tag{5.44}$$

where $k$'s are positive constants. In this model, we assume that the protein X synthesis is regulated by both MPF and the cytosolic calcium. Experiments by Mouillac et al [25] indicated that intracellular $Ca^{2+}$ modulates the production of IP$_3$. This suggests the possibility of a positive feedback mechanism of $Ca^{2+}$ on the production of IP$_3$. The stimulation of IP$_3$ production by $Ca^{2+}$ may even account for some observations of $Ca^{2+}$-induced $Ca^{2+}$ release (CICR) [18].

We propose that the induction of the channel by the protein X may be saturated and is given by

$$u_5 = \frac{[X]}{K_x + [X]}, \tag{5.45}$$

where $K_x$ is a positive constant.

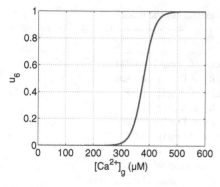 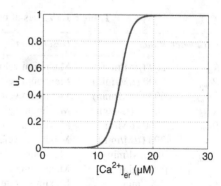

**Fig. 5.4.** *Left:* A feedback control of calcium homeostasis in Golgi. When the Golgi calcium level is above the resting level of 300 $\mu$M, vesicles in Golgi are activated and transport calcium out of Golgi. *Right:* A feedback control of calcium homeostasis in ER. When the ER calcium level is above the resting level of 10 $\mu$M, vesicles in ER are activated and transport calcium out of ER. Reproduced with permission from [41]

### 5.7.6 Control of Calcium Homeostasis in Golgi and ER

Since it is not clear how the calcium homeostasis in Golgi and ER is maintained, we propose mathematically the controllers $u_6$ and $u_7$ as follows

$$u_6 = \frac{1}{1 + a_1 \exp\left[a_2([\overline{Ca^{2+}}]_g - [Ca^{2+}]_g)\right]}, \tag{5.46}$$

$$u_7 = \frac{1}{1 + a_3 \exp\left[a_4([\overline{Ca^{2+}}]_{er} - [Ca^{2+}]_{er})\right]}, \tag{5.47}$$

where $[\overline{Ca^{2+}}]_g$ is the resting calcium level (300 $\mu$M) in Golgi, $[\overline{Ca^{2+}}]_{er}$ is the resting $Ca^{2+}$ level (10 $\mu$M) in the ER, and $k_8, k_9, a_1, a_2, a_3, a_4$ are positive constants. These two controllers regulate the transport of $Ca^{2+}$ out of Golgi and ER by vesicles, respectively. They are plotted in Fig. 5.4, which shows that when the Golgi (ER) calcium level is above the resting level of 300 $\mu$M (10 $\mu$M), vesicles in Golgi (ER) are activated and transport $Ca^{2+}$ out of Golgi (ER). Here the logistic function is used because a population of vesicles achieve the transport.

## 5.8 Simulation of Calcium Shocks

The model consisting of (5.7), (5.10), (5.15), (5.16), (5.23), (5.24), (5.34), (5.42), (5.43), and (5.44) is solved by using the function ode15s of MATLAB, the Math-Works, Inc. The values of parameters in the model are listed in Table 5.1. Initial conditions for numerical codes are listed in Table 5.2. These initial conditions are not the initial conditions shown in figures below because, in our numerical compu-

**Table 5.1.** Values of parameters of the model

| Parameter | Value | Description |
|---|---|---|
| $R_{ex}$ | 200 ($\mu$M/min) | Maximum velocity of pump X on PM |
| $R_{pmc}$ | 6000 ($\mu$M/min) | Maximum velocity of Pmc1p on vacuole |
| $R_{erpmc}$ | 6000 ($\mu$M/min) | Maximum velocity of Pmc1p on ER |
| $R_{vcx}$ | 10000 ($\mu$M/min) | Maximum velocity of Vcx1p |
| $R_{pmr}$ | 2000 ($\mu$M/min) | Maximum velocity of Pmr1p on Golgi |
| $R_{erpmr}$ | 200 ($\mu$M/min) | Maximum velocity of Pmr1p on ER |
| $R_{cod}$ | 10 ($\mu$M/min) | Maximum velocity of Cod1p |
| $R_{yvc}$ | 1.5 (/min) | Maximum rate of the channel Yvc1p |
| $R_x$ | 1.5 (/min) | Maximum rate of the channel X on vacuole stimulated by protein X |
| $K_{ex}$ | 500 ($\mu$M) | Michaelis-Menten constant of pump X on PM [21] |
| $K_{pmc}$ | 4.3 ($\mu$M) | Michaelis-Menten constant of Pmc1p [40] |
| $K_{erpmc}$ | 4.3 ($\mu$M) | Michaelis-Menten constant of Pmc1p |
| $K_{vcx}$ | 100 ($\mu$M) | Michaelis-Menten constant of Vcx1p [27] |
| $K_{yvc}$ | 100 ($\mu$M) | Constant of giving half maximum rate of Yvc1p |
| $K_{pmr}$ | 0.38 ($\mu$M) | Michaelis-Menten constant of Pmr1p on Golgi [37] |
| $K_x$ | 0.4 ($\mu$M) | Constant of giving half maximum rate of the channel on vacuole stimulated by protein X [4] |
| $K_{erpmr}$ | 0.38 ($\mu$M) | Michaelis-Menton constant of Pmr1p on ER [37] |
| $K_{cod}$ | 15 ($\mu$M) | Michaelis-Menton constant of Cod1p [8] |
| $[Ca^{2+}]_i$ | 0.06 ($\mu$M) | Steady state of $Ca^{2+}$ in the cytosol [1] |
| $[Ca^{2+}]_v$ | 30 ($\mu$M) | Steady state of $Ca^{2+}$ in the vacuole [16] |
| $[Ca^{2+}]_g$ | 300 ($\mu$M) | Steady state of $Ca^{2+}$ in Golgi [30] |
| $[Ca^{2+}]_{er}$ | 10 ($\mu$M) | Steady state of $Ca^{2+}$ in ER [1, 39] |
| $[CaM_0]$ | 25 ($\mu$M) | The total concentration of calmodulin [9] |
| $[CaN_0]$ | 25 ($\mu$M) | The total concentration of calcineurin [9] |
| $N$ | 13 | The number of relevant regulatory phosphorylation sites [36] |
| $L_0$ | $10^{-N/2}$ | The basic equilibrium constant [36] |
| $\lambda$ | 5 | The increment factor [36] |
| $a_1$ | 50 | Constant of determining the transport rate of calcium out of Golgi at the resting level |
| $a_2$ | 0.05 (1/$\mu$M) | Vesicle sensitivity to the calcium rise in Golgi |
| $a_3$ | 50 | Constant of determining the transport rate of calcium out of ER at the resting level |
| $a_4$ | 1 (1/$\mu$M) | Vesicle sensitivity to the calcium rise in ER |
| $p$ | 3/100 | Cell cycle scaling |
| $i$ | 1.2 | Cyclin input [26] |
| $k_1$ | 30 (1/(($\mu$M)$^3$ min)) | The rate of $Ca^{2+}$ binding to calmodulin |
| $k_2$ | 5 (1/($\mu$M min)) | The rate of calmodulin binding to calcineurin [5] |
| $k_3$ | 0.4 (1/min) | The import rate of Crz1p into nucleus [36] |
| $k_4$ | 28 (1/min) | The export rate of Crz1p from nucleus |
| $k_5$ | 0.1 (1/$\mu$M) | Calcineurin inhibition efficiency on Vcx1p |
| $k_6$ | 0.5 (1/min) | The rate of $Ca^{2+}$ binding to polyphosphate |
| $k_7$ | 0.3 | The percentage of $Ca^{2+}$ from Golgi into the vacuole |
| $k_8$ | 5 (1/min) | The maximum rate of $Ca^{2+}$ export from ER |

**Table 5.1** continued

| Parameter | Value | Description |
|-----------|-------|-------------|
| $k_9$ | 7 (1/min) | The maximum rate of $Ca^{2+}$ export from Golgi |
| $k_{10}$ | 0.1 | The percentage of $Ca^{2+}$ from Golgi into ER |
| $k_{11}$ | 3.5 | The rate of MPF production from cyclin [26] |
| $k_{12}$ | 1 | The rate of MPF production from cyclin under the activation of MPF [26] |
| $k_{13}$ | 10 | The degradation rate of MPF [26] |
| $k_{14}$ | 50000 ($\mu$M/min) | The maximum rate of production of the protein X |
| $k_{15}$ | 100 | The constant of giving half maximum impact of MPF on the production of the protein X |
| $k_{16}$ | 100 | The constant of giving half maximum impact of $Ca^{2+}$ on the production of the protein X |
| $k_{17}$ | 0.3 (1/min) | The degradation rate of the protein X |
| $k_{-1}$ | 10 (1/min) | The rate of $Ca^{2+}$ dissociation from calmodulin [9] |
| $k_{-2}$ | 5 (1/min) | The rate of calmodulin dissociation from calcineurin [9] |
| $l_{ex}$ | 0.15 (1/$\mu$M) | Calmodulin inhibition efficiency on the PM pump X |

**Table 5.2.** Initial conditions for the model

| Parameter | Value | Description |
|-----------|-------|-------------|
| $[Ca^{2+}]_i(0)$ | 0.089 ($\mu$M) | Initial cytosolic calcium |
| $[Ca^{2+}]_v(0)$ | 0.0 ($\mu$M) | Initial vacuolar calcium |
| $[Ca^{2+}]_g(0)$ | 0.0 ($\mu$M) | Initial Golgi calcium |
| $[Ca^{2+}]_{er}(0)$ | 0.0 ($\mu$M) | Initial ER calcium |
| $[CaM](0)$ | 0.0 ($\mu$M) | Initial $Ca^{2+}$-bound calmodulin |
| $[CaN](0)$ | $10^{-18}$ ($\mu$M) | Initial calmodulin-bound calcineurin |
| $h(0)$ | 0.0 | Initial nuclear fraction of Crz1p |
| $[C](0)$ | 0.8 | Initial cyclin [26] |
| $[M](0)$ | 0.4 | Initial maturation promoting factor [26] |
| $[X](0)$ | 0.0 ($\mu$M) | Initial protein X |

tations, we ran our MATLAB codes for some time so that the solution of the model reaches its steady state and then set that moment to 0.

The model can qualitatively reproduce calcium shocks observed in experiments. Fig. 5.5 shows that the calcium shock simulated by the model agrees qualitatively with the experimental data of Förster et al [17], although they do not match perfectly. It also shows that the $Ca^{2+}$ steady state in the cytosol is established within about 1 minute. This agrees with experimental observations of Dunn et al [16]. In simulating this calcium shock, we ran our MATLAB programs with the environmental calcium $[Ca^{2+}]_{ex}$ of 300 $\mu$M for 40 minutes so that the solution of the model reaches its steady state. We then set that moment to 0. 12 seconds later, $[Ca^{2+}]_{ex}$ was suddenly changed to 50000 $\mu$M in accordance with the experiments. Therefore, this simulation is independent of initial data.

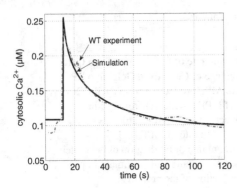

Fig. 5.5. Simulation of calcium shock. The environmental calcium is set to 300 $\mu$M during the initial 12 seconds and then is suddenly changed to 50000 $\mu$M, following the experiment of Förster and Kane [17]. The experimental data were provided by Förster and Kane. Reproduced with permission from [41]

## 5.9 Simulation of Calcium Accumulations

Kellermayer *et al* [19] observed experimentally that the total cellular Ca$^{2+}$ level in the *pmc1Δ* strain was roughly 2-fold lower than the WT strain. By contrast, the *pmr1Δ/pmc1Δ* strain contained 3.8-fold more Ca$^{2+}$ than the WT strain, and 2.2-fold more total cellular Ca$^{2+}$ than the *pmr1Δ* strain. Using our model, we reproduce this observation qualitatively in Fig. 5.6, which shows that our simulation is qualitatively close to the observation, Fig. 3B of [19]. In this simulation, the environmental calcium is set to 300 $\mu$M and calcium accumulates during a 30 minute time period, following the experiment of Kellermayer *et al* [19]. The inhibition constant $I_{ex}$ is set to 0. We found that the simulated Ca$^{2+}$ accumulation of *pmr1Δ* is very close to that of *pmr1Δ/pmc1Δ* if $I_{ex} = 0.15$. This may suggest that the inhibition of the pump X by calmodulin is very week. In simulating each mutant, its maximum velocity is set to 0.

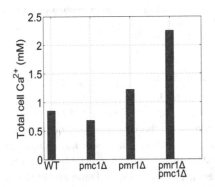
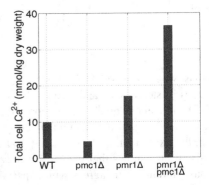

Fig. 5.6. Simulation of calcium accumulations. *Left:* Total cell Ca$^{2+}$ accumulations during a 30 minute time period simulated by the model with the extracellular calcium $[Ca^{2+}]_{ex}$ of 300 $\mu$M, following the experiment of Kellermayer *et al* [19]. *Right:* Modified from [19]. Data were estimated by eye [19]. Reproduced with permission from [41]

## 5.10 Prediction of Cell Cycle-dependent Oscillations of Calcium

Cell cycle-dependent calcium pulse was observed in *Xenopus* embryos [15]. This pulse is likely caused by the oscillation of cyclins and MPF during the cell cycle. Thus, we hypothesize that an unidentified protein X oscillates with MPF. This X releases calcium ions from the vacuole and could cause calcium oscillations in the cytosol, the Golgi lumen, and the ER lumen. Using our model, we simulate such oscillations as shown in Figs. 5.7 and 5.8. If we set the S-phase as the initial time 0, then Fig. 5.7 shows that as the cell cycle progresses from the S-phase to the G2-phase, cyclin B and MPF increase, causing $Ca^{2+}$ in the cytosol, ER, and Golgi to increase and $Ca^{2+}$ in the vacuole to decrease. The $Ca^{2+}$ levels in the cytosol, ER, and Golgi (in the vacuole) reach their maxima (minimum) in the M-phase and then decrease (increase) to the previous S-phase level during the G1-phase. This applies to the calmodulin, the calcineurin, the nuclear fraction $h$ of Crz1p, and the protein X, as shown in Fig. 5.8. These figures indicate that both oscillations and levels of the cytosolic calcium in *pmr1Δ* are higher than the oscillations and levels in wild type and *pmc1Δ* cells. This could suggest that Pmr1p plays a major role in controlling the cytosolic calcium dynamics. The figures also show that calmodulin, calcineurin, and

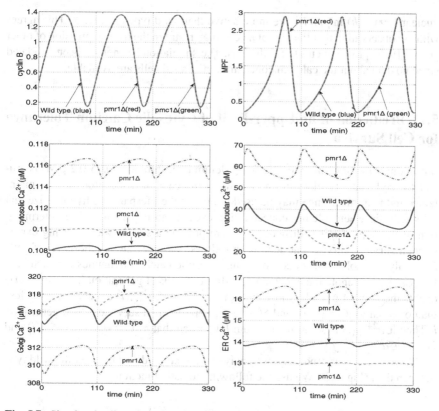

**Fig. 5.7.** Simulated cell-cycle-induced oscillations. Reproduced with permission from [41]

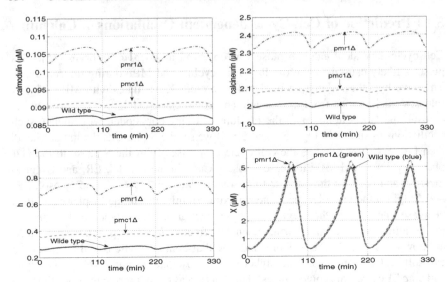

**Fig. 5.8.** Simulated cell-cycle-induced oscillations. Reproduced with permission from [41]

nuclear Crz1p ($h$) in *pmr1Δ* are more active than wild type or *pmc1Δ*. This agrees with the observation that the loss of a gene may result in the over-expression of other genes [19, 21] to compensate for the loss. The simulation has not yet been validated since no data about the calcium concentration in aged cells are available.

## 5.11 Prediction of an Upper-limit of Cytosolic Calcium Tolerance for Cell Survival

Since a narrow range of cytosolic free calcium concentration (0.05-0.2 $\mu$M) is crucial for cell viability, the time period when the calcium concentration was maintained in the normal range may be used to predict the lifespan. We hypothesize that cells will die when calcium concentration exceeds an upper-limit. The determination of the upper-limit is difficult. We here provide a simple way by using experimental survival curves. The idea is as follows. We generate a cohort (N) of cells randomly and solve our model to obtain cytosolic calcium dynamics for each cell. Let $[Ca^{2+}]_b(i)$ denote the calcium upper-limit of cells at generation $i$, $n([Ca^{2+}]_b(i))$ denote the number of cells whose cytosolic calcium $[Ca^{2+}]_i(i)$ at generation $i$ is less than or equal to $[Ca^{2+}]_b(i)$, and $S(i)$ be the experimental survival rate at generation $i$. Then $[Ca^{2+}]_b(i)$ is set to the calcium level such that $n([Ca^{2+}]_b(i))/N \leq S(i)$ and $(1+n([Ca^{2+}]_b(i)))/N > S(i)$.

To generate cells randomly, we analyze the random distribution of the wild type lifespan data (Fig. 5.2). We use the following logistic function

$$F(x) = \frac{1}{a+b\exp(-cx)}$$

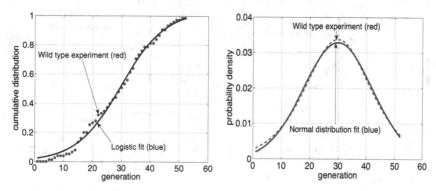

**Fig. 5.9.** Normal distribution of wild type lifespan. *Left:* A logistic function (blue solid curve) $F(x) = \frac{1}{a+b\exp(-cx)}$ fits into the experimental cumulative distribution of the wild type lifespan with $a = 1.059, b = 43.088$, and $c = 0.166$. *Right:* A normal probability density function (blue solid curve) $n(x) = \frac{1}{\sigma\sqrt{2\pi}}\exp\left(-\frac{(x-\mu)^2}{2\sigma^2}\right)$ with $\mu = 29.84$ and $\sigma = 12.13$ can be fitted well into the experimental probability density function (red dashed curve). Reproduced with permission from [41]

to fit the experimental cumulative distribution of the wild type lifespan (Fig. 5.9 (left)). This results in the parameter values $a = 1.059, b = 43.088$, and $c = 0.166$. We then obtain the probability density function

$$f(x) = F'(x) = \frac{bc\exp(-cx)}{(a+b\exp(-cx))^2}.$$

We find that the normal probability density function

$$n(x) = \frac{1}{\sigma\sqrt{2\pi}}\exp\left(-\frac{(x-\mu)^2}{2\sigma^2}\right) \tag{5.48}$$

with $\mu = 29.84$ and $\sigma = 12.13$ can be fitted well into $f(x)$ (Fig. 5.9 (right)). Thus the wild type lifespan follows the normal distribution. We then use the maximum likelihood estimator to obtain the mean $\mu = 28.43$ and the standard deviation $\sigma = 11.47$.

This analysis gives us a support to assume that maximum velocities of calcium pumps or channels follow a normal distribution. For instance, $R_{pmc}$ of Pmc1p can be randomized according to

$$\text{random } R_{pmc} = R_{pmc} + \frac{\sigma}{\mu}R_{pmc}\omega_n = R_{pmc} + \frac{11.47}{28.43}R_{pmc}\omega_n,$$

where $\omega_n$ is a random number generated from the standard normal distribution and $R_{pmc}$ is the mean given in Table 5.1.

By randomizing maximum velocities of calcium pumps or channels ($R_{pmc}, R_{vcx}, R_{pmr}, R_{erpmr}, R_{erpmc}, R_{cod}, R_{yvc}, R_x$), we generated a cohort (N=200) of wild type cells. The mutants $pmc1\Delta$ and $pmr1\Delta$ were simulated by setting $R_{pmc} = R_{erpmc} = 0$ or

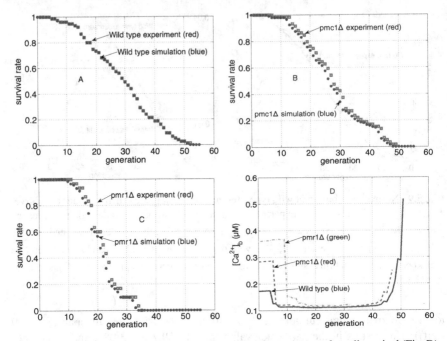

**Fig. 5.10.** Prediction of an upper-limit of cytosolic calcium tolerance for cell survival (Fig. D). The simulated survival rate is defined by $n([Ca^{2+}]_b(i))/N$, where $n([Ca^{2+}]_b(i))$ is the number of cells whose cytosolic calcium $[Ca^{2+}]_i(i)$ at generation $i$ is less than or equal to $[Ca^{2+}]_b(i)$, the upper-limit calcium $[Ca^{2+}]_b(i)$ is set to the calcium level such that $n([Ca^{2+}]_b(i))/N \leq S(i)$ and $(1 + n([Ca^{2+}]_b(i)))/N > S(i)$, $S(i)$ is the experimental survival rate at generation $i$, and $N$ is the number of total cells. In the simulation, the environmental calcium is set to 300 $\mu$M. Reproduced with permission from [41]

$R_{pmr} = R_{erpmr} = 0$, respectively. The generation time was assumed to be 120 minutes. We solve our model to obtain cytosolic calcium dynamics for each cell and then determined the upper-limit $[Ca^{2+}]_b(i)$ for wild type, $pmc1\Delta$, and $pmr1\Delta$, respectively (Fig. 5.10D). The simulated survival curves $n([Ca^{2+}]_b(i))/N$ and the experimental survival curves are plotted in Fig. 5.10 A, B, C.

The upper limits of Fig. 5.10D show that $pmc1\Delta$ and $pmr1\Delta$ have a high cytosolic calcium concentration of 0.35 $\mu$M for young ($< 10$ generations) cells. Examining Figs. 5.10B and C, this could suggest that almost all young mutants can tolerate such a high cytosolic calcium concentration. For aged cells ($> 35$ generations), no $pmr1\Delta$ can tolerate the concentration of 0.1 $\mu$M while a very small fraction (1%) of aged wild type cells ($> 50$ generations) can tolerate a high concentration of 0.5 $\mu$M.

Using the upper limit of cytosolic calcium tolerance for wild type cells, we simulate the life span of $pmc1\Delta$ and $pmr1\Delta$. Since a narrow range of calcium concentration (0.05-0.2 $\mu$M) is crucial for cell viability, the time period when the calcium concentration was maintained in the normal range may be used to predict the lifespan. We hypothesize that $pmc1\Delta$ and $pmr1\Delta$ will die when calcium concentration exceeds the upper limit of cytosolic calcium tolerance for wild type cells. This way we obtain

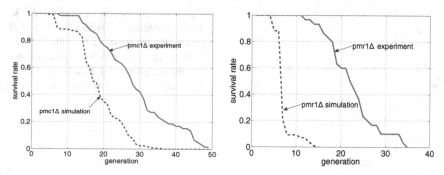

**Fig. 5.11.** Simulation of survival curves. The simulated survival curves were obtained by using the upper-limit of cytosolic calcium of wild type. The data are the same as in Fig. 5.10. Reproduced with permission from [41]

survival curves as shown in Fig. 5.11. The figure shows that the simulated survival curves are strikingly different from the experimental data. This difference could suggest that while the increased cytosolic calcium ion concentration is one factor of aging, some other factors in $pmc1\Delta$ and $pmr1\Delta$ may cancel the detrimental effect of high calcium ions. For example, unlike wild type cells, $pmr1\Delta$ cells do not increase intracellular hydrogen peroxide level after polychlorinated biphenyls treatment [35]. Thus, it is very likely that deletion of PMR1 triggers some protective responses that may extend the life span. Because our model did not consider these beneficial effects, our model either predicted a very high cytosolic calcium concentration (Fig. 5.10D) or a very short life span (Fig. 5.11(right)) for $pmr1\Delta$.

Note that the prediction of an upper-limit of cytosolic calcium tolerance for cell survival depends on the generation of random numbers of a software.

## 5.12 Model Limitation

Fig. 5.12 shows that the simulated shocks are qualitatively close to the experimental observation of Fig. 3A of Miseta $et$ $al$ [23], but $vcx1\Delta$ decays much faster than the observation after the shock. This indicates that our modeling about Vcx1p may not be accurate due to the lack of Vcx1p molecular regulation mechanisms. Since Vcx1p is an exchanger of $Ca^{2+}$ and $H^+$, the deletion of VCX1 may alter cytosolic pH and then may impact the function of calmodulin and calcineurin. Therefore the dynamics of $H^+$ may be needed to be included to refine the model.

In this simulation, we ran our MATLAB program with the environmental calcium of 300 $\mu M$ for 40 minutes so that the solution reaches its steady state. We then set that moment to 0. 12 seconds later, the environmental calcium $[Ca^{2+}]_{ex}$ is suddenly changed to 50000 $\mu M$, following the experiment of Miseta $et$ $al$ [23]. In simulating each mutant, its maximum velocity is set to 0. Because the loss of a gene may result in the over-expression of other genes [19, 21], the maximum velocity $R_{pmr}$ of Pmr1p in $pmc1\Delta$ is increased by 2-fold.

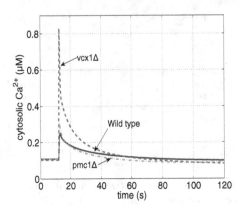

**Fig. 5.12.** Simulation of calcium shock. The environmental calcium is set to 300 $\mu$M during the initial 12 seconds and then is suddenly changed to 50000 $\mu$M, following the experiment of Miseta *et al* [23]. Reproduced with permission from [41]

## Exercises

**5.1.** The following are cell survival curve data [41]:

| $S$ | | | | | | | |
|---|---|---|---|---|---|---|---|
| 1 | 1 | 1 | 1 | 1 | 0.98667 | 0.98667 | 0.97333 |
| 0.96 | 0.96 | 0.96 | 0.94667 | 0.94667 | 0.93333 | 0.92 | 0.86667 |
| 0.84 | 0.8 | 0.8 | 0.74667 | 0.73333 | 0.72 | 0.68 | 0.66667 |
| 0.65333 | 0.62667 | 0.6 | 0.57333 | 0.56 | 0.52 | 0.49333 | 0.46667 |
| 0.44 | 0.38667 | 0.36 | 0.29333 | 0.26667 | 0.24 | 0.21333 | 0.21333 |
| 0.18667 | 0.18667 | 0.16 | 0.13333 | 0.093333 | 0.093333 | 0.066667 | 0.053333 |
| 0.04 | 0.026667 | 0.026667 | 0.013333 | 0.013333 | | | |
| $t\,(min)$ | | | | | | | |
| 0 | 120 | 240 | 360 | 480 | 600 | 720 | 840 |
| 960 | 1080 | 1200 | 1320 | 1440 | 1560 | 1680 | 1800 |
| 1920 | 2040 | 2160 | 2280 | 2400 | 2520 | 2640 | 2760 |
| 2880 | 3000 | 3120 | 3240 | 3360 | 3480 | 3600 | 3720 |
| 3840 | 3960 | 4080 | 4200 | 4320 | 4440 | 4560 | 4680 |
| 4800 | 4920 | 5040 | 5160 | 5280 | 5400 | 5520 | 5640 |
| 5760 | 5880 | 6000 | 6120 | 6240 | | | |

1. Construct a function to fit the cell survival curve data.
2. Define the cumulative distribution function by

$$cdf(t) = \text{the probability that cell lifespan is less than or equal to } t.$$

Then $cdf(t) = 1 - S(t)$. Fit the function

$$F(t) = \frac{1}{a + b\exp(-ct)}$$

into the experimental cumulative distribution $cdf(t)$.

3. The probability density function can be obtained by calculating the derivative of $F$ as follows:

$$f(t) = F'(t) = \frac{bc\exp(-ct)}{(a + b\exp(-ct))^2}.$$

Fit the normal probability density function

$$n(t) = \frac{1}{\sigma\sqrt{2\pi}} \exp\left(-\frac{(t-\mu)^2}{2\sigma^2}\right)$$

into $f(t)$.

**5.2.** Prove (5.32), (5.33), and (5.36).

**5.3.** Derive the dimensionless equations (5.42) and (5.43) from the equations (5.40) and (5.41).

**5.4.** Construct different calcium transport velocities for Yvc1p and the channel X on the vacuolar membrane and different feedback controllers for $u_4, u_5, u_6, u_7$ and then simulate the calcium shock as in Fig. 5.5.

**5.5.** Following the statistical analysis in Section 5.11, write a program with the model consisting of (5.7), (5.10), (5.15), (5.16), (5.23), (5.24), (5.34), (5.42), (5.43), and (5.44) to reproduce Fig. 5.10D.

**5.6.** Assume that cells can sense the cytosolic calcium concentration. Then an output equation can be set up:

$$y = [Ca^{2+}]_i.$$

1. Linearize the nonlinear control system (5.7), (5.10), (5.15), and (5.16) at a reasonable equilibrium of the states and controllers.
2. Examine the controllability and observability of the linearized control system.
3. Design a state feedback controller and an output feedback controller to stabilize the equilibrium point.

# References

1. Aiello D.P., Fu L., Miseta A., Bedwell D.M.: Intracellular Glucose 1-Phosphate and Glucose 6-Phosphate Levels Modulate $Ca^{2+}$ Homeostasis in Saccharomyces cerevisiae. J. Biol. Chem. **277**, 45751-45758 (2002).
2. Bagnat M., Keranen S., Shevchenko A., Shevchenko A., Simons K.: Lipid rafts function in biosynthetic delivery of proteins to the cell surface in yeast. Proc. Natl. Acad. Sci. USA. **97**, 3254-3259 (2000).
3. Baran I.: 1996. Calcium and cell cycle progression: possible effects of external perturbations on cell proliferation. Biophys J. **70**, 1198-1213 (1996).
4. Belde P.J., Vossen J.H., Borst-Pauwels G.W., Theuvenet A.P.: Inositol 1,4,5-trisphosphate releases $Ca^{2+}$ from vacuolar membrane vesicles of Saccharomyces cerevisiae. FEBS Lett. **323**, 113-118 (1993).
5. Bonilla M., Nastase K.K., Cunningham K.W.: Essential role of calcineurin in response to endoplasmic reticulum stress. EMBO J. **21**, 2343-2353 (2002).
6. Ciapa B, Pesando D, Wilding M, Whitaker M.: Cell-cycle calcium transients driven by cyclic change in inosital trisphosphate levels. Nature **368**, 875–878 (1994).

7. Courchesne W.E., Ozturk S.: Amiodarone induces a caffeine-inhibited, MID1 -depedent rise in free cytoplasmic calcium in Saccharomyces cerevisiae. Molecular Microbiology **47**, 223-234 (2003).

8. Cronin, S.R., Rao R., Hampton R.Y.: Cod1p/Spf1p is a P-type ATPase involved in ER function and $Ca^{2+}$ homeostasis. J. Cell Biology **157**, 1017-1028 (2002).

9. Cui J., Kaandorp J.A.: Mathematical modeling of calcium homeostasis in yeast cells. Cell Calcium. **39**, 337-348 (2006).

10. Cui J., Kaandorp J.A., Ositelu O.O., Beaudry V., Knight A., Nanfack Y.F., Cunningham K.W.: Simulating calcium influx and free calcium concentrations in yeast. Cell Calcium. **45**, 123-132 (2009).

11. Cunningham K.W., Fink G.R.: Calcineurin-dependent growth control in Saccharomyces cerevisiae mutants lacking PMC1, a homolog of plasma membrane $Ca^{2+}$ ATPases. J Cell Biol. **124**, 351-363 (1994).

12. Cunningham K.W., Fink G.R.: Calcineurin inhibits VCX1-dependent $H^+/Ca^{2+}$ exchange and induces $Ca^{2+}$ ATPases in Saccharomyces cerevisiae. Molecular and Cellular Biology **16**, 2226-2237 (1996).

13. Cyert M.S.: Genetic analysis of calmodulin and its targets in Sacchromyces cerevisiae. Annu. Rev. Genet. **35**, 647-672 (2001).

14. Denis V., Cyert M.S.: Internal $Ca^{2+}$ release in yeast is triggered by hypertonic shock and mediated by a TRP channel homologue. J. Cell Biology. **156**, 29-34 (2002).

15. Díaz J., Martínez-Mekler G.: Interaction of the IP3-$Ca^{2+}$ and MAPK signaling systems in the Xenopus blastomere: a possible frequency encoding mechanism for the control of the Xbra gene expression. Bull. Math. Biol. **67**, 433-465 (2005).

16. Dunn T., Gable K., Beeler T.: Regulation of Cellular $Ca^{2+}$ by Yeast Vacuoles. J. Biol. Chem. **269**, 7273-7278 (1994).

17. Förster C., Kane P.M.: Cytosolic $Ca^{2+}$ homeostasis is a constitutive function of the V-ATPase in Saccharomyces cerevisiae. J. Biol. Chem. **275**, 38245-38253 (2000).

18. Keizer J., De Young G.W.: Two roles for $Ca^{2+}$ in agonist stimulated $Ca^{2+}$ oscillations. Biophys. J. **61**, 649-660 (1992).

19. Kellermayer R., Aiello D.P., Miseta A., Bedwell D.M.: Extracellular $Ca^{2+}$ sensing contributes to excess $Ca^{2+}$ accumulation and vacuolar fragmentation in a pmr1$\Delta$ mutant of S. cerevisiae. J. Cell Science **116**, 1637-1646 (2003).

20. Klemm R.W., Ejsing C.S., Surma M.A., Kaiser H.J., Gerl M.J., Sampaio J.L., de Robillard Q., Ferguson C., Proszynski T.J., Shevchenko A., Simons K.: Segregation of sphingolipids and sterols during formation of secretory vesicles at the trans-Golgi network. J Cell Biol. **185**, 601-612 (2009).

21. Locke E.G., Bonilla M., Liang L., Takita Y., Cunningham K.W.: A Homolog of voltage-gated $Ca^{2+}$ channels stimulated by depletion of secretory $Ca^{2+}$ in yeast. Molecular and Cellular Biology **20**, 6686-6694 (2000).

22. Matsuura I., Kimura E., Tai K., Yazawa M.: Mutagenesis of the fourth calcium-binding domain of yeast calmodulin. J. Bio. Chem. **169**, 13267-13273 (1993).

23. Miseta A., Kellermayer R., Aiello D.P., Fu L., Bedwell D.M.: The vacuolar $Ca^{2+}/H^+$ exchanger Vcx1p/Hum1p tightly controls cytosolic $Ca^{2+}$ levels in *S. cerevisiae*. FEBS Letters **451**, 132-136 (1999).

24. Miseta A., Fu L., Kellermayer R., Buckley J., Bedwell D.M.: The Golgi apparatus Plays a Significant Role in the Maintenance of $Ca^{2+}$ Homeostasis in the vps33$\Delta$ Vacuolar Biogenesis Mutant of Saccharomyces cerevisiae. J. Biol. Chem. **274**, 5939-5947 (1999).

25. Mouillac B., Balestre M.N., Guillon G.: Positive feedback regulation of phospholipase C by vasopressin-induced calcium mobilization in WRK1 cells. Cellular Signaling **2**, 497-507 (1990).

26. Norel R., Agur Z.: A model for the adjustment of the mitotic clock by cyclin and MPF levels. Science **251**, 1076-1078 (1991).

27. Ohsumi Y., Anraku Y.: Calcium transport driven by a proton motive force in vacuolar membrane vesicles of Saccharomyces cerevisiae. J. Biol. Chem. **258**, 5614-5617 (1983).

28. Okamura H., Aramburu J., Garcia-Rodriguez C., Viola J.P.B., Raghavan A., Tahiliani M., Zhang X., Qin J., Hogan P.G., Rao A.: Concerted dephosphorylation of the transcription factor NFAT1 induces a conformational switch that regulates transcriptional activity. Molecular Cell **6**, 539-550 (2000).

29. Palmer C.P., Zhou X., Lin J., Loukin S.H., Kung C., Saimi Y.: A TRP homolog in Saccharomyces cerevisiae forms an intracellular $Ca^{2+}$-permeable channel in the yeast vacuolar membrane. Proc. Natl. Acad. Sci. U. S. A. **98**, 7801-7805 (2001).

30. Pinton P., Pozzan T., Rizzuto R.: The Golgi apparatus is an inositol 1,4,5-trisphosphate-sensitive $Ca^{2+}$ store, with functional properties distinct from those of the endoplasmic reticulum. EMBO Journal **17**, 5298-5308 (1998).

31. Pozos T.C., Sekler I., Cyert M.S.: The product of HUM1, a novel yeast gene, is required for vacuolar $Ca^{2+}/H^+$ exchange and is related to mammalian $Na^+/Ca^{2+}$ exchangers. Mol. Cell. Biol. **16**, 3730-3741 (1996).

32. Putney J.W. Jr.: A model for receptor-regulated calcium entry. Cell Calcium **7**, 1-12 (1986).

33. Putney J.W. Jr., McKay, R.R.: Capacitative calcium entry channels. Bioessays **21**, 38-46 (1999).

34. Putney J.W. Jr.: Recent breakthroughs in the molecular mechanism of capacitative calcium entry (with thoughts on how we got here), Cell Calcium **42**, 103-110 ( 2007).

35. Ryu J.H., Lee Y., Han S.K., Kim H.Y.: The role of hydrogen peroxide produced by polychlorinated biphenyls in PMR1-deficient yeast cells. J. Biochem. **134**, 137-142 (2003).

36. Salazar C., Höfer T.: Allosteric regulation of the transcription factor NFAT1 by multiple phosphorylation sites: a mathematical analysis. J. Mol. Biol. **327**, 31-45 (2003) .

37. Sorin A., Rosas G., Rao R.: PMR1, a $Ca^{2+}$-ATPase in yeast Golgi, has properties distinct from sarco/endoplasmic reticulum and plasma membrane calcium pumps. J. Biol. Chem. **272**, 9895-9901 (1997).

38. Starovasnik M.A., Davis T.N., Klevit, R.E.: Similarities and differences between yeast and vertebrate calmodulin: an examination of the calcium-binding and structural properties of calmodulin from the yeast Saccharomyces cerevisiae. Biochemistry **32**, 3261-3270 (1993).

39. Strayle J., Pozzan T., Rudolph H.K.: Steady-state free $Ca^{2+}$ in the yeast endoplasmic reticulum reaches only 10 mM and is mainly controlled by the secretory pathway pump Pmr1. EMBO Journal **18**, 4733-4743 (1999).

40. Takita Y., Engstrom L., Ungermann C., Cunningham K.W.: Inhibition of the $Ca^{2+}$-ATPase Pmc1p by the v-SNARE Protein Nyv1p. J. Biol. Chem. **276**, 6200-6206 (2001).

41. Tang F, Liu W.: An Age-dependent feedback control model for calcium in yeast cells. J. Math. Biol. **60**, 849-879 (2010).

42. Tang F., Watkins W, Bermudez M., Gray R., Gaban A., Portie K., Grace S., Kleve M., Craciun G.: A life span-extending form of autophagy employs the vacuole-vacuole fusion machinery. Autophagy **4**, 874-886 (2008).

43. Tanida I., Hasegawa A., Iida H., Ohya Y., Anraku Y.: Cooperation of calcineurin and vacuolar $H^+$-ATPase in intracellular $Ca^{2+}$ homeostasis of yeast cells. J. Biol. Chem. **270**, 10113-10119 (1995).

44. Ward J.P.T., Robertson T.P., Aaronson P.I.: Capacitative calcium entry: a central role in hypoxic pulmonary vasoconstriction? Am. J. Physiol. Lung. Cell. Mol. Physiol. **289**, L2-L4 (2005).

45. Wei Y., Marchi V., Wang R., Rao R.: An N-terminal EF hand-like motif modulates ion transprot by Pmr1, the yeast Golgi $Ca^{2+}/Mn^{2+}$-ATPase. Biochemistry **38**, 14534-14541 (1999).
46. Zamponi G.W.: The L-type calcium channe C-terminus: sparking interest beyond its role in calcium-dependent inactivation. J. Physiol. **552**, 333-333 (2003).
47. Zhou X.-L., Batiza A.F., Loukin S.H., Palmer C.P., Kung C., Saimi Y.: The transient receptor potential channel on the yeast vacuole is mechanosensitive. Proc. Natl. Acad. Sci. USA **100**, 7105-7110 (2003).

# 6

# Kinetics of Ion Pumps and Channels

Various ions, such as $Ca^{2+}$, $Na^+$, and $K^+$, enter and exit a cell through selective ion pumps and channels, such as potassium channels, voltage-gated sodium channels, voltage-gated calcium channels, sodium/calcium exchangers, and plasma membrane (PM) calcium ATPases, as demonstrated in Fig. 6.1. Calcium ions $Ca^{2+}$ enter the cytosol through store-operated channels (SOC) and voltage-gated calcium channels (VGCC). The sarcoplasmic or endoplasmic reticulum $Ca^{2+}$-ATPases (SERCA) pump $Ca^{2+}$ from the cytosol into the endoplasmic reticulum (ER) and $Ca^{2+}$ in ER are released to the cytosol through the inositol (1,4,5)-trisphosphate (IP$_3$)- and $Ca^{2+}$-mediated inositol (1,4,5)-trisphosphate receptors (IP$_3$R). $Ca^{2+}$ enter the mitochondrion through uniporters and exit through $Ca^{2+}/Na^+$ antiporters. $Ca^{2+}$ exit the cytosol through plasma membrane $Ca^{2+}$-ATPases (PMCA) and $Ca^{2+}/Na^+$ exchangers. Depletion of ER $Ca^{2+}$ stores causes STIM1 to move to ER-PM junctions, bind to Orai1, and activate store-operated channels for $Ca^{2+}$ entry [63]. Sodium ions $Na^+$ enter the cytosol through sodium channels (NC) and $Ca^{2+}/Na^+$ exchangers, and exit through $N^+/K^+$ ATPases (NKA). Potassium ions $K^+$ enter the cytosol through $N^+/K^+$ ATPases and exit through ATP-sensitive $K^+$ channels (ATPKC), delayed rectifying $K^+$ channels (DrKC), and calcium-activated $K^+$ channels (CAKC). $H^+$ ions in mitochondrion are ejected by the respiratory chain driven by the energy released from oxidation of NADP, which are produced from the tricarboxylic acid cycle (TAC, also called Krebs cycle). The established electrochemical gradient drives the electrogenic transport of ions, including ATP and ADP by adenine nucleotide translocators (ANT). $H^+$ ions in the cytosol flow back to the mitochondrion through $F_1F_0$-ATPases to power the ATP synthesis.

Mathematical models for ionic pumps and channels have been developed. We start with the Nernst-Planck equation that plays a key role in modeling the ionic pumps and channels.

Liu W.: Introduction to Modeling Biological Cellular Control Systems.
DOI 10.1007/978-88-470-2490-8_6, © Springer-Verlag Italia 2012

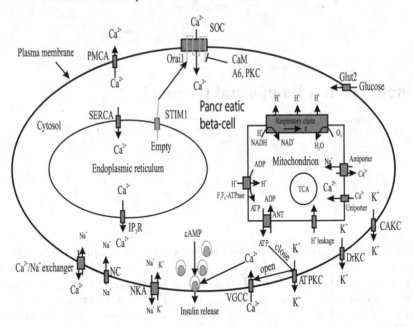

**Fig. 6.1.** Schematic diagram of ions fluxes in a pancreatic $\beta$-cell. The abbreviations used are: SOC, store-operated channel; CaM, calmodulin; A6, annexin 6; PKC, protein kinase C; Glut2, glucose transporter 2; CAKC, calcium-activated $K^+$ channel; DrKC, delayed rectifying $K^+$ channel; ATPKC, ATP-sensitive $K^+$ channel; VGCC, voltage-gated calcium channel; NKA, $N^+/K^+$ ATPase; NC, $Na^+$ channel; PMCA, plasma membrane $Ca^{2+}$ ATPase, SERCA, sarcoplasmic or endoplasmic reticulum $Ca^{2+}$-ATPase; IP$_3$R, inositol (1,4,5)-trisphosphate receptor; ANT, adenine nucleotide translocator; TCA, tricarboxylic acid cycle (also called Krebs cycle)

## 6.1 The Nernst-Planck Equation

The flow of ions across a membrane is driven by both a concentration gradient and an electric field. The relation among the flux of ions, the concentration gradient, and the electric field is governed by the Nernst-Planck equation. The flux of ions due to the concentration gradient is given by Fick's law [33]

$$J_c = -D\nabla c, \qquad (6.1)$$

where the positive constant $D$ is the diffusion constant and $c$ is the concentration of the ion. The flux of the ion due to the electric field is given by Planck's equation [26, 33]

$$J_e = -u\frac{z}{|z|}c\nabla\phi, \qquad (6.2)$$

where $u$ is the mobility of the ion, defined as the velocity of the ion under a constant unit electric field, $z$ is the valence of the ion, and $\phi$ is the electrical potential. The relationship between the ionic mobility $u$ and Fick's diffusion constant $D$ is given by (due to Einstein)

$$D = \frac{uRT}{|z|F},$$ (6.3)

where $R$ is the universal gas constant, $T$ is the absolute temperature, $F$ is Faraday's constant. Combining equations (6.1), (6.2) and (6.3), we obtain the Nernst-Planck equation

$$J = -D\left(\nabla c + \frac{zF}{RT}c\nabla\phi\right).$$ (6.4)

If the flow of ions and the electric field are transverse to the membrane, we can assume that $c$ and $\phi$ depends on $x$ only and then the equation (6.4) becomes

$$J = -D\left(\frac{dc}{dx} + \frac{zF}{RT}c\frac{d\phi}{dx}\right).$$ (6.5)

## 6.2  The Nernst Equilibrium Potential

At equilibrium, the ions do not move, and then $J = 0$. Thus we have

$$\frac{dc}{dx} + \frac{zF}{RT}c\frac{d\phi}{dx} = 0,$$

and then

$$\frac{1}{c}\frac{dc}{dx} + \frac{zF}{RT}\frac{d\phi}{dx} = 0.$$

Let the inside of the membrane be at $x = 0$ and the outside at $x = L$. Integrating the above equation from 0 to $L$, we obtain

$$\ln\left(\frac{c_o}{c_i}\right) = \frac{zF}{RT}(\phi_i - \phi_o),$$

where $c_o = c(L), c_i = c(0), \phi_o = \phi(L), \phi_i = c(0)$. Let $V_I = \phi_i - \phi_o$ denote the potential difference across the membrane due to the concentration difference of the ion. Then we obtain the Nernst equation

$$V_I = \frac{RT}{zF}\ln\left(\frac{c_o}{c_i}\right).$$ (6.6)

This potential is called the *Nernst potential*.

## 6.3  Current-voltage Relations

The relation between the electric current produced by the flow of ions across a membrane and the voltage potential across the membrane is generally governed by the Nernst-Planck equation (6.5). This general current-voltage model needs to be simplified so that it can be used conveniently in modeling. To this end, we assume that the electric field $\frac{d\phi}{dx}$ is constant across the membrane:

$$\frac{d\phi}{dx} = -\frac{V}{L},$$

where $V = \phi(0) - \phi(L)$ is the membrane voltage potential. The Nernst-Planck equation (6.5) becomes

$$\frac{dc}{dx} = \frac{zFV}{RTL}c - \frac{J}{D}. \tag{6.7}$$

At the steady state and without production of ions, the flux $J$ is constant. Solving (6.7), we obtain

$$\exp\left(\frac{-zVF}{RT}\right)c_o = \frac{JRTL}{zDVF}\left[\exp\left(\frac{-zVF}{RT}\right) - 1\right] + c_i,$$

and then

$$J = \frac{zDFV\left[c_i - c_o\exp\left(\frac{-zVF}{RT}\right)\right]}{LRT\left[1 - \exp\left(\frac{-zVF}{RT}\right)\right]}.$$

Because the ionic flux $J$ is related to the ionic current $I$ by the expression

$$I = zFJ, \tag{6.8}$$

multiplying $J$ by $zF$, we obtain the famous Goldman-Hodgkin-Katz (GHK) current equation

$$I_{ion} = \frac{Pz^2F^2V\left[c_i - c_o\exp\left(\frac{-zVF}{RT}\right)\right]}{RT\left[1 - \exp\left(\frac{-zVF}{RT}\right)\right]}, \tag{6.9}$$

where $P = D/L$ is the permeability of the membrane to the ion.

If the electric field is not constant across the membrane, we then need Poisson's equation for the potential $\phi$

$$\frac{d^2\phi}{dx^2} = -\frac{qc}{\varepsilon}, \tag{6.10}$$

where $q$ is the unit electric charge and $\varepsilon$ is the dielectric constant of the channel medium. Then the steady-state flux $J$ is governed by the system of equations (6.5) and (6.10). In general, it is impossible to solve the system exactly.

While the nonlinear GHK current-voltage model (6.9) is needed in some cases, such as the currents in open $Na^+$ and $K^+$ channels in vertebrate axons [33], a linear current-voltage equation is sufficient to approximately model the currents in many other cases, such as the currents in open $Na^+$ and $K^+$ channels in the squid giant axon.

To derive such a linear current-voltage model, we assume that the voltage potential $V$ across the plasma membrane consists of two components: the potential $V_I$ caused by the concentration difference of the ion and the potential $V_c$ caused by the electrical current $I_{ion}$ due to the movement of the ion. Then Ohm's law gives

$$V_c = rI_{ion},$$

where $r$ is a channel resistance. Thus we obtain

$$V = V_I + V_c = V_I + rI_{ion},$$

and then

$$I_{ion} = g(V - V_I), \tag{6.11}$$

where $g = 1/r$ is a membrane conductance. This is the *linear current-voltage equation*. Using the Nernst equation (6.6), we obtain

$$I_{ion} = gV - \frac{gRT}{zF} \ln\left(\frac{c_o}{c_i}\right).$$

## 6.4 The Potassium Channel

The potassium channel is a protein found in the plasma membrane of almost all cells [5]. It is a tube that links the cytosol with the extracellular fluid. Potassium ions, which cannot pass through the lipid bilayer of the plasma membrane, can pass through the potassium channel easily. The channel is selective for potassium and other ions cannot pass through. Potassium channels include the voltage-gated potassium channel, the calcium-activated potassium channel, and the ATP-sensitive potassium channel.

Potassium is much more concentrated in the cytosol than outside, typically 140 mmol/L in the cytosol but only 5 mmol/L in the extracellular medium. Thus there is a tendency for potassium ions to leave the cell down the concentration gradient. On the other hand, all cells have a voltage across their membrane when they are not being stimulated. This voltage is called the *resting voltage*, about -80 mV in a relaxed skeletal muscle cell (0 mV outside and -80 mV in the cytosol). Thus this negative voltage produces an electrical force pulling the positively charged potassium ions back inside. When the opposing electrical and concentration gradient are equal, the overall electrochemical gradient for potassium is zero and an equilibrium is established. The transmembrane voltage at the equilibrium is called the *equilibrium voltage* for potassium. For potassium, the equilibrium voltage is about -90 mV for a normal animal cell. Because the potassium channels are the major pathway by which ions can cross the plasma membrane of an *unstimulated* cell, the resting voltage has a value close to the potassium equilibrium voltage. In the following, we use the word *depolarization* to mean any positive increase in the transmembrane voltage.

### 6.4.1 The Voltage-Gated Potassium Channel

Voltage-gated potassium channels are potassium channels sensitive to voltage changes in the cell's membrane potential. Using the linear current-voltage equation (6.11), Hodgkin and Huxley [27] modeled the voltage-gated potassium ionic current of the squid giant axon as follows

$$I_k = G_k P_k (V - V_k), \tag{6.12}$$

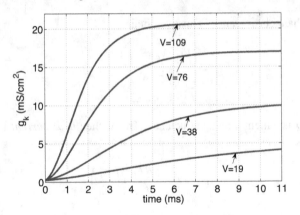

**Fig. 6.2.** Rise of potassium conductance associated with different depolarizations. This figure is modified from [27], using (6.13) with $g_{k0} = 0.24$ mS/cm$^2$, $g_{k\infty} = 20.7, 17, 10.29, 5$ mS/cm$^2$, $\tau = 1.05, 1.5, 2.6, 4.5$ ms for $V = 109, 76, 38, 19$ mV, respectively. In [27], the voltage potential was defined to be the difference between the outside and the inside of the membrane. Thus the depolarization potentials were negative. In this text, the voltage potential is defined to be the difference between the inside and the outside of the membrane. Hence, the negative depolarization potentials are changed to the positive ones

where $V$ is the potential difference between the inside and the outside of the membrane, $V_k$ is the equilibrium potential for the potassium ions, $G_k$ is a maximal ionic conductance, and $P_k$ is a channel open probability. The units of potential, current density, and conductance density are mV, $\mu$A/cm$^2$, mS/cm$^2$, respectively.

The experimental data showed that the conductance $g_k = G_k P_k$ is not constant and varies with time $t$ and voltage $V$, as shown in Fig. 6.2. Hodgkin and Huxley [27] observed that the time course data of $g_k$ can be well fitted by the function

$$g_k(t) = \left[ g_{k\infty}^{1/4} - \left( g_{k\infty}^{1/4} - g_{k0}^{1/4} \right) e^{-t/\tau} \right]^4 \tag{6.13}$$

with appropriate parameters $g_{k0}$, $g_{k\infty}$, and $\tau$, as shown in Fig. 6.2. Since $g_k$ depends on $V$, $g_{k\infty}$ should be a function of $V$. Thus we assume that

$$g_{k\infty}^{1/4} = \frac{G_k^{1/4} \alpha_n(V)}{\alpha_n(V) + \beta_n(V)},$$

where $\alpha_n$ and $\beta_n$ are constants which vary with voltage but not with time. Substituting it into (6.13), we obtain

$$g_k(t) = G_k \left[ \frac{\alpha_n(V)}{\alpha_n(V) + \beta_n(V)} - \left( \frac{\alpha_n(V)}{\alpha_n(V) + \beta_n(V)} - \left( \frac{g_{k0}}{G_k} \right)^{1/4} \right) e^{-t/\tau} \right]^4 . \tag{6.14}$$

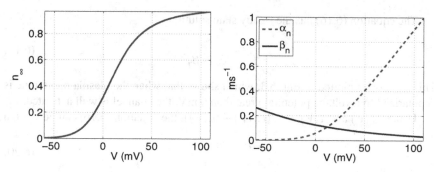

**Fig. 6.3.** Dependence of the potassium channel activation on voltage potential. $\alpha_n$, $\beta_n$, and $n_\infty$ are defined by (6.17), (6.18), and (6.19), respectively

Note that $n = \dfrac{\alpha_n(V)}{\alpha_n(V)+\beta_n(V)} - \left(\dfrac{\alpha_n(V)}{\alpha_n(V)+\beta_n(V)} - \left(\dfrac{g_{k0}}{G_k}\right)^{1/4}\right) e^{-t/\tau}$ is the solution of

$$\frac{dn}{dt} = -\frac{n}{\tau} + \frac{1}{\tau}\frac{\alpha_n(V)}{\alpha_n(V)+\beta_n(V)}, \quad n(0) = \left(\frac{g_{k0}}{G_k}\right)^{1/4}.$$

Selecting $\tau = 1/(\alpha_n(V)+\beta_n(V))$, we obtain from (6.14) that

$$P_k = n^4, \tag{6.15}$$

$$\frac{dn}{dt} = \alpha_n(1-n) - \beta_n n, \tag{6.16}$$

where $\alpha_n$ and $\beta_n$ have the dimension of [time]$^{-1}$ and $n$ is a dimensionless variable which can vary between 0 and 1. Using the experimental data at the temperature of 6°C, the rates $\alpha_n$ and $\beta_n$ were estimated as follows (see Exercise 6.2)

$$\alpha_n = \frac{0.01(10-V)}{\exp((10-V)/10)-1}, \tag{6.17}$$

$$\beta_n = 0.125\exp\left(\frac{-V}{80}\right). \tag{6.18}$$

Hodgkin and Huxley [27] gave a physical basis of the equation (6.16). If it is assumed that potassium ions can only cross the membrane when four similar particles occupy a certain region of the membrane, then $n$ represents the proportion of the particles in a certain position (for example at the outside of the membrane) and $1-n$ represents the proportion that are somewhere else (for example at the inside of the membrane). Thus, $\alpha_n$ determines the rate of transfer from inside to outside, while $\beta_n$ determines the transfer in the opposite direction, as demonstrated below:

$$n \underset{\alpha_n}{\overset{\beta_n}{\rightleftharpoons}} 1-n.$$

When the membrane potential $V$ increases, more K$^+$ flow out of cells and then $\alpha_n$ should increase and $\beta_n$ should decrease, as shown in Fig. 6.3.

The equation (6.16) has the steady state solution

$$n_\infty = \frac{\alpha_n}{\alpha_n + \beta_n}. \tag{6.19}$$

This solution is plotted in Fig. 6.3, which shows that when the plasma membrane is depolarized to a voltage potential great than 0 mV, the channel is well activated.

Using $n_\infty(V)$ and $\tau(V) = 1/(\alpha_n(V) + \beta_n(V))$, the equation (6.16) can be written as

$$\frac{dn}{dt} = \frac{n_\infty - n}{\tau}. \tag{6.20}$$

Here $\tau$ is called a *time constant*. This form clearly specifies the steady state $n_\infty$ and the time constant $\tau$, and it is often used in the literature.

In summary, the voltage-gated potassium current can be modeled by

$$I_k = G_k n^l (V - V_k), \tag{6.21}$$

$$\frac{dn}{dt} = \frac{n_\infty - n}{\tau}, \tag{6.22}$$

where $l$ is a positive number, $n_\infty = n_\infty(V)$ is a steady state of activation probability, and $\tau = \tau(V)$ is a time constant. Different $l, n_\infty, \tau$ were proposed in the literature. Based on the experimental data of delayed potassium currents in mouse pancreatic $\beta$-cell obtained by Rorsman and Trube [56] (Fig. 6.4), Sherman et al [58] proposed the following first-order activation model:

$$l = 1, \tag{6.23}$$

$$n_\infty(V) = \frac{1}{1 + \exp[(V_{kh} - V)/C_{kh}]}, \tag{6.24}$$

$$\tau(V) = \frac{c}{\exp[(V - \bar{V})/a] + \exp[(\bar{V} - V)/b]}, \tag{6.25}$$

where $V_{kh} = -19$ mV, $C_{kh} = 5.6$ mV (Exercise 6.3), $a = 65$ mV, $b = 20$ mV, $\bar{V} = -75$ mV, and $c = 60$ ms. Bertram et al [3] used the following

$$l = 1, \tag{6.26}$$

$$n_\infty(V) = \frac{1}{1 + \exp[(-9 - V)/10]}, \tag{6.27}$$

$$\tau(V) = \frac{8.3}{1 + \exp[(V + 9)/10]}. \tag{6.28}$$

Based on the experimental result of Smith et al [59], Fridlyand et al [19] proposed the following second-order activation model

$$l = 2, \tag{6.29}$$

$$n_\infty = \frac{1}{1 + \exp[(-9 - V)/5]}, \tag{6.30}$$

$$\tau = 25 \text{ ms} \tag{6.31}$$

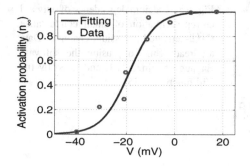

**Fig. 6.4.** Voltage dependence of outward current activation of delayed rectifier potassium channels. Data are read from [56] using the software Engauge Digitizer 4.1, and fitted by the function (6.24)

with $G_k = 45000$ pS and $V_k = -75$ mV. The steady $n_\infty$ in (6.30) is similar to Smith et al's activation curve [59]:

$$n_\infty = \frac{1}{1 + \exp[(-19 - V)/8]}.$$

Note that the kinetic model (6.12) derived from the squid giant axon is not universal. For other types of cells, other models may be more appropriate. For instance, for vertebrate axons, the nonlinear GHK current-voltage model (6.9) is required [6, 15, 16, 17, 33].

### 6.4.2 The Calcium-Activated Potassium Channel

The slowly varying potassium current through the calcium-activated potassium channel can be modeled by the linear current-voltage equation (6.11) as follows

$$I_{kca} = G_{kca} P_{kca} (V - V_k), \tag{6.32}$$

where $G_{kca}$ is the maximum conductance and $P_{kca}$ is the probability of the channel opening. Different models for the channel open probability $P_{kca}$ were proposed. One of such models was proposed by Plant [51]. $Ca^{2+}$ were assumed to interact with the potassium channel (denoted by $E$) by a first-order process

$$E + Ca^{2+} \underset{k_2}{\overset{k_1}{\rightleftharpoons}} O,$$

where $E$ and $O$ represent the closed and open states of the potassium channel, respectively. At equilibrium, the open and closed states satisfy

$$k_1[E][Ca^{2+}]_i = k_2[O],$$

and then

$$[E][Ca^{2+}]_i = K_{kca}[O], \tag{6.33}$$

where $K_{kca} = k_2/k_1$ and $[Ca^{2+}]_i$ is the concentration of free intracellular $Ca^{2+}$. Let $[E] + [O] = [E_0]$, where $[E_0]$ is the concentration of total potassium channels. We

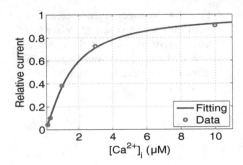

**Fig. 6.5.** Calcium dependence of outward current activation of small conductance $Ca^{2+}$-activated potassium channels. Data are read from [65] using the software Engauge Digitizer 4.1, and fitted by the Hill function (6.35)

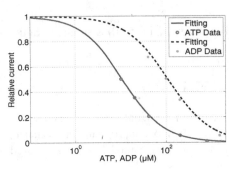

**Fig. 6.6.** ATP or ADP dependence of outward current activation of ATP-sensitive potassium channels in HEK293 cells. Data are read from [29] using the software Engauge Digitizer 4.1, and fitted by the Hill function (6.37) with $K_{katp} = 11$ $\mu$M and $n = 0.97$ for ATP, and $K_{katp} = 95$ $\mu$M and $n = 0.94$ for ADP

deduce from (6.33) that the channel open probability $P_{kca}$ is

$$P_{kca} = \frac{[O]}{[E_0]} = \frac{[Ca^{2+}]_i}{K_{kca} + [Ca^{2+}]_i}. \qquad (6.34)$$

Later on, the experimental data [36, 38, 65] showed that the activation probability of the $Ca^{2+}$-activated potassium channels can be modeled by the Hill function (Exercise 6.4)

$$P_{kca} = \frac{[Ca^{2+}]_i^n}{K_{kca}^n + [Ca^{2+}]_i^n} \qquad (6.35)$$

with $K_{kca} = 1.5$ $\mu$M and $n = 1.3$ (Fig. 6.5). Consequently, different Hill exponents were used in the literature, for example, 3 in [10], 4 in [18], and 5 in [2, 48] with $G_{kca} = 130$ pS and $K_{kca}$ varying in the range of 0.05-0.9 $\mu$M.

### 6.4.3 The ATP-Sensitive Potassium Channel

An ATP-sensitive potassium channel is a type of potassium channel that is inhibited by ATP. The current through ATP-sensitive potassium channel can be modeled by the linear I-V model:

$$I_{katp} = G_{katp}P_{katp}(V - V_k), \qquad (6.36)$$

where $G_{katp}$ (7500 pS used by Keizer *et al* [34] and 3000 pS used by Magnus *et al* [42]) is a maximum conductance and $P_{katp} = P_{katp}([ATP], [ADP])$ is a channel open probability to be found. There are two ways to determine the channel open probability $P_{katp}$. One way is to fit a function like the Hill function into data. The other way is to propose a molecular model of multiple channel states and test the model with data.

The experimental data obtained by John *et al* [29] showed that $P_{katp}$ can be described by the Hill function

$$P_{katp} = \frac{K_{katp}^n}{K_{katp}^n + [ATP]^n} \tag{6.37}$$

with $K_{katp} = 11\ \mu M$ and $n = 0.97$ (Fig. 6.6).

The experiments conducted by Kakei *et al* [31] showed that ADP shifts the ATP-dose-dependent activation curve of current to the right as shown in Fig. 6.7. Thus ADP promotes the activation of the ATP-sensitive potassium channel. Based on this experimental result, Keizer *et al* [34] assumed that ADP and ATP bind to the channel competitively:

$$E + ADP \underset{k_2}{\overset{k_1}{\rightleftharpoons}} EADP,$$

$$E + ATP \underset{k_4}{\overset{k_3}{\rightleftharpoons}} EATP,$$

where $E$ stands for the channel. At equilibrium, we derive that

$$[EADP] = \frac{[E][ADP]}{K_1}, \quad [EATP] = \frac{[E][ATP]}{K_2},$$

where $K_1 = k_2/k_1$ and $K_2 = k_4/k_3$. If only the unbound state and ADP-bound state are conducting, then the activation fraction of the channel is given by

$$\begin{aligned}
P_{katp} &= \frac{[E] + [EADP]}{[E] + [EADP] + [EATP]} \\
&= \frac{[E] + [E][ADP]/K_1}{[E] + [E][ADP]/K_1 + [E][ATP]/K_2} \\
&= \frac{1 + [ADP]/K_1}{1 + [ADP]/K_1 + [ATP]/K_2}. \tag{6.38}
\end{aligned}$$

Based on their experiments, Hopkins *et al* [28] proposed an ATP-sensitive $K^+$ channel model as demonstrated in Fig. 6.8. They assumed that the channel has three binding sites: one competitive site for $ATP^{4-}$ and $ADP^{3-}$ and two sites for $MgADP^-$. When the channel is free or bound by a single $MgADP^-$, its open probability is 0.08. When it is bound by two $MgADP^-$, its open probability is 0.89. When it is bound by $ATP^{4-}$ or $ADP^{3-}$ in all cases, it is closed.

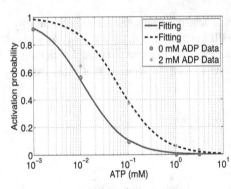

**Fig. 6.7.** ATP or ADP dependence of outward current activation of ATP-sensitive potassium channels in rat pancreatic $\beta$-cells. Data are read from [31] using the software Engauge Digitizer 4.1, and fitted by the competitive binding kinetics (6.38) with $K_1 = 0.45$ mM and $K_2 = 0.012$ mM. This figure is modified from [34]

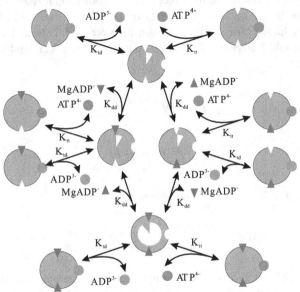

**Fig. 6.8.** ATP-sensitive $K^+$ channel model proposed by Hopkins *et al* [28]. It was assumed that the channel has three binding sites: one competitive site for $ATP^{4-}$ and $ADP^{3-}$ and two sites for $MgADP^-$. When the channel is free or bound by a single $MgADP^-$, its open probability is 0.08. When it is bound by two $MgADP^-$, its open probability is 0.89. When it is bound by $ATP^{4-}$ or $ADP^{3-}$ in all cases, its open probability is 0

The binding reactions in Fig. 6.8 can be described by

$$E + MgADP^- \underset{K_{dd}}{\Longleftrightarrow} E_1 MgADP,$$

$$E + MgADP^- \underset{K_{dd}}{\Longleftrightarrow} E_2 MgADP,$$

$$E_1 MgADP + MgADP^- \underset{K_{dd}}{\Longleftrightarrow} E2MgADP,$$

$$E_2 MgADP + MgADP^- \underset{K_{dd}}{\Longleftrightarrow} E2MgADP,$$

$$E + ATP^{4-} \underset{K_{tt}}{\Longleftrightarrow} EATP,$$

$$E_1MgADP + ATP^{4-} \underset{K_{tt}}{\Longleftrightarrow} E_1MgADPATP,$$

$$E_2MgADP + ATP^{4-} \underset{K_{tt}}{\Longleftrightarrow} E_2MgADPATP,$$

$$E2MgADP + ATP^{4-} \underset{K_{tt}}{\Longleftrightarrow} E2MgADPATP,$$

$$E + ADP^{3-} \underset{K_{td}}{\Longleftrightarrow} EADP,$$

$$E_1MgADP + ADP^{3-} \underset{K_{td}}{\Longleftrightarrow} E_1MgADPADP,$$

$$E_2MgADP + ADP^{3-} \underset{K_{td}}{\Longleftrightarrow} E_2MgADPADP,$$

$$E2MgADP + ADP^{3-} \underset{K_{td}}{\Longleftrightarrow} E2MgADPADP,$$

where $E$ denotes the free K$^+$ channel, $E_1MgADP$ denotes the K$^+$ channel with a single $MgADP^-$ bound at the above site, $E_2MgADP$ denotes the K$^+$ channel with a single $MgADP^-$ bound at the site below, and $K_{dd}, K_{td}, K_{tt}$ are dissociation constants. For a general reaction

$$AB \rightleftharpoons mA + nB,$$

the dissociation constant is defined as

$$K_d = \frac{[A]^m [B]^n}{[AB]}.$$

At equilibrium, we derive from the above reaction that

$$[E][MgADP^-] = K_{dd}[E_1MgADP],$$
$$[E][MgADP^-] = K_{dd}[E_2MgADP],$$
$$[E_1MgADP][MgADP^-] = K_{dd}[E2MgADP],$$
$$[E_2MgADP][MgADP^-] = K_{dd}[E2MgADP],$$
$$[E][ATP^{4-}] = K_{tt}[EATP],$$
$$[E_1MgADP][ATP^{4-}] = K_{tt}[E_1MgADPATP],$$
$$[E_2MgADP][ATP^{4-}] = K_{tt}[E_2MgADPATP],$$
$$[E2MgADP][ATP^{4-}] = K_{tt}[E2MgADPATP],$$
$$[E][ADP^{3-}] = K_{td}[EADP],$$
$$[E_1MgADP][ADP^{3-}] = K_{td}[E_1MgADPADP],$$
$$[E_2MgADP][ADP^{3-}] = K_{td}[E_2MgADPADP],$$
$$[E2MgADP][ADP^{3-}] = K_{td}[E2MgADPADP].$$

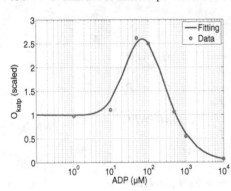

**Fig. 6.9.** ADP dependence of outward current activation of ATP-sensitive potassium channels in mouse pancreatic $\beta$-cells. Data are read from [28] using the software Engauge Digitizer 4.1, and fitted by the function (6.40) with $K_{dd} = 5.7$ $\mu$M, $K_{td} = 8.2$ $\mu$M, $K_{tt} = 0.95$ $\mu$M, and $[ATP] = 0.00001$ $\mu$M

Let $[E_T]$ denote the concentration of total K$^+$ channels. Then we have

$$
\begin{aligned}
[E_T] &= [E] + [E_1 MgADP] + [E_2 MgADP] + [E2MgADP] + [EATP] \\
&\quad + [E_1 MgADPATP] + [E_2 MgADPATP] + [E2MgADPATP] \\
&\quad + [EADP] + [E_1 MgADPADP] \\
&\quad + [E_2 MgADPADP] + [E2MgADPADP] \\
&= [E] + 2\frac{[E][MgADP^-]}{K_{dd}} + \frac{[E][MgADP^-]^2}{K_{dd}^2} \\
&\quad + \frac{[E][ATP^{4-}]}{K_{tt}} + 2\frac{[E][MgADP^-][ATP^{4-}]}{K_{dd}K_{tt}} + \frac{[E][MgADP^-]^2[ATP^{4-}]}{K_{dd}^2 K_{tt}} \\
&\quad + \frac{[E][ADP^{3-}]}{K_{td}} + 2\frac{[E][MgADP^-][ADP^{3-}]}{K_{dd}K_{td}} + \frac{[E][MgADP^-]^2[ADP^{3-}]}{K_{dd}^2 K_{td}} \\
&= [E]\left(1 + \frac{[MgADP^-]}{K_{dd}}\right)^2 \left(1 + \frac{[ATP^{4-}]}{K_{tt}} + \frac{[ADP^{3-}]}{K_{td}}\right).
\end{aligned}
$$

It then follows that the channel open probability is given by

$$
\begin{aligned}
P_{katp} &= \frac{0.08([E] + [E_1 MgADP] + [E_2 MgADP])}{[E_T]} + \frac{0.89[E2MgADP]}{[E_T]} \\
&= \frac{0.08\left(1 + 2\frac{[MgADP^-]}{K_{dd}}\right) + 0.89\frac{[MgADP^-]^2}{K_{dd}^2}}{\left(1 + \frac{[MgADP^-]}{K_{dd}}\right)^2 \left(1 + \frac{[ATP^{4-}]}{K_{tt}} + \frac{[ADP^{3-}]}{K_{td}}\right)}.
\end{aligned} \tag{6.39}
$$

The ions $MgADP^-$, $ATP^{4-}$, $ADP^{3-}$ are related to ADP and ATP as follows (see, e.g., [43])

$$
[MgADP^-] = 0.165[ADP], \quad [ADP^{3-}] = 0.135[ADP], \quad [ATP^{4-}] = 0.05[ATP].
$$

It then follows from (6.39) that

$$
P_{katp} = \frac{0.08\left(1+\frac{0.33[ADP]}{K_{dd}}\right)+0.89\left(\frac{0.165[ADP]}{K_{dd}}\right)^2}{\left(1+\frac{0.165[ADP]}{K_{dd}}\right)^2\left(1+\frac{0.135[ADP]}{K_{td}}+\frac{0.05[ATP]}{K_{tt}}\right)}.
\tag{6.40}
$$

This function after scaled to 1 at $[ADP] = 1$ $\mu$M can be well fitted into the data of Fig. 2 of Hopkins et al [28] with $K_{dd} = 5.7$ $\mu$M, $K_{td} = 8.2$ $\mu$M, $K_{tt} = 0.95$ $\mu$M, and $[ATP] = 0.00001$ $\mu$M, as shown in Fig. 6.9.

## 6.5 The Voltage-Gated Sodium Channel

The cells in multicellular animals that are specialized for rapid conduction have a voltage-gated sodium channel. When the transmembrane voltage is -70 mV, the voltage-gated sodium channel is gated shut. When the plasma membrane is depolarized, the channel opens rapidly and then, after about 1 ms, inactivates. After the channel has gone through this cycle, it must spend at least 1 ms with the transmembrane voltage at the resting voltage before it can be opened by a second depolarization.

Like the potassium channel, the sodium ionic current of the squid giant axon was modeled by Hodgkin and Huxley [27] as follows

$$
I_{na} = G_{na}P_{na}(V - V_{na}),
\tag{6.41}
$$

where $V$ is the potential difference between the inside and the outside of the membrane, $V_{na}$ is the equilibrium potential for the sodium ions, $G_{na}$ is a maximal ionic conductance, and $P_{na}$ is a channel open probability.

As in the case of the potassium conductance, the experimental data showed that the conductance $g_{na} = G_{na}P_{na}$ is not constant and varies with time $t$ and voltage $V$, as shown in Fig. 6.10. Hodgkin and Huxley [27] observed that the time course data of $g_{na}$ can be well fitted by the function

$$
g_{na}(t) = G_{na}\left[m_\infty - (m_\infty - m_0)e^{-t/\tau_m}\right]^3\left[h_\infty - (h_\infty - h_0)e^{-t/\tau_h}\right]
\tag{6.42}
$$

with appropriate parameters $G_{na}$, $m_\infty$, $m_0$, $\tau_m$, $h_\infty$, $h_0$, and $\tau_h$ (Fig. 6.10), where $G_{na}$ is a constant with the dimension of conductance/cm$^2$. In the resting state, the sodium conductance is very small compared with the value attained during a large depolarization. Therefore, $m_0$ can be neglected if the depolarization is greater than 30 mV. Further, inactivation is very nearly complete if $V > 30$ mV so that $h_\infty$ may also be neglected. The expression for the sodium conductance then becomes

$$
g_{na}(t) = G_{na}^*\left(1 - e^{-t/\tau_m}\right)^3 e^{-t/\tau_h},
\tag{6.43}
$$

where $G_{na}^* = G_{na}m_\infty^3 h_0$. The function (6.43) was fitted into experimental data as shown in Fig. 6.10.

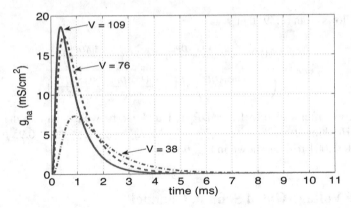

**Fig. 6.10.** Changes of sodium conductance associated with different depolarizations. This figure is modified from [27], using (6.43) with $g_{nam}^* = 40.3, 39.5, 20$ mS/cm$^2$, $\tau_m = 0.14, 0.189, 0.382$ ms, and $\tau_h = 0.67, 0.84, 1.27$ ms for $V = 109, 76, 38$ mV, respectively. In [27], the voltage potential was defined to be the difference between the outside and the inside of the membrane. Thus the depolarization potentials were negative. In this text, the voltage potential is defined to be the difference between the inside and the outside of the membrane. Hence, the negative depolarization potentials are changed to the positive ones

Note that $h = h_\infty - (h_\infty - h_0)e^{-t/\tau_h}$ and $m = m_\infty - (m_\infty - m_0)e^{-t/\tau_m}$ are the solutions of

$$\frac{dh}{dt} = \frac{h_\infty - h}{\tau_h}, \quad h(0) = h_0, \tag{6.44}$$

$$\frac{dm}{dt} = \frac{m_\infty - m}{\tau_m}, \quad m(0) = m_0. \tag{6.45}$$

We then have that

$$P_{na} = hm^3. \tag{6.46}$$

Since $g_{na}$ depends on $V$, the parameters $h_\infty, \tau_h, m_\infty, \tau_m$ are functions of $V$. To construct these functions, we define $\alpha_h, \beta_h, \alpha_m, \beta_m$ as follows

$$\tau_h = 1/(\alpha_h + \beta_h),$$
$$h_\infty = \alpha_h/(\alpha_h + \beta_h),$$
$$\tau_m = 1/(\alpha_m + \beta_m),$$
$$m_\infty = \alpha_m/(\alpha_m + \beta_m).$$

$\alpha$'s and $\beta$'s have the dimension of [time]$^{-1}$. Using the experimental data at the temperature of 6°C [27], the rates $\alpha_h, \alpha_m, \beta_h,$ and $\beta_m$ were estimated as follows (see Exercise 6.7)

$$\alpha_h = 0.07 \exp\left(\frac{-V}{20}\right), \tag{6.47}$$

$$\beta_h = \frac{1}{\exp\left(\frac{30-V}{10}\right) + 1}, \tag{6.48}$$

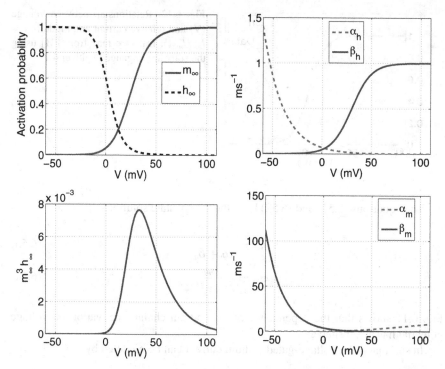

**Fig. 6.11.** Dependence of the sodium channel activation on voltage potential. Functions $\alpha_h$, $\beta_h$, $\alpha_m$, $\beta_m$, $h_\infty$, and $m_\infty$ are defined by (6.47), (6.48), (6.49), (6.50), (6.53), and (6.54), respectively

$$\alpha_m = \frac{0.1(25 - V)}{\exp\left(\frac{25-V}{10}\right) - 1}, \tag{6.49}$$

$$\beta_m = 4\exp\left(\frac{-V}{18}\right). \tag{6.50}$$

The equations (6.44) and (6.45) may be given a physical basis if sodium conductance is assumed to be proportional to the number of sites on the inside of the membrane which are occupied simultaneously by three activating molecules but are not blocked by an inactivating molecule. For this, we rewrite the equations (6.44) and (6.45) as

$$\frac{dh}{dt} = \alpha_h(1-h) - \beta_h h, \tag{6.51}$$

$$\frac{dm}{dt} = \alpha_m(1-m) - \beta_m m. \tag{6.52}$$

Then it can be explained that $m$ represents the proportion of activating molecules on the inside and $1 - m$ the proportion on the outside; $h$ is the proportion of inactivating molecules on the outside and $1 - h$ the proportion on the inside. $\alpha_m$ or $\beta_h$ and $\beta_m$ or $\alpha_h$ represent the transfer rate constants in the two directions.

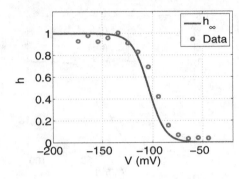

**Fig. 6.12.** Relationship between voltage (V) and relative current amplitude ($h_\infty = I_{na}/I_{max}$). Data are read from [21] using the software Engauge Digitizer 4.1

The equations (6.51) and (6.52) have the steady state solutions

$$h_\infty = \frac{\alpha_h}{\alpha_h + \beta_h}, \tag{6.53}$$

$$m_\infty = \frac{\alpha_m}{\alpha_m + \beta_m}. \tag{6.54}$$

Fig. 6.11 shows that the dependence of the sodium channel activation on voltage potential has a bell shape.

In summary, the voltage-gated sodium current can be modeled by

$$I_{na} = G_{na} m^j h^k (V - V_{na}), \tag{6.55}$$

$$\frac{dh}{dt} = \frac{h_\infty - h}{\tau_h}, \tag{6.56}$$

$$\frac{dm}{dt} = \frac{m_\infty - m}{\tau_m}, \tag{6.57}$$

where $j, k$ are positive numbers, $h_\infty = h_\infty(V)$ and $m_\infty = m_\infty(V)$ are steady states of inactivation probability and activation probability, respectively, and $\tau_h = \tau_h(V)$, $\tau_m = \tau_m(V)$ are time constants.

The other model of sodium ionic current used in the literature for pancreatic $\beta$-cells [18, 21] is given by

$$I_{na} = G_{na} \frac{V - V_{na}}{1 + \exp[(104 + V)/8]}, \tag{6.58}$$

where $G_{na}$ is the maximum whole cell conductance. The relationship between voltage (V) and relative current amplitude

$$h_\infty(V) = I_{na}/I_{max} = \frac{1}{1 + \exp[(104 + V)/8]}$$

is shown in Fig. 6.12, reproduced from Fig. 5B of Göpel *et al* [21].

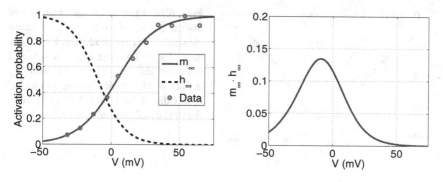

**Fig. 6.13.** Voltage dependence of inward current activation of voltage-gated calcium channels. Data are read from [56] using the software Engauge Digitizer 4.1, and fitted by the function (6.67). The function $h_\infty$ is defined by (6.68)

## 6.6  The Voltage-Gated Calcium Channel

The voltage-gated calcium channel is a protein in the plasma membrane [5]. It forms a tube that opens to the extracellular medium, but closes at the end of the cytosol side when the transmembrane voltage is at the resting voltage. If the membrane depolarizes so that the cytosol is less negative, the channel is open and then calcium ions can pass through it to enter the cytosol. After about 100 ms, an inactivation plug can bind to the inside of the open channel, resulting in the inactivation of the channel. If the membrane is repolarized, the channel quickly recloses.

Like the potassium and sodium channels, the calcium ionic current through the voltage-gated calcium channel can be described by the linear current-voltage equation (6.11) as follows

$$I_{ca} = G_{ca}P_{ca}(V - V_{ca}),  \tag{6.59}$$

where $V$ is the potential difference between the inside and the outside of the membrane, $V_{ca}$ is the equilibrium potential for the calcium ions, $G_{ca}$ is a maximal ionic conductance, and $P_{ca}$ is a channel open probability. A number of models for $P_{ca}$ were proposed in the literature.

Since the calcium channel gating in the $\beta$-cells is similar to the sodium channel gating in the squid giant axon, Chay and Keizer [9] adopted the usual Hodgkin-Huxley model (6.46) for the calcium current as follows

$$P_{ca} = hm^3,  \tag{6.60}$$

$$\frac{dh}{dt} = \alpha_h(1 - h) - \beta_h h,  \tag{6.61}$$

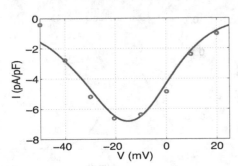

**Fig. 6.14.** Current-voltage relation of inward current of voltage-gated calcium channels. Data are read from [56] using the software Engauge Digitizer 4.1, and fitted by the function (6.59) with $P_{ca}$ defined by (6.69) with scaling it to pA/pF by dividing it by $1000 \times 5.31$ (5.31 pF is the total membrane capacitance)

$$\frac{dm}{dt} = \alpha_m(1-m) - \beta_m m, \tag{6.62}$$

$$\alpha_h = 0.07 \exp\left(\frac{-V-V_{ca}^*}{20}\right), \tag{6.63}$$

$$\beta_h = \frac{1}{\exp\left(\frac{30-V-V_{ca}^*}{10}\right)+1}, \tag{6.64}$$

$$\alpha_m = 0.1\frac{25-V-V_{ca}^*}{\exp\left(\frac{25-V-V_{ca}^*}{10}\right)-1}, \tag{6.65}$$

$$\beta_m = 4\exp\left(\frac{-V-V_{ca}^*}{18}\right). \tag{6.66}$$

They used the exact $\alpha_h, \beta_h, \alpha_m$, and $\beta_m$ given by Hodgkin and Huxley except that the voltage $V$ was shifted to the left by $V_{ca}^* = 50$ mV. This shift is plausible because Fig. 6.11 and Fig. 6.13 show that the voltage-gated calcium channel is activated at a lower voltage than the voltage-gated sodium channel.

Based on the experimental data of voltage-gated calcium current in mouse pancreatic $\beta$-cell obtained by Rorsman and Trube [56], Sherman *et al* [58] proposed a model for the calcium current. The data showed that the activation probability of the voltage-gated calcium channel at steady state can be modeled by (Fig. 6.13)

$$m_\infty(V) = \frac{1}{1+\exp[(V_{mh}-V)/C_{mh}]} \tag{6.67}$$

with $V_{mh} = 4$ mV and $C_{mh} = 14$ mV (Exercise 6.8). They found that the current-voltage relation (6.59) with $P_{ca} = m_\infty(V)$ and $V_{ca} = 110$ mV was inadequate to describe all of Rorsman and Trube's data for the calcium current (Fig. 6.14). Thus, they introduced the following reverse sigmoidal factor

$$h_\infty(V) = \frac{1}{1+\exp[(V-V_{hh})/C_{hh}]}. \tag{6.68}$$

This factor is very similar to $h_\infty$ defined in (6.53) for the voltage-gated sodium channel (Fig. 6.11). Then they obtained the following channel open probability

$$P_{ca} = m_\infty(V)h_\infty(V). \tag{6.69}$$

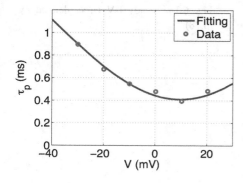

**Fig. 6.15.** Time constant of activation of voltage-gated calcium channels in mouse pancreatic $\beta$-cells. Data are read from [21] using the software Engauge Digitizer 4.1, and fitted by the function (6.74)

The current-voltage relation (6.59) with this channel open probability can be well fitted into Rorsman and Trube's data for the calcium current (Fig. 6.14) with $V_{hh} = -13$ mV, $C_{hh} = 9$ mV, and $G_{ca} = 2548$ pS (Exercise 6.9).

Another model for the calcium ionic current in pancreatic $\beta$-cells was given by Bertram *et al* [3] as follows

$$I_{ca} = G_{ca}m_\infty(V)(V - V_{ca}), \qquad (6.70)$$

$$m_\infty(V) = \frac{1}{1 + \exp[(-22 - V)/7.5]}, \qquad (6.71)$$

where $G_{ca} = 280$ pS is the maximum whole cell conductance and $V_{ca} = 100$ mV is the $Ca^{2+}$ resting potential. This model was used by Fridlyand *et al* [18] with different parameter values.

While $I_{ca}$ is rapidly activated by membrane depolarization, its inactivation is controlled by both $Ca^{2+}$ and the membrane potential [22]. Binding of intracellular $Ca^{2+}$ to calmodulin causes a channel conformational change to occlude the channel pore [7, 44]. This $Ca^{2+}$-dependent inactivation was not considered in the above models.

To account for the $Ca^{2+}$-dependent inactivation, Fridlyand *et al* [19] proposed the following model

$$P_{ca} = p_o h_v h_{ca}, \qquad (6.72)$$

where

$$p_\infty = 0.002 + \frac{1}{1 + \exp[(-2 - V)/8.8]}, \qquad (6.73)$$

$$\tau_p = 2.2 - 1.79 \exp\left[-((V - 9.7)/70.2)^2\right], \qquad (6.74)$$

$$h_v = \frac{1}{1 + \exp[(9 + V)/8]}, \qquad (6.75)$$

$$\frac{dp_o}{dt} = \frac{p_\infty - p_o}{\tau_p}, \qquad (6.76)$$

$$\frac{dh_{ca}}{dt} = 0.007(1 - h_{ca}) - 0.0025(-I_{ca}/G_{ca})h_{ca}, \qquad (6.77)$$

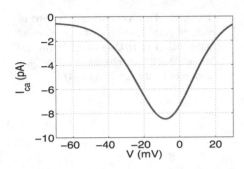

**Fig. 6.16.** Steady I-V curve defined by (6.78)

where $G_{ca} = 1500\,\text{pS}$ and $V_{ca} = 100\,\text{mV}$. In this model, $p_o$ is the variable of activation of the channel upon membrane depolarization, $h_v$ is the voltage-sensitive inactivation variable, and $h_{ca}$ is the $Ca^{2+}$-sensitive inactivation variable. Note that $p_\infty$ is close to $m_\infty$ defined by (6.67) and $h_v$ is close to $h_\infty$ defined by (6.68). The equation for the time constant $\tau_p$ was obtained by fitting the data from Fig. 4B of Göpel *et al* [21], as shown in Fig. 6.15.

To model the $Ca^{2+}$-sensitive channel inactivation, Fridlyand *et al* suggested that the rate of inactivation is proportional to a $Ca^{2+}$ current through a unitary channel (the term $I_{ca}/G_{ca}$) rather than $Ca^{2+}$ concentration as has been suggested in other models [12, 40]. The dynamics of $h_{ca}$ is determined by the relative rates of activation and inactivation as described in (6.77).

Solving the steady state equation of (6.77) for the steady state $h_{ca\infty}$, we can obtain the steady I-V relationship:

$$I_{ca} = G_{ca}p_\infty h_v h_{ca\infty}(V - V_{ca}). \tag{6.78}$$

We plot the I-V curve in Fig. 6.16, which shows that the U-shape of the I-V curve agrees well with experimental results obtained in [21, 49, 52, 56].

## 6.7 The IP$_3$ Receptor

The action of an extracellular agonist, such as hormones, growth factors, and neuro-transmitters, on its specific receptor typically activates the phosphoinositide-specific phospholipase C (PLC). PLC breaks down the phosphatidylinositol 4,5 bisphosphate (PIP$_2$) to generate two second messengers, the inositol 1,4,5 trisphosphate (IP$_3$) and diacylglycerol (DAG) [53]. DAG is known to activate protein kinase C (PKC) iso-forms, but can also regulate ion channels in a PKC-independent manner. IP$_3$ diffuses through the cytoplasm until it binds and activates its receptor to release $Ca^{2+}$ from the endoplasmic reticulum (ER) into the cytosol.

This IP$_3$ receptor (IP$_3$R) is a universal intracellular $Ca^{2+}$-release channel predominantly located on the endoplasmic reticulum (ER). The release of $Ca^{2+}$ from the ER into the cytoplasm controlled by IP$_3$ is crucial for setting up complex spatio-temporal $Ca^{2+}$ signals, which control cellular processes as different as fertilization, cell divi-

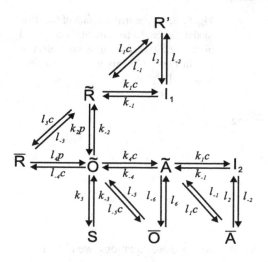

Fig. 6.17. An IP$_3$R model proposed by Sneyd et al [60]. R, receptor; O, open; A, activated; S, shut; I, inactivated. c is [Ca$^{2+}$]; p is [IP$_3$]

sion, cell migration, differentiation, metabolism, muscle contraction, secretion, neuronal processing, and ultimately cell death [64]. Since the ER calcium concentration is much higher than the cytosolic one, we assume that the Ca$^{2+}$ transport through the receptor is driven by the calcium concentration gradient:

$$r_{ipr} = R_{ipr}P_{ipr}\left([Ca^{2+}]_{er} - [Ca^{2+}]_i\right) \tag{6.79}$$

where $R_{ipr}$ is a maximal rate, $P_{ipr}$ is a channel open probability, and $[Ca^{2+}]_{er}$ denotes the concentration of Ca$^{2+}$ in the ER.

IP$_3$R is a tetramer. Its four subunits have a similar general structure. Each subunit consists of about 2700 a.a. (amino acids). The linear sequence of the IP$_3$R consists of three large regions, an N-terminally located IP$_3$-binding region of about 600 a.a., a large modulatory and transducing region (about 1600 a.a.) and a small C-terminal region (about 500 a.a.) containing the 6 transmembrane domains. More recently, it has been shown that the N-terminal IP$_3$-binding region is composed of a suppressor domain and an IP$_3$-binding core, while the C-terminal region is composed of a channel region and a coupling region [64]. In addition, the IP$_3$R structure undergoes major conformational changes under influence of Ca$^{2+}$. Today at least 12 different protein kinases are known to directly phosphorylate the IP$_3$R. Several mathematical models for the IP$_3$R kinetics have been established.

Sneyd et al [60] constructed an IP$_3$R model as shown in Fig. 6.17. Although it appears to contain a multiplicity of states, there are specific reasons for each one. The background structure is simple. Ignoring the various tildes, hats, and primes, we see that a receptor, R, can bind Ca$^{2+}$ and inactivate to state I$_1$, or it can bind IP$_3$ and open to state O. State O can then shut (state S) or bind Ca$^{2+}$ and activate to state A. State A can then bind Ca$^{2+}$ and inactivate to state I$_2$. This structure is clearer in Fig. 6.18.

To simplify the model, Sneyd et al [60] assumed that the transitions $\tilde{R} \rightleftharpoons \bar{R}$, $\tilde{O} \rightleftharpoons \bar{O}$, $\tilde{A} \rightleftharpoons \bar{A}$, and $\tilde{R} \rightleftharpoons R'$ are fast and in instantaneous equilibrium. They then obtained that $c\tilde{R} = L_3\bar{R}$, $c\tilde{R} = L_1R'$, $c\tilde{O} = L_5\bar{O}$, and $c\tilde{A} = L_1\bar{A}$, where $c = [Ca^{2+}]$, $p = [IP_3]$, and $L_i = l_{-i}/l_i$ for every appropriate integer $i$. They introduced the new variables

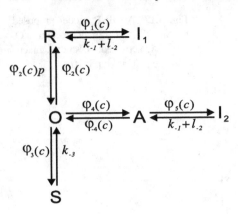

**Fig. 6.18.** Simplified diagram of the IP$_3$R model. Given the fast equilibria described in the text, this diagram is equivalent to that in Fig. 6.17. This figure is modified from [60]

$R = \tilde{R} + \bar{R} + R'$, $O = \tilde{O} + \bar{O}$, and $A = \tilde{A} + \bar{A}$. Solving these equations, we obtain

$$\tilde{R} = \frac{L_1 R}{L_1 + c(1 + L_1/L_3)}, \tag{6.80}$$

$$R' = \frac{cR}{L_1 + c(1 + L_1/L_3)}, \tag{6.81}$$

$$\bar{R} = \frac{cL_1 R/L_3}{L_1 + c(1 + L_1/L_3)}, \tag{6.82}$$

$$\bar{A} = \frac{cA}{L_1 + c}, \tag{6.83}$$

$$\tilde{O} = \frac{L_5 O}{L_5 + c}. \tag{6.84}$$

To derive the forward rate from $R$ to $I_1$, we focus on the reactions among $\tilde{R}$, $\bar{R}$, $R'$, and $I_1$ of Fig. 6.17, and obtain

$$k_1 c \tilde{R} + l_2 R' = \frac{(k_1 L_1 + l_2)cR}{L_1 + c(1 + L_1/L_3)},$$

where we have used (6.80) and (6.81). This implies that the forward rate from $R$ to $I_1$ is equal to

$$\phi_1(c) = \frac{(k_1 L_1 + l_2)c}{L_1 + c(1 + L_1/L_3)}.$$

It is evident that the forward rate $\phi_1(c)$ is saturable with respect to $c$, that is, the limit of $\phi_1(c)$ as $c \to \infty$ is a constant. In the same way, we can derive other reaction rates indicated in Fig. 6.18 as follows:

$$\phi_2(c) = \frac{k_2 L_3 + l_4 c}{L_3 + c(1 + L_3/L_1)}, \tag{6.85}$$

$$\phi_{-2}(c) = \frac{k_{-2} + l_{-4} c}{1 + c/L_5}, \tag{6.86}$$

$$\phi_3(c) = \frac{k_3 L_5}{L_5 + c}, \tag{6.87}$$

$$\phi_4(c) = \frac{(k_4 L_5 + l_6)c}{L_5 + c}, \tag{6.88}$$

$$\phi_{-4}(c) = \frac{L_1(k_{-4} + l_{-6})}{L_1 + c}, \tag{6.89}$$

$$\phi_5(c) = \frac{(k_1 L_1 + l_2)c}{L_1 + c}. \tag{6.90}$$

From Fig. 6.18, we derive the governing differential equations

$$\frac{dR}{dt} = \phi_{-2} O - \phi_2[IP_3]R + (k_{-1} + l_{-2})I_1 - \phi_1 R, \tag{6.91}$$

$$\frac{dO}{dt} = \phi_2[IP_3]R - (\phi_{-2} + \phi_4 + \phi_3)O + \phi_{-4}A + k_{-3}S, \tag{6.92}$$

$$\frac{dA}{dt} = \phi_4 O - \phi_{-4}A - \phi_5 A + (k_{-1} + l_{-2})I_2, \tag{6.93}$$

$$\frac{dI_1}{dt} = \phi_1 R - (k_{-1} + l_{-2})I_1, \tag{6.94}$$

$$\frac{dI_2}{dt} = \phi_5 A - (k_{-1} + l_{-2})I_2, \tag{6.95}$$

where $R + O + A + S + I_1 + I_2 = 1$. The parameters were determined by fitting to experimental data from the type II hepatocyte IP$_3$R [13] as follows: $k_1 = 0.64 \, \mu M^{-1} \cdot s^{-1}$, $k_{-1} = 0.04 \, s^{-1}$, $k_2 = 37.4 \, \mu M^{-1} \cdot s^{-1}$, $k_{-2} = 1.4 \, s^{-1}$, $k_3 = 0.11 \, \mu M^{-1} \cdot s^{-1}$, $k_{-3} = 29.8 \, s^{-1}$, $k_4 = 4 \, \mu M^{-1} \cdot s^{-1}$, $k_{-4} = 0.54 \, s^{-1}$, $L_1 = 0.12 \, \mu M$, $L_3 = 0.025 \, \mu M$, $L_5 = 54.7 \, \mu M$, $l_2 = 1.7 \, s^{-1}$, $l_4 = 1.7 \, \mu M^{-1} \cdot s^{-1}$, $l_6 = 4707 \, s^{-1}$, $l_{-2} = 0.8 \, s^{-1}$, $l_{-4} = 2.5 \, \mu M^{-1} \cdot s^{-1}$, $l_{-6} = 11.4 \, s^{-1}$.

The model assumes that the binding of IP$_3$ and Ca$^{2+}$ is sequential, not independent. So, for instance, Ca$^{2+}$ can bind to the activating site only after IP$_3$ has bound. Experimental data [46] indicated that the binding of IP$_3$ and Ca$^{2+}$ are not independent events, with Ca$^{2+}$ being unable to bind to the activating site until IP$_3$ has first bound.

Sneyd et al [60] assumed that the IP$_3$ receptor allows Ca$^{2+}$ current when all four subunits are in state O, or all four are in state A, or some intermediate combination

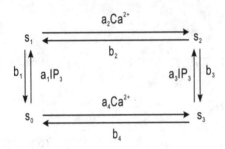

**Fig. 6.19.** A kinetic diagram of the IP$_3$-receptor proposed in [35]

(for instance, when three are in state O, and one is in state A). Furthermore, they assumed that the more subunits there are in state A, the greater the open probability of the receptor. With these assumptions, the open probability of the receptor is most conveniently written as

$$P_{ipr} = (0.1O + 0.9A)^4. \tag{6.96}$$

The numbers 0.1 and 0.9 are not crucial. Others, for instance, $(0.1O + A)^4$ or $(0.05O + 0.9A)^4$ can be used as the open probability.

Another channel open probability model was established by Keizer *et al* [35], who kinetically modeled the IP$_3$-receptor as consisting of four independent and equivalent subunits, as demonstrated in Fig. 6.19. Each subunit is endowed with an IP$_3$ and a Ca$^{2+}$ binding site that interact with each other, such that when the Ca$^{2+}$ site is occupied, the affinity $K_d$ for binding of IP$_3$ is increased. Thus a subunit can exist in four states: state $s_0$ consists of a subunit with neither IP$_3$ nor Ca$^{2+}$ bound; $s_1$ has only IP$_3$ bound; $s_2$ has both IP$_3$ and Ca$^{2+}$ bound; and $s_3$ has only Ca$^{2+}$ bound. An open channel is assumed to result only when each one of the four subunits is in the state $s_1$. All other states of the tetramer are assumed to be closed. Thus a rise in [Ca$^{2+}$]$_i$ shifts the channel into a blocked state. Another way to incorporate the inhibitory effect of Ca$^{2+}$ would be to assume that the tetramer remains open even with subunits in the $s_2$ state but that the rate of Ca$^{2+}$ flux is reduced. For simplicity this possibility is ignored.

Assuming mass action kinetics, the kinetic equations governing the state of a subunit are

$$\frac{dx_0}{dt} = -a_1[IP_3]x_0 + b_1x_1 - a_4[Ca^{2+}]_ix_0 + b_4x_3, \tag{6.97}$$

$$\frac{dx_1}{dt} = a_1[IP_3]x_0 - b_1x_1 - a_2[Ca^{2+}]_ix_1 + b_2x_2, \tag{6.98}$$

$$\frac{dx_2}{dt} = a_3[IP_3]x_3 - b_3x_2 + a_2[Ca^{2+}]_ix_1 - b_2x_2, \tag{6.99}$$

$$\frac{dx_3}{dt} = -a_3[IP_3]x_3 + b_3x_2 + a_4[Ca^{2+}]_ix_0 - b_4x_3, \tag{6.100}$$

where $x_i$ denotes the fraction of subunits in state $s_i$, $a_1 = 50$, $a_2 = 1$, $a_3 = 20$, $a_4 = 0.9$ $\mu$M$^{-1}$s$^{-1}$, and $b_1 = 6.5$, $b_2 = 0.5$, $b_3 = 14.5$, $b_4 = 0.0806$ s$^{-1}$. The open probability of the receptor is equal to

$$P_{ipr} = x_1^4. \tag{6.101}$$

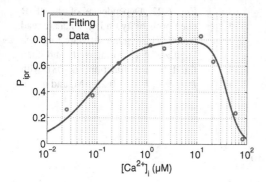

**Fig. 6.20.** Dependence of the IP$_3$R open probability on cytosolic Ca$^{2+}$. Data are read from [45] using the software Engauge Digitizer 4.1, and fitted by the function (6.102) with $P_{max} = 0.8$, $K_{rca} = 0.077$ $\mu$M, and $K_{inh} = 39$ $\mu$M

Other static open probability models of the receptor were proposed in the literature. The dependence of the IP$_3$R open probability on cytosolic Ca$^{2+}$ obtained by Mak *et al* [45] (Fig. 6.20) was well fitted to a biphasic Hill equation as follows

$$P_{ipr} = P_{max} \frac{[Ca^{2+}]_i}{K_{rca} + [Ca^{2+}]_i} \frac{K_{inh}^3}{K_{inh}^3 + [Ca^{2+}]_i^3}. \qquad (6.102)$$

The dependence of the IP$_3$R open probability on IP$_3$ was also well fitted to a Hill equation but with conflicting Hill exponents:

$$P_{ipr} = P_{max} \frac{[IP_3]^n}{K_{ip3}^n + [IP_3]^n}. \qquad (6.103)$$

According to Hagar and Ehrlich [25] (Fig. 6.21), the Hill exponent $n$ was found to be 1.9 in vitro. On the other hand, the Hill exponent $n$ was estimated to 1 (Fig. 6.22) in vivo for rat basophilic leukemia cells according to Meyer *et al* [47] (it was estimated to be about 4 in their paper [47] because the function $P_{ipr} = P_{max} \frac{[IP_3]^n}{(K_{ip3} + [IP_3])^n}$ was used in their estimation). On the basis of these data, the following model was proposed [18, 39]:

$$P_{ipr} = \frac{[Ca^{2+}]_i}{K_{rca} + [Ca^{2+}]_i} \frac{K_{inh}^3}{K_{inh}^3 + [Ca^{2+}]_i^3} \frac{[IP_3]^n}{K_{ip3}^n + [IP_3]^n}. \qquad (6.104)$$

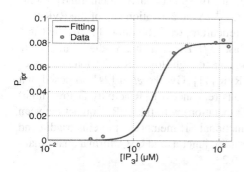

**Fig. 6.21.** Dependence of the IP$_3$R open probability on IP$_3$. Data are read from [25] using the software Engauge Digitizer 4.1, and fitted by the function (6.103) with $P_{max} = 0.08$, $K_{ip3} = 3.2$ $\mu$M, and $n = 1.9$

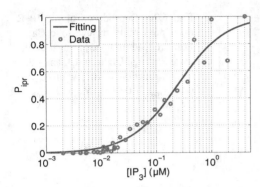

**Fig. 6.22.** Dependence of the IP$_3$R open probability on IP$_3$. Data are read from [47] using the software Engauge Digitizer 4.1, normalized, and then fitted by the function (6.103) with $P_{max} = 1$, $K_{ip3} = 0.25$ $\mu$M, and $n = 1$

## 6.8 The Ryanodine Receptor

Similar to the IP$_3$ receptors, ryanodine receptors (RyR) localized on the membrane of sarcoplasmic reticulum (SR) transport Ca$^{2+}$ from the sarcoplasmic reticulum into the cytosol by recognizing Ca$^{2+}$ on their cytosolic side, thus establishing a positive feedback mechanism: a small amount of Ca$^{2+}$ in the cytosol near the receptors causes them to release even more Ca$^{2+}$ from the sarcoplasmic reticulum into the cytosol, a process called calcium-induced calcium release (CICR). Since the SR calcium concentration is much higher than the cytosolic one, it was assumed that the Ca$^{2+}$ transport through the receptor is driven by the calcium concentration gradient [37]:

$$r_{ryr} = R_{ryr}P_{ryr}\left([Ca^{2+}]_{sr} - [Ca^{2+}]_i\right) \tag{6.105}$$

where $R_{ryr}$ is a maximal rate, $P_{ryr}$ is a channel opening probability, and $[Ca^{2+}]_{sr}$ denotes the concentration of Ca$^{2+}$ in the SR.

The plant alkaloid ryanodine, for which this receptor was named, has become an invaluable investigative tool. At low concentrations ($< 10$ $\mu$M), ryanodine binding to RyRs locks the RyRs into a long-lived sub-conductance (half-open) state and eventually depletes the store, while higher concentrations (about $100$ $\mu$M) irreversibly inhibit channel opening.

Ryanodine receptors are controlled by cytosolic Ca$^{2+}$, SR lumenal Ca$^{2+}$, the Ca$^{2+}$-binding protein calsequestrin (CSQ), and two junctional SR membrane proteins triadin and junctin. Binding of cytosolic Ca$^{2+}$ to the activation site of the RyR triggers Ca$^{2+}$ release from the sarcoplasmic reticulum, while binding of cytosolic Ca$^{2+}$ to the inactivation site terminates the release [57]. Experimental studies demonstrate that Ca$^{2+}$-sensing sites exist in the SR lumen and high (low) SR Ca$^{2+}$ load enhances (decreases) the open probability of the RyR [11]. Gyorke *et al* [24] proposed that CSQ, the SR lumenal Ca$^{2+}$ buffer, inhibits (enhances) RyR activity at low (high) SR Ca$^{2+}$ load through its association (dissociation) with the RyR. Contact between the RyR and CSQ is mediated by two junctional SR membrane proteins triadin and junctin [23]. These four proteins, RyR, triadin, junctin, and CSQ, form a quarternary "RyR complex", as demonstrated in Fig. 6.23.

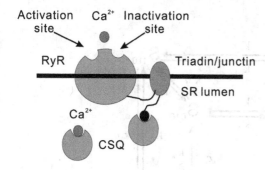

Activation    Ca²⁺  Inactivation
site                        site

RyR                              Triadin/junctin

SR lumen

Ca²⁺

CSQ

**Fig. 6.23.** Schematic description of RyR, CSQ, junctin, and triadin. The RyR has two cytosolic binding sites (activation and inactivation) and one junctional SR lumenal CSQ binding site. CSQ can bind either with $Ca^{2+}$ or the RyR mediated by the linkage of triadin and junctin. This figure is modified from [37]

Based on the above control mechanisms, Lee *et al* [37] constructed a model of the RyR channel opening probability by modeling it as having three binding sites, one for CSQ binding from the junctional SR lumen and two for $Ca^{2+}$ binding from the diadic space, one for activation and one for inactivation of the receptor. Thus, the RyR can be in eight states: $S_{ijk}$, where $i, j, k = 0$ or 1, and 0 and 1 represent the unoccupied and occupied binding sites, respectively. The index $i$ represents the CSQ binding site in the junctional SR lumen, the index $j$ denotes the cytosolic $Ca^{2+}$ activation site, and the index $k$ indicates the cytosolic $Ca^{2+}$ inactivation site. The open (i.e., conducting) states of the RyR channel are those states $S_{010}$ and $S_{110}$ for which the activation site is bound by $Ca^{2+}$ and $Ca^{2+}$ is not bound to the inactivation site. The rate constants of the RyR kinetics depend on the concentration of free CSQ (denoted as $[CSQ]$ or $q$) and free $Ca^{2+}$ concentration ($[Ca^{2+}]_i$ or $c$) in the diadic space. The diagram of a general eight-state model of the RyR channel is shown in Fig. 6.24.

We let $x_{ijk}$ denote the fraction of subunits in state $S_{ijk}$. To reduce the eight-state model to a four-state model, it was assumed that CSQ binding to the RyR occurs at a faster time scale than the activation of the RyR. Thus, it was assumed that the states $S_{0jk}$ and $S_{1jk}$ are in equilibrium, which leads to

$$k_1 q x_{000} = k_{-1} x_{100}, \quad k_3 q x_{010} = k_{-3} x_{110}, \tag{6.106}$$

$$k_{10} q x_{011} = k_{-10} x_{111}, \quad k_8 q x_{001} = k_{-8} x_{101}. \tag{6.107}$$

Define the new state variables

$$x_{00} = x_{000} + x_{100}, \quad x_{10} = x_{010} + x_{110}, \quad x_{11} = x_{011} + x_{111}, \quad x_{01} = x_{001} + x_{101}. \tag{6.108}$$

Solving (6.106), (6.107), and (6.108) gives

$$x_{100} = \frac{q}{K_1^d + q} x_{00}, \quad x_{000} = \frac{K_1^d}{K_1^d + q} x_{00}, \quad x_{110} = \frac{q}{K_3^d + q} x_{10}, \tag{6.109}$$

$$x_{010} = \frac{K_3^d}{K_3^d + q} x_{10}, \quad x_{111} = \frac{q}{K_{10}^d + q} x_{11}, \quad x_{011} = \frac{K_{10}^d}{K_{10}^d + q} x_{11}, \tag{6.110}$$

$$x_{101} = \frac{q}{K_8^d + q} x_{01}, \quad x_{001} = \frac{K_8^d}{K_8^d + q} x_{01}, \tag{6.111}$$

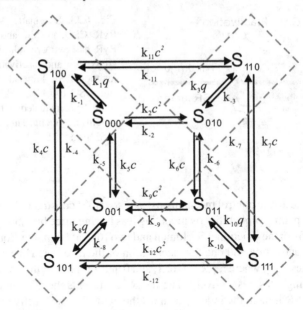

**Fig. 6.24.** A diagram of general eight-state RyR kinetic model. $c$ and $q$ represent the cytosolic $[Ca^{2+}]$ and free CSQ concentration ($[CSQ]$), respectively. $S_{ijk}$ denotes a kinetic state of the RyR. The two states enclosed within each of the dashed-line boxes have the same cytosolic binding states. This figure is modified from [37]

where

$$K_1^d = \frac{k_{-1}}{k_1}, \; K_3^d = \frac{k_{-3}}{k_3}, \; K_3^d = \frac{k_{-3}}{k_3}, \; K_{10}^d = \frac{k_{-10}}{k_{10}}. \tag{6.112}$$

Using the law of mass action, we derive a system of differential equations governing the four states:

$$\frac{dx_{00}}{dt} = k_{-11}x_{110} + k_{-2}x_{010} + k_{-5}x_{001} + k_{-4}x_{101}$$
$$- (k_{11}c^2 + k_4 c)x_{100} - (k_2 c^2 + k_5 c)x_{000},$$

$$\frac{dx_{10}}{dt} = -(k_{-11} + k_7 c)x_{110} - (k_{-2} + k_6 c)x_{010} + k_{11}c^2 x_{100}$$
$$+ k_2 c^2 x_{000} + k_{-7}x_{111} + k_{-6}x_{011},$$

$$\frac{dx_{01}}{dt} = -(k_{12}c^2 + k_{-4})x_{101} - (k_9 c^2 + k_{-5})x_{001} + k_{-12}x_{111}$$
$$+ k_{-9}x_{011} + k_5 c x_{000} + k_4 c x_{100},$$

$$\frac{dx_{11}}{dt} = k_7 c x_{110} + k_6 c x_{010} + k_{12}c^2 x_{101} + k_9 c^2 x_{001}$$
$$- (k_{-7} + k_{-12})x_{111} - (k_{-6} + k_{-9})x_{011}.$$

Substituting (6.109), (6.110), and (6.111) into the above system gives

$$\frac{dx_{00}}{dt} = \frac{k_{-2}K_3^d + k_{-11}q}{K_3^d + q}x_{10} + \frac{k_{-5}K_8^d + k_{-4}q}{K_8^d + q}x_{01}$$
$$- \frac{(k_2K_1^d + k_{11}q)c^2 + (k_5K_1^d + k_4q)c}{K_1^d + q}x_{00}, \qquad (6.113)$$

$$\frac{dx_{10}}{dt} = -\frac{k_{-2}K_3^d + k_{-11}q + (k_6K_3^d + k_7q)c}{K_3^d + q}x_{10}$$
$$+ \frac{(k_2K_1^d + k_{11}q)c^2}{K_1^d + q}x_{00} + \frac{k_{-6}K_{10}^d + k_{-7}q}{K_{10}^d + q}x_{11}, \qquad (6.114)$$

$$\frac{dx_{01}}{dt} = \frac{k_{-9}K_{10}^d + k_{-12}q}{K_{10}^d + q}x_{11} + \frac{(k_5K_1^d + k_4q)c}{K_1^d + q}x_{00}$$
$$- \frac{(k_9K_8^d + k_{12}q)c^2 + k_{-5}K_8^d + k_{-4}q}{K_8^d + q}x_{01}, \qquad (6.115)$$

$$\frac{dx_{11}}{dt} = \frac{(k_6K_3^d + k_7q)c}{K_3^d + q}x_{10} + \frac{(k_9K_8^d + k_{12}q)c^2}{K_8^d + q}x_{01}$$
$$- \frac{k_{-6}K_{10}^d + k_{-7}q + k_{-9}K_{10}^d + k_{-12}q}{K_{10}^d + q}x_{11}. \qquad (6.116)$$

In this scheme, since the states $S_{010}$ and $S_{110}$ represent the open state of the RyR, the open probability is given by

$$P_{ryr} = x_{10}. \qquad (6.117)$$

To reduce the number parameters, it was assumed that dissociation rates for the three binding sites are independent of $Ca^{2+}$ or CSQ binding:

$$k_{-2} = k_{-11} = k_{-9} = k_{-12}, \qquad (6.118)$$

$$k_{-4} = k_{-5} = k_{-6} = k_{-7}, \qquad (6.119)$$

$$k_{-1} = k_{-3} = k_{-8} = k_{-10}. \qquad (6.120)$$

It was further assumed that the rate of $Ca^{2+}$ binding to the activation site is independent of whether the $Ca^{2+}$ inactivation site is occupied or not, and the rate of $Ca^{2+}$ binding to the $Ca^{2+}$ inactivation site is independent of whether the $Ca^{2+}$ activation site is occupied or not:

$$k_2 = k_9, \ k_{11} = k_{12}, \ k_4 = k_7, \ k_5 = k_6. \qquad (6.121)$$

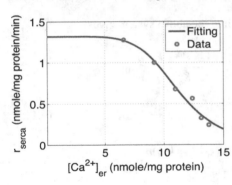

**Fig. 6.25.** Inhibition of ER $Ca^{2+}$ on SERCA $Ca^{2+}$ uptake. Data are read from [14] using the software Engauge Digitizer 4.1, and fitted by the Hill-type function (6.129) with $R_1 = 1.32$ nmole $Ca^{2+}$/mg protein/min, $K_{er} = 11.19$ nmole $Ca^{2+}$/mg protein, and $n = 6.14$

Moreover, it follows from the detailed balance in each cycle that

$$k_{11}k_{-3}k_{-2}k_1 = k_{-11}k_{-1}k_2k_3, \tag{6.122}$$

$$k_{-1}k_5k_8k_{-4} = k_1k_4k_{-8}k_{-5}, \tag{6.123}$$

$$k_{-8}k_9k_{10}k_{-12} = k_8k_{12}k_{-10}k_{-9}, \tag{6.124}$$

$$k_{-6}k_3k_7k_{-10} = k_6k_{10}k_{-7}k_{-3}, \tag{6.125}$$

$$k_{-2}k_5k_9k_{-6} = k_2k_6k_{-9}k_{-5}. \tag{6.126}$$

Under the conditions (6.118), (6.119), (6.120), and (6.121), these balance constraints are not independent and they can be reduced to the following three independent constraints:

$$k_{11}k_1 = k_2k_3, \quad k_5k_8 = k_1k_4, \quad k_3k_4 = k_5k_{10}. \tag{6.127}$$

Furthermore, detailed balance requires that

$$K_3^d = \frac{K_1^d k_{11}}{k_2}, \quad K_8^d = \frac{K_1^d k_6}{k_7}. \tag{6.128}$$

Therefore, the model parameters have been reduced to seven independent parameters, the unbinding rate constants $k_{-2}$ and $k_{-6}$, the binding rate constants $k_2, k_6, k_7$, and $k_{11}$, and the equilibrium constant $K_1^d$. The values of these parameters were estimated by Lee *et al* [37] as follows: $k_{-2} = 60$ s$^{-1}$, $k_{-6} = 5$ s$^{-1}$, $k_2 = 0.045$ $\mu$M$^{-2}$s$^{-1}$, $k_6 = 0.3$ $\mu$M$^{-1}$s$^{-1}$, $k_7 = 0.47$ $\mu$M$^{-1}$s$^{-1}$, $k_{11} = 0.0045$ $\mu$M$^{-2}$s$^{-1}$, and $K_1^d = 1000$ $\mu$M.

## 6.9 The Sarcoplasmic or Endoplasmic Reticulum Calcium ATPase

The sarcoplasmic or endoplasmic reticulum $Ca^{2+}$-ATPase (SERCA) resides on the membrane of intracellular sarcoplasmic or endoplasmic reticulum organelles and pumps $Ca^{2+}$ from the cytosol into the organelles. The enzymatic cycle of $Ca^{2+}$ transport for SERCA is schematically described in Fig. 2.8.

Based on their experimental data, Favre *et al* [14] found that the ER $Ca^{2+}$ inhibit SERCA $Ca^{2+}$ uptake into the ER. Moreover, they found that the rate of $Ca^{2+}$ uptake depends on the ER $Ca^{2+}$ concentration and can be well fitted by the following Hill-type function (Fig. 6.25)

$$r_{serca} = \frac{R_1 K_{er}^n}{K_{er}^n + [Ca^{2+}]_{er}^n} \tag{6.129}$$

with $R_1 = 1.32$ nmole $Ca^{2+}$/mg protein/min, $K_{er} = 11.19$ nmole $Ca^{2+}$/mg protein, and $n = 6.14$. Their experimental data further showed that the maximal rate $R_1$ depends on the cytosolic $Ca^{2+}$ concentration and can be well fitted by the Michaelis-Menten function (Fig. 6.26)

$$R_1 = \frac{R_2 [Ca^{2+}]_i}{K_i + [Ca^{2+}]_i} \tag{6.130}$$

with $R_2 = 1.74$ nmole $Ca^{2+}$/mg protein/min and $K_i = 0.54$ $\mu$M. In addition, their experimental data showed that the constant $K_{er}$ also depends on the cytosolic $Ca^{2+}$ concentration and can be well fitted by the Hill function (Fig. 6.27)

$$K_{er} = \frac{K[Ca^{2+}]_i^n}{K_m^n + [Ca^{2+}]_i^n} \tag{6.131}$$

with $K = 10.78$ nmole $Ca^{2+}$/mg protein, $K_m = 0.33$ $\mu$M, and $n = 1.3$. Substituting (6.130) into (6.129) gives

$$r_{serca} = \frac{R_2 [Ca^{2+}]_i}{K_i + [Ca^{2+}]_i} \frac{K_{er}^n}{K_{er}^n + [Ca^{2+}]_{er}^n}. \tag{6.132}$$

Other SERCA $Ca^{2+}$ uptake rates were constructed. The experimental data of Lytton $et\ al$ [41] (Fig. 2.9) showed that the $Ca^{2+}$ uptake rate can be described by

$$r_{serca} = \frac{R_{serca}[Ca^{2+}]_i^n}{K_{1/2}^n + [Ca^{2+}]_i^n}, \tag{6.133}$$

where $R_{serca}$ is a maximum rate, $K_{1/2}$ is the concentration $[Ca^{2+}]_i$ which gives half of $R_{serca}$, and $n = 2$ is the Hill exponent. Their experimental data showed that SERCA1 expressed in COS cell microsomes, one of SERCA family members, has a $K_{1/2}$ of 0.4

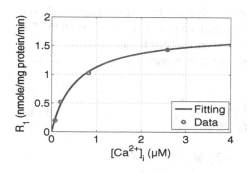

**Fig. 6.26.** Dependence of SERCA $Ca^{2+}$ uptake rate on cytosolic $Ca^{2+}$. Data are read from [14] using the software Engauge Digitizer 4.1, and fitted by the Michaelis-Menten function (6.130) with $R_2 = 1.74$ nmole $Ca^{2+}$/mg protein/min and $K_i = 0.54$ $\mu$M

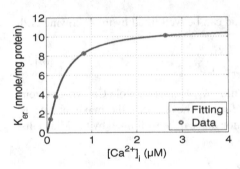

**Fig. 6.27.** Dependence of $K_{er}$ on cytosolic $Ca^{2+}$. Data are read from [14] using the software Engauge Digitizer 4.1, and fitted by the Hill function (6.131) with $K = 10.78$ nmole $Ca^{2+}$/mg protein, $K_m = 0.33$ $\mu$M, and $n = 1.3$

$\mu$M. SERCA2a expressed in COS cells also had a $K_{1/2}$ of 0.4 $\mu$M, contrasting with its apparent affinity in cardiac muscle sarcoplasmic reticulum ($K_{1/2}$ of about 0.9 $\mu$M). SERCA2b seemed to display a slightly higher apparent affinity for calcium ($K_{1/2}$ of 0.27 $\mu$M). SERCA3 had a much lower apparent calcium affinity, $K_{1/2}$ of about 1.1 $\mu$M, very similar to that observed for the $Ca^{2+}$-ATPase of cardiac sarcoplasmic reticulum. In all cases, the data for calcium dependence of enzyme activity were best fitted by the Hill function with an exponent of about 2, exhibiting two highly cooperative calcium-binding sites for activity.

Other different rates were also used in the literature. Atri *et al* [1] used the following rate:

$$r_{serca} = \frac{R_{serca}[Ca^{2+}]_i}{K_{serca} + [Ca^{2+}]_i}, \tag{6.134}$$

where $R_{serca} = 2.0\ \mu M \cdot s^{-1}$ and $K_{serca} = 0.1\ \mu$M. Sneyd at al [62] used the following rate:

$$r_{serca} = \left( \frac{R_{serca}[Ca^{2+}]_i}{K_{serca} + [Ca^{2+}]_i} \right) \left( \frac{1}{[Ca^{2+}]_{er}} \right), \tag{6.135}$$

where $R_{serca} = 120\ \mu M^2 \cdot s^{-1}$ and $K_{serca} = 0.18\ \mu$M. The factor $1/[Ca^{2+}]_{er}$ was introduced as a negative feedback control of ER $Ca^{2+}$ as observed by Favre *et al* [14].

## 6.10 The Plasma Membrane Calcium ATPase

The plasma membrane calcium ATPase hydrolyzes ATP and the energy released from ATP drives calcium ions out of the cell and the proton $H^+$ into the cell [5]. For every ATP hydrolyzed, one $Ca^+$ ion is moved out and one $H^+$ is moved in.

The calcium extrusion rate across the plasma membrane by the calcium ATPase was modeled by (see, e.g., [48, 61, 62])

$$r_{pmca} = \frac{R_{pmca}[Ca^{2+}]_i^2}{K_{pmca}^2 + [Ca^{2+}]_i^2}, \tag{6.136}$$

where $R_{pmca} = 28\ \mu M \cdot s^{-1}$ and $K_{pmca} = 0.42\ \mu M$.

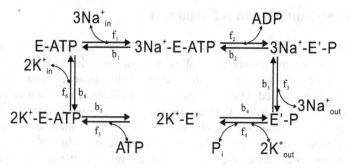

**Fig. 6.28.** A six-step biochemical reaction scheme of Na$^+$/K$^+$-ATPase (denoted by E in the figure) proposed in [8]

## 6.11 The Sodium/Potassium ATPase

The sodium/potassium ATPase is a single protein in the plasma membrane. It hydrolyzes ATP and the energy released from ATP drives sodium ions out of the cell and potassium ions into the cell. For every ATP hydrolyzed, three Na$^+$ ions are moved out and two K$^+$ ions are moved in, generating a net outward flow of cations [5]. Thus, Na$^+$/K$^+$-ATPase is electrogenic.

Based on a six-step biochemical reaction scheme (Fig. 6.28) proposed by Chapman *et al* [8], a model for the Na$^+$/K$^+$ pump current was constructed by Miwa *et al* [50] and was used with slight modifications by Fridlyand *et al* [18] as follows:

$$I_{nka} = \frac{P_{nka}(F_1 f_2 f_3 F_4 F_5 f_6 - b_1 B_2 B_3 B_4 b_5 B_6)}{D}, \tag{6.137}$$

where

$$D = f_2 f_3 F_4 F_5 f_6 + b_1 f_3 F_4 F_5 f_6 + b_1 B_2 F_4 F_5 f_6 + b_1 B_2 B_3 F_5 f_6$$
$$+ b_1 B_2 B_3 B_4 f_6 + b_1 B_2 B_3 B_4 b_5,$$

$$F_1 = f_1 [Na^+]_i^3, \; F_4 = f_4 [K^+]_o^2, \; F_5 = f_5 [ATP]_i,$$

$$B_2 = b_2 [ADP]_i, \; B_3 = b_3 [Na^+]_o^3, \; B_4 = b_4 [P], \; B_6 = b_6 [K^+]_i^2,$$

$$f_5 = f_5^* \exp(VF/(2RT)), \; b_5 = b_5^* \exp(-VF/(2RT)).$$

In these expressions, $P_{nka}$ is a coefficient for resulting current in the presence of saturating level of ATP and $[P]$ is an inorganic phosphate concentration. All others are the rate constants. The parameter values used by Fridlyand *et al* [18] are as follows: $[P] = 4{,}950 \; \mu M$, $f_1 = 2.5 \cdot 10^{-10} \; \mu M^{-3} ms^{-1}$, $f_2 = 10 \; ms^{-1}$, $f_3 = 0.172 \; ms^{-1}$, $f_4 = 1.5 \cdot 10^{-8} \; \mu M^{-2} \, ms^{-1}$, $f_5^* = 0.002 \; \mu M^{-1} \, ms^{-1}$, $f_6 = 11.5 \; ms^{-1}$, $b_1 = 100 \; ms^{-1}$, $b_2 = 0.0001 \; \mu M^{-1} \, ms^{-1}$, $b_3 = 1.72 \cdot 10^{-17} \; \mu M^{-3} \, ms^{-1}$, $b_4 = 0.0002 \; \mu M^{-1} \, ms^{-1}$, $b_5^* = 0.03 \; ms^{-1}$, and $b_6 = 6 \cdot 10^{-7} \; \mu M^{-1} \, ms^{-1}$.

## 6.12 The Sodium/Calcium Exchanger

Like channels, the sodium/calcium exchanger is an integral membrane protein that forms a tube across the membrane to allow sodium and calcium ions to cross the membrane [5, 66]. Unlike channels, the exchanger tube is never open all the way through; it is always closed at one or other end. The exchange can exist in two shapes, one open to the extracellular medium and one open to the cytosol. Inside the tube there are three sites that can bind sodium ions and one site that can bind a calcium ion.

When the exchanger is open to the extracellular medium, three $Na^+$ bind to the sodium binding sites. It then switches its shape and opens to the cytosol so that the three $Na^+$ is released to the cytosol. After the $Na^+$ release, one $Ca^{2+}$ binds to the calcium binding site and the exchanger switches its shape and opens to the extracellular medium to release the $Ca^{2+}$. The overall effect of one cycle is to carry three sodium ions into the cell and one calcium ion out of the cell.

A number of mathematical models for the exchanger have been constructed. We first present a simplified version (see [33]) of the four-state model constructed by Kang et al [32]. Let $E_i$ denote the conformation of the exchanger whose binding sites are exposed to the interior of the cell and $E_o$ denote the conformation of the exchanger whose binding sites are exposed to the outside. Starting at the state $E_i 3Na^+$ (the exchanger bound with three $Na^+$), the exchanger can bind $Ca^{2+}$ inside the cell, simultaneously releasing three $Na^+$ to the interior. A change of conformation to $E_o$ allows the exchanger to release the $Ca^{2+}$ to the outside and bind three external $Na^+$. A return to the $E_i$ conformation completes the cycle. This binding cycle is summarized as follows:

$$E_i 3Na_i^+ + Ca_i^{2+} \underset{k_{-1}}{\overset{k_1}{\rightleftharpoons}} E_i Ca_i^{2+} + 3Na_i^+, \tag{6.138}$$

$$E_i Ca_i^{2+} \underset{k_{-2}}{\overset{k_2}{\rightleftharpoons}} E_o Ca_o^{2+}, \tag{6.139}$$

$$E_o Ca_o^{2+} + 3Na_o^+ \underset{k_{-3}}{\overset{k_3}{\rightleftharpoons}} E_o 3Na_o^+ + Ca_o^{2+}, \tag{6.140}$$

$$E_o 3Na_o^+ \underset{k_{-4}}{\overset{k_4}{\rightleftharpoons}} E_i 3Na_i^+. \tag{6.141}$$

Let $x_1, x_2, x_3$, and $x_4$ denote the fraction of the exchangers in the state $E_i 3Na_i^+$, $E_i Ca_i^{2+}, E_o Ca_o^{2+}$, and $E_o 3Na_o^+$, respectively. It follows from the reactions (6.138)-(6.141) that the governing equations for $x_1, x_2, x_3$, and $x_4$ are

$$\frac{dx_1}{dt} = k_{-1} x_2 [Na_i^+]^3 + k_4 x_4 - (k_1 [Ca_i^{2+}] + k_{-4}) x_1, \tag{6.142}$$

$$\frac{dx_2}{dt} = k_{-2} x_3 + k_1 [Ca_i^{2+}] x_1 - (k_2 + k_{-1} [Na_i^+]^3) x_2, \tag{6.143}$$

$$\frac{dx_3}{dt} = k_2 x_2 + k_{-3}[Ca_o^{2+}]x_4 - (k_{-2} + k_3[Na_o^+]^3)x_3, \tag{6.144}$$

$$1 = x_1 + x_2 + x_3 + x_4. \tag{6.145}$$

The steady-state solution of these equations can be found by using a mathematical software such as Maple. Then the $Na^+$ inward flux and the $Ca^{2+}$ outward flux can be found as follows

$$r_{cana} = k_4 x_4 - k_{-4} x_1$$

$$= \frac{k_1 k_2 k_3 k_4 \left([Ca_i^{2+}][Na_o^+]^3 - K_1 K_2 K_3 K_4 [Ca_o^{2+}][Na_i^+]^3\right)}{16 \text{ positive terms}}, \tag{6.146}$$

where $K_i = k_{-i}/k_i$.

Since each cycle of the $3Na^+/Ca^{2+}$ exchanger transports two positive charges out and three positive charges in, an electric current is generated. Thus the reaction constants depend on the membrane potential difference. To find such dependence, we look at the chemical potential of the reactions in the cycle. The overall reaction in the cycle begins with three $Na^+$ outside the cell and one $Ca^{2+}$ inside the cell, and ends with three $Na^+$ inside the cell and one $Ca^{2+}$ outside the cell. It can be written as

$$3Na_o^+ + Ca_i^{2+} \longrightarrow 3Na_i^+ + Ca_o^{2+}.$$

It then follows from (2.54) and (B.2) that the change in electrochemical potential of this reaction is

$$\Delta\mu = \Delta\mu^\circ + RT \ln\left(\frac{[Ca_o^{2+}][Na_i^+]^3}{[Ca_i^{2+}][Na_o^+]^3}\right) + FV,$$

where $V = V_i - V_o$ is the transmembrane potential. At equilibrium, we must have $\Delta\mu = 0$. Assuming that the standard free energy for the reaction is the same on both sides of the membrane ($\Delta\mu^\circ = 0$), it then follows that

$$\frac{[Ca_o^{2+}][Na_i^+]^3}{[Ca_i^{2+}][Na_o^+]^3} = \exp\left(-\frac{FV}{RT}\right).$$

Around any closed reaction loop, the product of the forward rates must be equal to the product of the reverse rates. This gives

$$k_1[Ca_i^{2+}]k_2 k_3[Na_o^+]^3 k_4 = [Na_i^+]^3 k_{-1} k_{-4}[Ca_o^{2+}]k_{-3}k_{-2},$$

and then

$$K_1 K_2 K_3 K_4 = \frac{[Ca_i^{2+}][Na_o^+]^3}{[Ca_o^{2+}][Na_i^+]^3} = \exp\left(\frac{FV}{RT}\right).$$

It then follows from (6.146) that

$$r_{cana} = \frac{k_1 k_2 k_3 k_4 \left([Ca_i^{2+}][Na_o^+]^3 - \exp\left(\frac{FV}{RT}\right)[Ca_o^{2+}][Na_i^+]^3\right)}{16 \text{ positive terms}}. \tag{6.147}$$

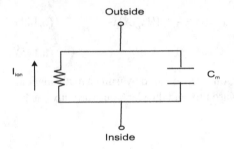

**Fig. 6.29.** An electrical circuit model of cell membrane

For the pancreatic $\beta$-cells, a detailed study by Gall *et al* [20] showed that the electrogenic $3Na^+/Ca^{2+}$ exchanger current can be modeled by

$$I_{cana} = G_{cana} \frac{[Ca^{2+}]_i^5}{K_{cana}^5 + [Ca^{2+}]_i^5} (V - V_{cana})$$

$$= G_{cana} \frac{[Ca^{2+}]_i^5}{K_{cana}^5 + [Ca^{2+}]_i^5} \left( V - \frac{RT}{F} \left( 3\ln \frac{[Na^+]_o}{[Na^+]_i} - \ln \frac{[Ca^{2+}]_o}{[Ca^{2+}]_i} \right) \right), \quad (6.148)$$

where $[Na^+]_o$ ($[Ca^{2+}]_o$) and $[Na^+]_i$ ($[Ca^{2+}]_i$) denote the $Na^+$ ($Ca^{2+}$) concentration outside the cell and the concentration inside the cell, respectively. This model was used by Fridlyand *et al* [18] with the parameter values: $G_{cana} = 271$ pS and $K_{cana} = 0.75\ \mu M$.

## 6.13 Membrane Potential Models

Since a cell membrane can act as an insulator to separate charge, it can be viewed as a capacitor parallel to a resistor (Fig. 6.29). The capacitance $C_m$ of any insulator is defined as the ratio of the charge $Q$ across the capacitor to the voltage potential $V$ necessary to hold that charge:

$$C_m = \frac{Q}{V}.$$

If we assume that $C_m$ is constant, then the capacitive current is given by

$$\frac{dQ}{dt} = C_m \frac{dV}{dt}.$$

Assume that there is no net buildup of charge on either side of the membrane. Then, according to Kirchhoff's current law, the sum of the ionic and capacitive currents must be zero:

$$C_m \frac{dV}{dt} + I_{ion} = 0.$$

If there are $N$ different ions moving across the membrane, then we have

$$C_m \frac{dV}{dt} + \sum_{n=1}^{N} I_n = 0, \quad (6.149)$$

where $I_n$ denotes the ionic current of the ion $c_n$.

Substituting the linear current-voltage equation (6.11) for $I_n$ in (6.149), we obtain

$$C_m \frac{dV}{dt} + \sum_{n=1}^{N} g_n(V - V_{I_n}) = 0,$$

where $g_n$ denotes the membrane conductance of the ion $c_n$. Using the Nernst equation (6.6), we obtain a linear voltage potential model:

$$C_m \frac{dV}{dt} + V \sum_{n=1}^{N} g_n - \sum_{n=1}^{N} \frac{g_n RT}{z_n F} \ln\left(\frac{c_{no}}{c_{ni}}\right) = 0, \qquad (6.150)$$

where $c_{no}$ and $c_{ni}$ denote the concentrations of the outside and inside ion $c_n$, respectively.

Substituting the nonlinear GHK current-voltage equation (6.9) for $I_n$ in (6.149), we obtain a nonlinear voltage potential model:

$$C_m \frac{dV}{dt} + \sum_{n=1}^{N} \frac{P_n z_n^2 F^2 V \left[c_{ni} - c_{no} \exp\left(\frac{-z_n VF}{RT}\right)\right]}{RT \left[1 - \exp\left(\frac{-z_n VF}{RT}\right)\right]} = 0. \qquad (6.151)$$

Due to its simplicity, the linear model (6.150) is used in modeling the membrane voltage potentials more frequently than (6.151).

## 6.14 The Hodgkin-Huxley Model

A mathematical model for the electrical behavior of membrane of the squid giant axon was established by Hodgkin and Huxley [27]. The electrical current was represented by the network shown in Fig. 6.30. The current can be carried through the membrane either by charging the membrane capacity or by movement of ions through channels in parallel with the capacity. The ionic channels can open and close in response to changes in the membrane potential.

The ionic current can be split into components carried by sodium ions $I_{na}$, potassium ions $I_k$, and other ions such as the chloride $I_l$. These ionic currents can be mod-

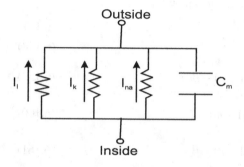

**Fig. 6.30.** An electrical circuit model of cell plasma membrane of the squid giant axon proposed in [27]

**Table 6.1.** Parameters of the model (6.156)-(6.167) from [27]

| | |
|---|---|
| $C_m = 1\ \mu F/cm^2$ | $G_{na} = 120\ mS/cm^2$ |
| $G_k = 36\ mS/cm^2$ | $T = 6°\ C$ |
| $G_l = 0.3\ mS/cm^2$ | $V_k = -12\ mV$ |
| $V_{na} = 115\ mV$ | $V_l = 10.613\ mV$ |
| $V(0) = 15\ mV$ | $h(0) = 0.6$ |
| $m(0) = 0.053$ | $n(0) = 0.32$ |

eled by the relations

$$I_{na} = G_{na}P_{na}(V - V_{na}), \tag{6.152}$$

$$I_k = G_kP_k(V - V_k), \tag{6.153}$$

$$I_l = G_l(V - V_l), \tag{6.154}$$

where $V$ is the potential difference between the inside and the outside of the membrane, $V_{na}$ and $V_k$ are the equilibrium potentials for the sodium and potassium ions, $V_l$ is the potential at which the "leakage current" due to chloride and other ions is zero, $G_{na}, G_k, G_l$ are maximal ionic conductances, and $P_{na}, P_k$ are channel open probabilities to be found.

The total membrane current is divided into a capacity current and the ionic current. It then follows from Kirchhoff's current law and the equations (6.152), (6.153), and (6.154) that

$$C_m\frac{dV}{dt} = -I_{na} - I_k - I_l + I_{in}$$
$$= -G_{na}P_{na}(V - V_{na}) - G_kP_k(V - V_k) - G_l(V - V_l) + I_{in}, \tag{6.155}$$

where $I_{in}$ is an input current and the outward current is assumed to be positive. The potassium channel open probability $P_k$ was modeled by (6.15), (6.16), (6.17), and (6.18). The sodium channel open probability $P_{na}$ was modeled by (6.46), (6.47), (6.48), (6.49), (6.50), (6.51), and (6.52).

We now collect all equations to form a complete feedback control system

$$C_m\frac{dV}{dt} = -G_{na}P_{na}(V - V_{na}) - G_kP_k(V - V_k) - G_l(V - V_l) + I_{in}, \tag{6.156}$$

where the feedback controllers $P_{na}$ and $P_k$ are given by

$$P_{na} = hm^3, \tag{6.157}$$

$$P_k = n^4, \tag{6.158}$$

$$\frac{dh}{dt} = \alpha_h(1 - h) - \beta_h h, \tag{6.159}$$

$$\frac{dm}{dt} = \alpha_m(1 - m) - \beta_m m, \tag{6.160}$$

$$\frac{dn}{dt} = \alpha_n(1 - n) - \beta_n n, \tag{6.161}$$

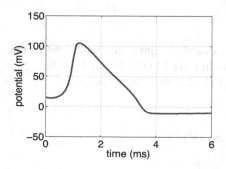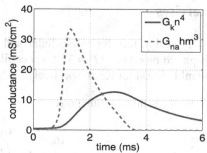

**Fig. 6.31.** Numerical simulation of sodium and potassium channel gating with the Hodgkin-Huxley model (6.156)-(6.167). The initial depolarization of 15 mV results in an increase of the potential and the opening of potassium and sodium channels. After a few milli-seconds, the potential returns to its resting level and the channels return to their closing state

$$\alpha_h = 0.07 \exp\left(\frac{-V}{20}\right), \tag{6.162}$$

$$\beta_h = \frac{1}{\exp\left(\frac{30-V}{10}\right)+1}, \tag{6.163}$$

$$\alpha_m = \frac{0.1(25-V)}{\exp\left(\frac{25-V}{10}\right)-1}, \tag{6.164}$$

$$\beta_m = 4 \exp\left(\frac{-V}{18}\right), \tag{6.165}$$

$$\alpha_n = \frac{0.01(10-V)}{\exp\left(\frac{10-V}{10}\right)-1}, \tag{6.166}$$

$$\beta_n = 0.125 \exp\left(\frac{-V}{80}\right). \tag{6.167}$$

The units of potential, current density, conductance density, and capacitance density are mV, $\mu A/cm^2$, $mS/cm^2$, $\mu F/cm^2$, respectively. The system of equations (6.156)-(6.167) is called the Hodgkin-Huxley model for the sodium and potassium channel gating.

A numerical solution of the model is presented in Fig. 6.31 and the parameter values for the numerical solution are given in Table 6.1. The figure shows that the initial depolarization of 15 mV results in an increase of the potential and the opening of potassium and sodium channels. After a few milli-seconds, the potential returns to its resting level and the channels return to their closing state.

## Exercises

**6.1.** Typical values for intracellular and extracellular ionic concentrations of the human red blood cell ($T = 37°C$) at equilibrium are 19 mM, 155 mM for $Na^+$, and 136 mM, 5 mM for $K^+$, respectively [33]. Calculate their Nernst potentials.

**6.2.** Fit the following functions

$$\alpha_n = \frac{\alpha_1 (\alpha_2 + V)}{\exp\left(\frac{\alpha_2 + V}{\alpha_2}\right) - \alpha_3},$$

$$\beta_n = \beta_1 \exp\left(\frac{V}{\beta_2}\right)$$

into the data of potassium conductance of the squid giant axon from Table 1 of [27]:

| $V$ | −109 | −100 | −88 | −76 | −63 | −51 | −38 | −32 | −26 | −19 | −10 | −6 |
|---|---|---|---|---|---|---|---|---|---|---|---|---|
| $\alpha_n$ | 0.915 | 0.866 | 0.748 | 0.61 | 0.524 | 0.419 | 0.31 | 0.241 | 0.192 | 0.15 | 0.095 | 0.085 |
| $\beta_n$ | 0.037 | 0.043 | 0.052 | 0.057 | 0.064 | 0.069 | 0.075 | 0.071 | 0.072 | 0.072 | 0.096 | 0.105 |

The units of $V$, $\alpha_n$, and $\beta_n$ are mV, ms$^{-1}$, and ms$^{-1}$, respectively.

**6.3.** Fit the following functions

$$n_\infty(V) = \frac{1}{1 + \exp[(V_{kh} - V)/C_{kh}]},$$

$$n_\infty(V) = \frac{a_k \exp((V - V_{ah})/C_{ah})}{a_k \exp((V - V_{ah})/C_{ah}) + b_k \exp((V_{bh} - V)/C_{bh})}$$

into the activation probability data of delayed rectifier potassium channels in mouse pancreatic $\beta$-cells obtained by Rorsman and Trube [56]:

| $V(mV)$ | −40.80 | −31.32 | −20.98 | −20.40 | −11.49 | −10.92 | −2.012 | 6.61 | 17.24 |
|---|---|---|---|---|---|---|---|---|---|
| $n_\infty$ | 0.02 | 0.22 | 0.29 | 0.50 | 0.77 | 0.95 | 0.91 | 0.99 | 1 |

The data are read from Fig. 10B of [56] using the software Engauge Digitizer 4.1.

**6.4.** Fit the Hill function

$$P_{kca} = \frac{[Ca^{2+}]_i^n}{K_{kca}^n + [Ca^{2+}]_i^n}$$

into the activation probability data of small conductance $Ca^{2+}$-activated potassium channels in smooth muscle cells of the mouse obtained by Vogalis et al [65]:

| $[Ca^{2+}]_i(\mu M)$ | 0.15 | 0.3 | 1 | 3 | 10 |
|---|---|---|---|---|---|
| $P_{kca}$ | 0.0420769 | 0.101626 | 0.383386 | 0.72532 | 0.904607 |

The data are read from Fig. 5D of [65] using the software Engauge Digitizer 4.1.

**6.5.** Let

$$P_{katp}([ADP]) = \frac{0.08\left(1+\frac{0.33[ADP]}{K_{dd}}\right)+0.89\left(\frac{0.165[ADP]}{K_{dd}}\right)^2}{\left(1+\frac{0.165[ADP]}{K_{dd}}\right)^2\left(1+\frac{0.135[ADP]}{K_{td}}+\frac{0.05\times0.00001}{K_{tt}}\right)}.$$

Fit the function $f([ADP]) = P_{katp}([ADP])/P_{katp}(1)$ into the data of ADP dependence of outward current activation of ATP-sensitive potassium channels in mouse pancreatic $\beta$-cells:

| $[ADP](\mu M)$ | 1.03197 | 10.088 | 50.0852 | 101.996 | 494.734 | 1004.35 | 10005.6 |
|---|---|---|---|---|---|---|---|
| $f$ | 0.97685 | 1.10743 | 2.61544 | 2.49791 | 1.0622 | 0.548409 | 0.0707625 |

The data are read from Fig. 2 of Hopkins *et al* [28] using the software Engauge Digitizer 4.1.

**6.6.** Fit the Hill function

$$O = \frac{K^n}{K^n+x^n}$$

into the activation probability data of ATP or ADP dependence of outward current activation of ATP-sensitive potassium channels in HEK293 cells:

| $ATP(M)$ | $1.04\times10^{-5}$ | $2.05\times10^{-5}$ | $4.13\times10^{-5}$ | 0.000201687 | 0.000771298 |
|---|---|---|---|---|---|
| $O$ | 0.502291 | 0.354306 | 0.206343 | 0.0558664 | 0.00988113 |

and

| $ADP(M)$ | $4.01\times10^{-5}$ | 0.000100331 | 0.000198058 | 0.00146967 |
|---|---|---|---|---|
| $O$ | 0.675907 | 0.504508 | 0.336254 | 0.0544307 |

The data obtained by John *et al* [29] are read from Fig. 1F of [29] using the software Engauge Digitizer 4.1.

**6.7.** Fit the following functions

$$\alpha_h = \alpha_{h1}\exp\left(\frac{V}{\alpha_{h2}}\right),$$

$$\beta_h = \frac{1}{\exp\left(\frac{\beta_{h1}+V}{\beta_{h2}}\right)+1},$$

$$\alpha_m = \frac{\alpha_{m1}(\alpha_{m2}+V)}{\exp\left(\frac{\alpha_{m2}+V}{\alpha_{m3}}\right)-1},$$

$$\beta_m = \beta_{m1}\exp\left(\frac{V}{\beta_{m2}}\right)$$

into the data of sodium conductance of the squid giant axon from Table 2 of [27]:

| $V(mV)$ | $-109$ | $-100$ | $-88$ | | $-76$ | $-63$ | $-51$ | $-38$ | $-32$ | $-26$ | $-19$ | $-10$ | $-6$ |
|---|---|---|---|---|---|---|---|---|---|---|---|---|---|
| $\alpha_h(ms^{-1})$ | 0 | 0 | 0 | | 0 | 0 | 0 | 0 | 0 | 0.02 | 0.03 | 0.05 | 0.06 |
| $\beta_h(ms^{-1})$ | 1.5 | 1.5 | 1.5 | | 1.19 | 1.19 | 0.94 | 0.79 | 0.75 | 0.65 | 0.4 | 0.13 | 0.09 |
| $\alpha_m(ms^{-1})$ | 7 | 6.2 | 5.15 | | 5.15 | 3.82 | 2.82 | 2.03 | 1.36 | 0.95 | 0.81 | 0.66 | 0.51 |
| $\beta_m(ms^{-1})$ | 0.14 | 0.02 | $(-0.14)$ | | 0.13 | 0.15 | 0.33 | 0.58 | 0.56 | 0.72 | 1.69 | 3.9 | 4.5 |

**6.8.** Fit the following function

$$m_\infty(V) = \frac{1}{1+\exp[(V_{mh}-V)/C_{mh}]}$$

into the activation probability data of voltage-gated calcium channels in mouse pancreatic $\beta$-cells obtained by Rorsman and Trube [56]:

| $V(mV)$ | $-32.14$ | $-22.68$ | $-13.03$ | $-3.14$ | 4.73 | 15.73 | 25.64 | 34.11 | 44.23 | 53.66 | 64.45 |
|---|---|---|---|---|---|---|---|---|---|---|---|
| $m_\infty$ | 0.07 | 0.12 | 0.23 | 0.37 | 0.53 | 0.66 | 0.79 | 0.93 | 0.92 | 1.00 | 0.93 |

The data are read from Fig. 6B of [56] using the software Engauge Digitizer 4.1.

**6.9.** Fit the following function

$$I_{ca}(V) = \frac{G_{ca}(V-110)}{5310(1+\exp[(4-V)/14])(1+\exp[(V-V_{hh})/C_{hh}])}$$

into the current-voltage data of voltage-gated calcium channels in mouse pancreatic $\beta$-cells obtained by Rorsman and Trube [56]:

| $V(mV)$ | $-50$ | $-40.06$ | $-30.11$ | $-20.58$ | $-10.22$ | 0.12 | 9.67 | 20.03 |
|---|---|---|---|---|---|---|---|---|
| $I_{ca}(pA/pF)$ | $-0.42$ | $-2.78$ | $-5.19$ | $-6.63$ | $-6.37$ | $-4.86$ | $-2.39$ | $-1.01$ |

The data are read from Fig. 3C of [56] using the software Engauge Digitizer 4.1.

**6.10.** Fit the following function

$$\tau = a - b\exp\left[-((V-c)/d)^2\right]$$

into the data of time constant of activation of voltage-gated calcium channels in mouse pancreatic $\beta$-cells from Fig. 4B of Göpel et al [21]:

| $V(mV)$ | $-29.9562$ | $-19.8898$ | $-10.0097$ | 0.207763 | 10.2546 | 20.2763 |
|---|---|---|---|---|---|---|
| $\tau(ms)$ | 0.895937 | 0.675499 | 0.547328 | 0.478959 | 0.394263 | 0.483318 |

The data are read from Fig. 4B of Göpel et al [21] using the software Engauge Digitizer 4.1.

**6.11.** Solve the steady state equation of (6.77) for the steady state $h_{ca\infty}$, derive the steady I-V relationship from (6.72), and plot the I-V curve.

**6.12.** Find the steady state of the system (6.91)-(6.95). Plot the channel open probability $P = (0.1O + 0.9A)^4$ against the calcium concentration $c$ for a given IP$_3$ concentration and against the IP$_3$ concentration $[IP_3]$ for a given calcium concentration. Fit the probability function into the data of Figs. 6B and 7B of Ramos-Franco *et al* [55] and the data of Figs. 2B and 3C of Dufour *et al* [13].

**6.13.** Find the steady state of the system (6.97)-(6.100). Plot the channel open probability $P = x_1^4$ against the calcium concentration $c$ for a given IP$_3$ concentration and against the IP$_3$ concentration $[IP_3]$ for a given calcium concentration. Fit the probability function into the data of Figs. 4a and 5 of Joseph *et al* [30] and the data of Figs. 2b and 3b of Bezprozvanny *et al* [4].

**6.14.** Fit the IP$_3$ receptor open probability function (6.104) into the data of Fig. 1B of Hagar *et al* [25], the data of Fig. 4 of Mak *et al* [45], and the data of Fig. 2a of Meyer *et al* [47].

**6.15.** Solve (6.106), (6.107), and (6.108) to obtain (6.109), (6.110), and (6.111).

**6.16.** Derive the equations (6.113)-(6.116).

**6.17.** Derive the equation (6.127) from the equations (6.118), (6.119), (6.120), (6.121), (6.122), and (6.123).

**6.18.** The sarcoplasmic or endoplasmic reticulum Ca$^{2+}$-ATPase (SERCA) is an enzyme that resides on the membrane of intracellular sarcoplasmic or endoplasmic reticulum organelles and pumps Ca$^{2+}$ from the cytosol into the organelles. The calcium uptake cycle by SERCA, as demonstrated in Fig. 2.8, is thought to include a binding of two Ca$^{2+}$ ions to the cytosolic portion of the Ca$^{2+}$-ATPase (Ca$_2^{2+}$-E$_1$), an ATP-dependent phosphorylation (Ca$_2^{2+}$-E$_1$-P), a translocation of Ca$^{2+}$ to the lumenal portion of the Ca$^{2+}$-ATPase (Ca$_2^{2+}$-E$_2$-P), a presumably sequential dissociation of Ca$^{2+}$ to the Ca$^{2+}$ store lumen (E$_2$-P), a dephosphorylation of the enzyme (E$_2$), and finally a regain of the original conformation (E$_1$) (see [14, 54]). Use this enzymatic cycle and equilibrium analysis to construct a model for the rate of Ca$^{2+}$ transport by the SERCA pump.

**6.19.** Use an equilibrium analysis to derive the steady-state fluxes for Na$^+$ and K$^+$ from the six-step biochemical reaction scheme of Na$^+$/K$^+$-ATPase described in Fig. 6.28.

**6.20.** ([33]) Simplify the model (6.138)-(6.141) of the Na$^+$/Ca$^{2+}$ exchanger by assuming that the binding and unbinding of Na$^+$ and Ca$^{2+}$ are fast compared to the exchange processes between the inside and the outside of the cell. This assumption of fast equilibrium gives

$$k_1[Ca_i^{2+}]x_1 = k_{-1}[Na_i^+]^3x_2, \quad k_{-3}[Ca_o^{2+}]x_4 = k_3[Na_o^+]^3x_3.$$

Calculate the steady-state flux, introduce the new variables $x = x_1 + x_2$ and $y = x_3 + x_4$, and derive the equations for $x$ and $y$.

# References

1. Atri A., Amundson J., Clapham D., Sneyd J.: A Single-pool model for Intracelilular calcium oscillations and waves in the Xenopus laevis oocyte. Biophys. J. **65**, 1727-1739 (1993).
2. Bertram R., Smolen P., Sherman A., Mears D., Atwater I., Martin F., Soria, B.: A role for calcium release-activated current (CRAC) in cholinergic modulation of electrical activity in pancreatic $\beta$-Cells. Biophysical J. **68**, 2323-2332 (1995).
3. Bertram R., Previte J., Sherman A., Kinard T.A., Satin L.S.: The phantom burster model for pancreatic $\beta$-cells, Biophysical J. **79**, 2880-2892 (2000).
4. Bezprozvanny I., Watras J., Ehrlich B.E.: Bell-shaped calcium-response curve of Ins(1,4,5)P3- and calcium-gated channels from endoplasmic reticulum of cerebellum. Nature **351**, 751-754 (1991).
5. Bolsover S.R., Hyams J.S., Shephard E.A., White H.A., Wiedemann C.G.: Cell Biology, A Short Course, Seond Edition. John Wiley & Sons, Inc., Hoboken, New Jersey (2004).
6. Campbell E.T., Hille B.: Kinetic and pharmacological properties of the sodium channel of frog skeletal muscle. J. General Physiology **67**, 309-323 (1976).
7. Cens T., Rousset M., Leyris J. P., Fesquet P., Charnet P.: Voltage- and calcium-dependent inactivation in high voltage-gated $Ca^{2+}$ channels. Prog. Biophys. Mol. Biol. **90**, 104-117 (2006).
8. Chapman J.B., Johnson E.A., Kootsey J.M.: Electrical and biochemical properties of an enzyme model of the sodium pump. J. Membr. Biol. **74**, 139-153 (1983).
9. Chay T.R., Keizer J.: Minimal model for membrane oscillations in the pancreatic $\beta$-cell. Biophysical J. **42**, 181-190 (1983).
10. Chay T.R.: Effects of extracellular calcium on electrical bursting and intracellular and luminal calcium oscillations in insulin secreting pancreatic $\beta$-cells. Biophys. J. **73**, 1673-1688 (1997).
11. Ching L.L., Williams A.J., Sitsapesan R.: Evidence for $Ca^{2+}$ activation and inactivation sites on the luminal side of the cardiac ryanodine receptor complex. Circ. Res. **87**, 201-206 (2000).
12. DiFrancesco D., Noble D.: A model of cardiac electrical activity incorporating ionic pumps and concentration changes. Philos. Trans. R. Soc. Lond. B Biol. Sci. **307**, 353-398 (1985).
13. Dufour J.-F., Arias I.M., Turner T.J.: Inositol 1,4,5-trisphosphate and calcium regulate the calcium channel function of the hepatic inositol 1,4,5-trisphosphate receptor. J. Biol. Chem. **272**, 2675-2681 (1997).
14. Favre C.J., Schrenzel J., Jacquet J., Lew D.P., Krause K.-H.: Highly supralinear feedback inhibition of $Ca^{2+}$ uptake by the $Ca^{2+}$ load of intracellular stores. J. Biol. Chem. **271**, 14925-14930 (1996).
15. Frankenhaeuser B.: Quantitative description of sodium currents in myelinated nerve fibres of *Xenopus Laevis*. J. Physiol. **151**, 491-501 (1960).
16. Frankenhaeuser B.: Sodium permeability in toad nerve and in squid nerve. J. Physiol. **152**, 159-166 (1960).
17. Frankenhaeuser B.: A quantitative description of potassium currents in myelinated nerve fibres of *Xenopus Laevis*. J. Physiol. **169**, 424-430 (1963).
18. Fridlyand L.E., Tamarina N., Philipson L.H.: Modeling of $Ca^{2+}$ flux in pancreatic $\beta$-cells: role of the plasma membrane and intracellular stores. Am. J. Physiol. Endocrinol. Metab. **285**, E138-E154 (2003).
19. Fridlyand L.E., Jacobson D.A., Kuznetsov A., Philipson L.H.: A model of action potentials and fast $Ca^{2+}$ dynamics in pancreatic beta-cells. Biophys J. **96**, 3126-3139 (2009).

20. Gall D., Gromada J., Susa I., Rorsman P., Herchuelz A., Bokvist K.: Significance of Na/Ca exchange for $Ca^{2+}$ buffering and electrical activity in mouse pancreatic $\beta$-cells. Biophys. J. **76**, 2018-2028 (1999).

21. Göpel S., Kanno T., Barg S., Galvanovskis J., Rorsman P.: Voltage-gated and resting membrane currents recorded from $\beta$-cells in intact mouse pancreatic islets. J. Physiology, **521**, 717-728 (1999).

22. Grandi E., Morotti S., Ginsburg K.S., Severi S., Bers D.M.: Interplay of voltage and Ca-dependent inactivation of L-type Ca current. Progress in Biophysics and Molecular Biology **103**, 44-50 (2010).

23. Guo W., Campbell K.P.: Association of triadin with the ryanodine receptor and calsequestrin in the lumen of the sarcoplasmic reticulum. J. Biol. Chem. **270**, 9027-9030 (1995).

24. Gyorke I., Hester N., Jones L.R., Gyorke S.: The role of calsequestrin, triadin, and junctin in conferring cardiac ryanodine receptor responsiveness to luminal calcium. Biophys. J. **86**, 2121-2128 (2004).

25. Hagar R.E., Ehrlich B.E.: Regulation of the type III InsP(3) receptor by InsP(3) and ATP. Biophys J. **79**, 271-278 (2000).

26. Hille B.: Ion Channels of Excitable Memebranes. Sinauer Associates, INC., Sunderland, Massachusetts (2001).

27. Hodgkin A.L., Huxley A.F.: A quantitativfe description of membrane current and its application to conduction and excitation in nerve. J. Physicol. **117**, 500-544 (1952).

28. Hopkins W.F., Fatherazi S., Peter-Riesch B., Corkey B.E., Cook D.L.: Two sites for adenine-nucleotide regulation of ATP-sensitive potassium channels in mouse pancreatic $\beta$-cells and HIT cells. J. Membr. Biol. **129**, 287-295 (1992).

29. John S.A., Weiss J.N., Ribalet B.: ATP sensitivity of ATP-sensitive $K^+$ channels: role of the $\gamma$ phosphate group of ATP and the R50 residue of mouse Kir6.2. J. Physiol. **568**, 931-940 (2005).

30. Joshph S.K., Rice H.L., Williamson J.R.: The effect of external calcium and pH on inositol trisphosphate-mediated calcium release from cerebellum microsomal fractions. Biochem. J. **258**, 261-265 (1989).

31. Kakei M., Kelly R.D., Ashcroft S., Ashcroft F.: The ATP-sensitivity of $K^+$ channels in rat pancreatic $\beta$-cells is modulated by ADP. FEBS Lett. **208**, 63-66 (1986).

32. Kang T.M., Hilgemann D.W.: Multiple transport modes of the cardiac $Na^+/Ca^{2+}$ exchanger. Nature **427**, 544-548 (2004).

33. Keener J., Sneyrd J.: Mathematical Physiology I: Cellular Physiology, II: Systems Physiology, Second Edition. Springer, New York (2009).

34. Keizer J., Magnus G.: ATP-sensitive potassium channel and bursting in the pancreatic beta cell. A theoretical study. Biophys. J. **56**, 229-242 (1989).

35. Keizer J., De Young G.W.: Two roles for $Ca^{2+}$ in agonist stimulated $Ca^{2+}$ oscillations. Biophys. J. **61**, 649-660 (1992).

36. Köhler M., Hirschberg B., Bond C.T., Kinzie J.M., Marrion N.V., Maylie J., Adelman J.P.: Small-conductance, calciuma-ctivated potassium channels from mammalian brain. Science **273**, 1709-1714 (1996).

37. Lee Y., Keener J.P.: A calcium-induced calcium release mechanism mediated by calsequestrin. J. Theor. Biol. **253**, 668-679 (2008).

38. Leinders T., Vijverberg H.P.M.: $Ca^{2+}$ dependence of small $Ca^{2+}$-activated $K^+$ channels in cultured N1E-115 mouse neuroblastoma cells. Pflügers Arch. **422**, 223-232 (1992).

39. Liu W., Tang F., Chen J.: Designing dynamical output feedback controllers for store-operated $Ca^{2+}$ entry. Math. Biosci. **228**, 110-118 (2010).

40. Luo C.H., Rudy Y.: A dynamic model of the cardiac ventricular action potential. I. Simulations of ionic currents and concentration changes. Circ. Res. **74**, 1071-1096 (1994).

41. Lytton J., Westlin M., Burk S.E., Shull G.E., MacLennan D.H.: Functional comparisons between isoforms of the sarcoplasmic or endoplasmic reticulum family of calcium pumps. J. Biol. Chem. **267**, 14483-14489 (1992).
42. Magnus G., Keizer J.: Model of $\beta$-cell mitochondrial calcium handling and electrical activity. I. Mitochondrial variables, Am. J. Physiol. Cell Physiol. **274**, C1158-C1173 (1998).
43. Magnus G., Keizer J.: Model of $\beta$-cell mitochondrial calcium handling and electrical activity. II. Cytoplasmic variables. Am. J. Physiol. Cell Physiol. **274**, C1174-C1184 (1998).
44. Mahajan A., Shiferaw Y., Sato D., Baher A., Olcese R., Xie L.-H., Yang M.-J., Chen P.-S., Restrepo J.G., Karma, A., Garfinkel A., Qu Z., Weiss J.N.: A rabbit ventricular action potential model replicating cardiac dynamics at rapid heart rates. Biophysical J. **94**, 392-410 (2008).
45. Mak D.O., McBride S., Foskett J.K.: Regulation by $Ca^{2+}$ and inositol 1,4,5-trisphosphate (InsP$_3$) of single recombinant type 3 InsP$_3$ receptor channels. $Ca^{2+}$ activation uniquely distinguishes types 1 and 3 insp$_3$ receptors. J. Gen. Physiol. **117**, 435-446 (2001).
46. Marchant J.S., Taylor C.W.: Rapid activation and partial inactivation of inositol trisphosphate receptors by inositol trisphosphate. Biochemistry **37**, 11524-11533 (1998).
47. Meyer T., Wensel T., Stryer L.: Kinetics of calcium channel opening by inositol 1,4,5-trisphosphate. Biochemistry **29**, 32-37 (1990).
48. Mears D., Sheppard N.F. Jr, Atwater I., Rojas E., Bertram R., Sherman A.: Evidence that calcium release-activated current mediates transient glucose-induced electrical activity in the pancreatic $\beta$-cell. J. Membr. Biol. **155**, 47-55 (1997).
49. Mears D., Rojas E.: Properties of voltage-gated $Ca^{2+}$ currents measured from mouse pancreatic $\beta$-cells in situ. Biol. Res. **39**, 505-520 (2006).
50. Miwa Y., Imai Y.: Simulation of spike-burst generation and $Ca^{2+}$ oscillation in pancreatic $\beta$-cells. Jpn. J. Physiol. **49**, 353-364 (1999).
51. Plant R.E.: The effects of calcium++ on bursting neurons. A modeling study. Biophys. J. **21**, 217-237 (1978).
52. Plant T.D.: Properties and calcium-dependent inactivation of calcium currents in cultured mouse pancreatic $\beta$-cells. J. Physiol. **404**, 731-747 (1988).
53. Potier M., Trebak M.: New developments in the signaling mechanisms of the store-operated calcium entry pathway. Pflugers Arch. **457**, 405-415 (2008).
54. Raeymaekers L., Vandecaetsbeek I., Wuytack F., Vangheluwe P.: Modeling $Ca^{2+}$ dynamics of mouse cardiac cells points to a critical role of SERCA's affinity for $Ca^{2+}$. Biophys. J. **100**, 1216-1225 (2011).
55. Ramos-Franco J., Bare D., Caenepeel S., Nani A., Fill M., Mignery G.: Single-channel function of recombinant type 2 inositol 1,4,5-trisphosphate receptor. Biophy. J. **79**, 1388-1399 (2000).
56. Rorsman P., Trube G.: Calcium and delayed potassium currents in mouse pancreatic $\beta$-cells under voltage clamp conditions. J. Physiol. (Lond.) **375**, 531-550 (1986).
57. Sham J.S., Song L.S., Chen Y., Deng L.H., Stern M.D., Lakatta E.G., Cheng H.: Termination of $Ca^{2+}$ release by a local inactivation of ryanodine receptors in cardiac myocytes. Proc. Natl. Acad. Sci. USA **95**, 15096-15101 (1998).
58. Sherman A., Rinzel J., Keizer J.: Emergence of organized bursting in clusters of pancreatic $\beta$-cells by channel sharing. Biophys. J. **54**, 411-425 (1988).
59. Smith P.A., Bokvist K., Arkhammar P., Berggren P.-O., Rorsman, P.: Delayed rectifying and calcium-activated $K^+$ channels and their significance for action potential repolarization in mouse pancreatic $\beta$-cells. J. Gen. Physiol. **95**, 1041-1059 (1990).

60. Sneyd J., Dufour J.F.: A dynamic model of the type-2 inositol trisphosphate receptor. Proc. Natl. Acad. Sci. USA **99**, 2398-2403 (2002).

61. Sneyd J., Tsaneva-Atanasova K., Bruce J.I., Straub S.V., Giovannucci D.R., Yule D.I.: A Model of calcium waves in pancreatic and parotid acinar cells. Biophys. J. **85**, 1392-1405 (2003).

62. Sneyd J., Tsaneva-Atanasova K., Yule D.I., Thompson J.L., Shuttleworth T.J.: Control of calcium oscillations by membrane fluxes. PNAS. **101**, 1392-1396 (2004).

63. Tamarina N.A., Kuznetsov A., Philipson L.H.: Reversible translocation of EYFP-tagged STIM1 is coupled to calcium infux in insulin secreting $\beta$-cells. Cell Calcium **44**, 533-544 (2008).

64. Vanderheyden V., Devogelaere B., Missiaen L., De Smedt H., Bultynck G., Parys J.B.: Regulation of inositol 1,4,5-trisphosphate-induced $Ca^{2+}$ release by reversible phosphorylation and dephosphorylation. BBA - Molecular Cell Research **1793**, 959-970 (2009).

65. Vogalis F., Goyal R.K.: Activation of small conductance $Ca^{2+}$-dependent $K^+$ channels by purinergic agonists in smooth muscle cells of the mouse. J. Physiology **502**, 497-508 (1997).

66. Yang Y.C., Fann M.J., Chang W.H., Tai L.H., Jiang J.H., Kao L.S.: Regulation of sodium-calcium exchanger activity by creatine kinase under energy-compromised conditions. J. Biol. Chem. **285**, 28275-28285 (2010).

# Control of Intracellular Calcium Oscillations

The voltage-gated calcium channel is closed at the end of the cytosol side when the transmembrane voltage is at the resting voltage. If the membrane depolarization exceeds a threshold, most of the voltage-gated calcium channels stay shut, but some open, and calcium ions rush into the cell down their electrochemical gradient. The flow of positive charge inward on the calcium ions outweighs the outward flow of positive charge on potassium ions, so the cytosol is gaining positive charge and its voltages is moving in the positive direction. As a result, more voltage-gated calcium channels open, and this allows more calcium ions to flood in to further ele-

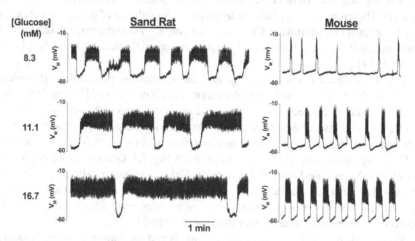

**Fig. 7.1.** Glucose-dependent action potentials in pancreatic $\beta$-cells. Membrane potential was recorded using intracellular microelectrodes. Sand rat islet (*left*): Traces are segments of a continuous recording, each showing the final 6 minutes of a 20-minute exposure to the glucose concentration indicated at the left. Mouse islet (*right*): Traces are segments of a continuous recording, each showing the final 5 minutes of a 10-minute exposure to the glucose concentration indicated at the left. The time scale applies to all traces. Reproduced with permission from [20]

Liu W.: Introduction to Modeling Biological Cellular Control Systems.
DOI 10.1007/978-88-470-2490-8_7, © Springer-Verlag Italia 2012

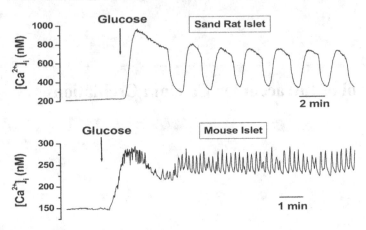

**Fig. 7.2.** Intracellular Ca$^{2+}$ ([Ca$^{2+}$]$_i$) dynamics in sand rat and mouse islets. Time course of [Ca$^{2+}$]$_i$ measured in (A) a sand rat islet and (B) a mouse islet. Islets were loaded with the fluorescent Ca$^{2+}$ indicator fura-2. The extracellular glucose concentration was raised from 2.8 mM to 11 mM as indicated by the arrows. Traces are representative of 6 experiments with islets from each species. Reproduced with permission from [20]

vate the voltage. This elevated voltage is called an *action potential*, as shown in Fig. 7.1. After about 100 ms, the voltage-gated calcium channels begin to inactivate and inward flow of calcium ions stops. As potassium leaves the cytosol through its voltage-gated potassium channels, the transmembrane voltage returns to its resting level. Therefore, the depolarizing and repolarizing phases of the action potentials have been attributed to inward Ca$^{2+}$ and outward K$^+$ currents, respectively, which seem sufficient for the generation of the action potentials starting from the plateau potential [14].

The intracellular calcium [Ca$^{2+}$]$_i$ and the membrane potential in pancreatic $\beta$-cells exhibit a bursting and oscillatory dynamics. Zimliki *et al* [20] made the first intracellular electrical recordings from sand rat islets, which showed that sand rat $\beta$-cells display bursting electrical activity over a wide range of glucose concentrations, depolarizing the plasma membrane to a plateau potential ($\approx$ -40 mV) and initiate Ca$^{2+}$-dependent action potentials, as shown in Fig. 7.1. Glucose with a concentration of 11 mM produced a regular pattern of [Ca$^{2+}$]$_i$ oscillations (Fig. 7.2). Overall, the pattern of glucose-induced [Ca$^{2+}$]$_i$ changes in sand rat islets resembled the bursting electrical activity. However, the [Ca$^{2+}$]$_i$ oscillations, with a period range of 98 to 195 seconds, were somewhat slower than the electrical bursts.

Mathematical models for the action potentials and oscillatory calcium dynamics have been developed. In this chapter, we first present a simple model constructed by Atwater *et al* [1] and Chay *et al* [3], and then other comprehensive models.

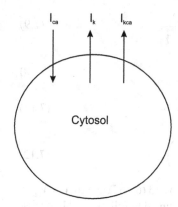

**Fig. 7.3.** Schematic representation of ionic currents through the plasma membrane of a $\beta$-cell. $I_{ca}$: voltage-gated $Ca^{2+}$ current; $I_k$: voltage-gated $K^+$ current; $I_{kca}$: $Ca^{2+}$-activated $K^+$ current

## 7.1 The Chay-Keizer Feedback Control System

One of the first mathematical models for intracellular calcium oscillations was developed by Atwater *et al* [1] and Chay *et al* [3]. Since inward $Ca^{2+}$ and outward $K^+$ currents seem sufficient for the generation of the action potentials [14], they considered only the voltage-gated $Ca^{2+}$ current $I_{ca}$, the voltage-gated $K^+$ current $I_k$, and the $Ca^{2+}$-activated $K^+$ current $I_{kca}$ in their model as shown in Fig. 7.3.

Since the potassium and calcium channel gating in the $\beta$-cell could be similar to the gating of potassium and sodium channels in the squid giant axon, the Hodgkin-Huxley model (6.156)-(6.167) was used with the sodium channel replaced by the calcium channel. Thus the membrane potential can be modeled by the following system [3, 8]

$$C_{pm}\frac{dV}{dt} = -G_{ca}P_{ca}(V - V_{ca}) - (G_{kca}P_{kca} + G_kP_k)(V - V_k) - G_l(V - V_l), \quad (7.1)$$

where the feedback controllers $P_{ca}$ and $P_k$ are given by

$$P_{ca} = hm^3, \tag{7.2}$$

$$P_k = n^4, \tag{7.3}$$

$$\frac{dh}{dt} = \alpha_h(1-h) - \beta_h h, \tag{7.4}$$

$$\frac{dm}{dt} = \alpha_m(1-m) - \beta_m m, \tag{7.5}$$

$$\frac{dn}{dt} = \alpha_n(1-n) - \beta_n n, \tag{7.6}$$

$$\alpha_h = 0.07 \exp\left(\frac{-V - V_{ca}^*}{20}\right), \tag{7.7}$$

$$\beta_h = \frac{1}{\exp\left(\frac{30-V-V_{ca}^*}{10}\right) + 1}, \tag{7.8}$$

$$\alpha_m = 0.1 \frac{25 - V - V_{ca}^*}{\exp\left(\frac{25-V-V_{ca}^*}{10}\right) - 1},\tag{7.9}$$

$$\beta_m = 4\exp\left(\frac{-V - V_{ca}^*}{18}\right),\tag{7.10}$$

$$\alpha_n = 0.01\frac{10 - V - V_k^*}{\exp\left(\frac{10-V-V_k^*}{10}\right) - 1},\tag{7.11}$$

$$\beta_n = 0.125\exp\left(\frac{-V - V_k^*}{80}\right).\tag{7.12}$$

Note that the voltage $V$ is replaced by $V + V_k^*$ in (6.166) and (6.167) and by $V + V_{ca}^*$ in (6.162)-(6.165). The $Ca^{2+}$-activated $K^+$ channel is assumed to follow the Michaelis-Menten law. Then the feedback controller $P_{kca}$ is given by

$$P_{kca} = \frac{[Ca^{2+}]_i}{K_{kca} + [Ca^{2+}]_i},\tag{7.13}$$

where $K_{kca}$ is a positive constant.

To complete the model, an equation for intracellular $Ca^{2+}$ in the cytosol is needed. It was assumed that the calcium leaves the cytosol at a rate proportional to its concentration and flows in at a rate proportional to the $Ca^{2+}$ current $I_{ca}$. Hence the equation for the intracellular $Ca^{2+}$ is

$$\frac{d[Ca^{2+}]_i}{dt} = f(-k_1 G_{ca}P_{ca}(V - V_{ca}) - k_{ca}[Ca^{2+}]_i),\tag{7.14}$$

where $k_1, k_{ca}$ are constants. The constant $f$ is a scalar factor. The units of potential, current density, conductance density, capacitance density, and calcium are mV, $\mu A/cm^2$, $mS/cm^2$, $\mu F/cm^2$, $\mu M$, respectively.

A numerical solution of the model is presented in Fig. 7.4 and the parameter values for the numerical solution are given in Table 7.1. The figure shows that the model can simulate the experimental action potential and calcium oscillations (Figs. 7.1 and 7.2).

**Table 7.1.** Parameters of the model (7.1)-(7.14) from [8]

| | |
|---|---|
| $C_{pm} = 1\ \mu F/cm^2$ | $G_{kca} = 0.02\ mS/cm^2$ |
| $G_k = 3\ mS/cm^2$ | $G_{ca} = 3.2\ mS/cm^2$ |
| $G_l = 0.012\ mS/cm^2$ | $V_k = -75\ mV$ |
| $V_{ca} = 100\ mV$ | $V = -40\ mV$ |
| $V_k^* = 30\ mV$ | $V_{ca}^* = 50\ mV$ |
| $K_{kca} = 1\ \mu M$ | $f = 0.007$ |
| $k_1 = 0.0275\ \mu M\ cm^2/nC$ | $k_{ca} = 0.02/ms$ |
| $V(0) = -55\ mV$ | $h(0) = 0.6$ |
| $m(0) = 0.05$ | $n(0) = 0.32$ |
| $[Ca^{2+}](0) = 0.8\ \mu M$ | |

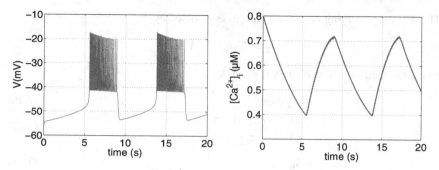

**Fig. 7.4.** Numerical simulation of action potential and calcium oscillation by the model (7.1)-(7.14)

## 7.2 The BSSMAMS Feedback Control System

Bertram *et al* [2] constructed a calcium oscillation model by adding an ATP-sensitive potassium current $I_{katp}$ and a store-operated calcium entry (SOCE) current $I_{soce}$ to the Chay-Keizer's model, which considered only the voltage-gated $Ca^{2+}$ current $I_{ca}$, voltage-gated $K^+$ current $I_k$, and $Ca^{2+}$-activated $K^+$ current $I_{kca}$. The resulted membrane potential equation was then given by

$$C_{pm}\frac{dV}{dt} = -I_{caf} - I_{cas} - I_k - I_{kca} - I_{katp} - I_{soce}, \tag{7.15}$$

where the various currents are modeled as follows. The voltage-gated $Ca^{2+}$ current is classified into a fast inactivation current $I_{caf}$ and a slow inactivation current $I_{cas}$, which are modeled as follows

$$I_{caf} = G_{caf}P_{caf}(V - V_{ca}), \tag{7.16}$$
$$I_{cas} = G_{cas}P_{cas}(V - V_{ca}), \tag{7.17}$$

where the feedback controllers $P_{caf}$ and $P_{cas}$ are given by

$$P_{caf} = m_f, \tag{7.18}$$
$$P_{cas} = m_s h, \tag{7.19}$$
$$\frac{dh}{dt} = \frac{h_\infty(V) - h}{\tau_h(V)}, \tag{7.20}$$
$$m_f = \frac{1}{1 + \exp[(-20 - V)/7.5]}, \tag{7.21}$$
$$m_s = \frac{1}{1 + \exp[(-16 - V)/10]}, \tag{7.22}$$
$$h_\infty = \frac{1}{1 + \exp[(53 + V)/2]}, \tag{7.23}$$

$$\tau_h = 1500 + \frac{50000}{\exp[(53 + V)/4] + \exp[-(53 + V)/4]}. \tag{7.24}$$

The delayed rectifying potassium current is modeled by

$$I_k = G_k P_k (V - V_k), \tag{7.25}$$

where the feedback controller $P_k$ is given by

$$P_k = n, \tag{7.26}$$

$$\frac{dn}{dt} = \frac{n_\infty(V) - n}{\tau_n(V)}, \tag{7.27}$$

$$n_\infty = \frac{1}{1 + \exp[(-15 - V)/6]}, \tag{7.28}$$

$$\tau_n = \frac{4.86}{1 + \exp[(15 + V)/6]}. \tag{7.29}$$

The model for the $Ca^{2+}$-activated $K^+$ current $I_{kca}$ is a little bit different from Chay-Keizer's model (7.13) and is given by

$$I_{kca} = G_{kca} P_{kca} (V - V_k), \tag{7.30}$$

where the feedback controller $P_{kca}$ is given by

$$P_{kca} = \frac{[Ca^{2+}]_i^5}{(0.55)^5 + [Ca^{2+}]_i^5}.$$

The ATP-sensitive potassium current is modeled by

$$I_{katp} = G_{katp}(V - V_k). \tag{7.31}$$

The SOCE current is modeled by

$$I_{soce} = G_{soce} P_{soce} (V - V_{soce}), \tag{7.32}$$

where the feedback controller $P_{soce}$ is given by

$$P_{soce} = \frac{1}{1 + \exp([Ca^{2+}]_{er} - 3)}.$$

The calcium sequestration into and release from the ER plays a key role in controlling calcium oscillations and are included in the model. Taking this into account, the equations for the intracellular and ER $Ca^{2+}$ are given by

$$\frac{d[Ca^{2+}]_i}{dt} = \frac{1}{\lambda}\left[\left(\frac{P_l}{P_{ip3}} + O_\infty\right)([Ca^{2+}]_{er} - [Ca^{2+}]_i) - \frac{r_{serca}}{P_{ip3}}\right] + \frac{r_{pm}}{V_{ieff}}, \tag{7.33}$$

$$\frac{d[Ca^{2+}]_{er}}{dt} = -\frac{1}{\sigma\lambda}\left[\left(\frac{P_l}{P_{ip3}} + O_\infty\right)([Ca^{2+}]_{er} - [Ca^{2+}]_i) - \frac{r_{serca}}{P_{ip3}}\right]. \tag{7.34}$$

In these equation, $P_l$ is the calcium leakage permeability through the ER membrane, $P_{ip3}$ is the calcium permeability through IP$_3$-activated calcium channels in

**Table 7.2.** Parameter values of the model (7.15)-(7.37) from [2]

| | | |
|---|---|---|
| $C_{pm} = 6158$ fF | $G_{caf} = 810$ pS | $G_{cas} = 510$ pS |
| $G_k = 3900$ pS | $G_{kca} = 1200$ pS | $G_{soce} = 75$ pS |
| $V_{ca} = 100$ mV | $V_k = -70$ mV | $V_{soce} = 0$ mV |
| $R_{serca} = 0.24\ \mu$M | $\lambda = 250$ ms | $P_l/P_{ip3} = 0.02$ |
| $f_i = 0.01$ | $V_{ieff} = 7.19 \times 10^6\ \mu$M$^3$ | $\alpha/V_{ieff} = 3.6 \times 10^{-8}$ fA$^{-1}\ \mu$M ms$^{-1}$ |
| $\sigma = 5$ | $k_{ca}/V_{ieff} = 0.0007$ ms$^{-1}$ | $G_{katp} = 150$ pS |
| $[IP_3] = 0\ \mu$M | $V(0) = -65$ mV | $[Ca^{2+}]_i(0) = 0.12\ \mu$M |
| $[Ca^{2+}]_{er}(0) = 4\ \mu$M | $h(0) = 0.68$ | $n(0) = 0.2$ |

the ER membrane, $V_{ieff}$ and $V_{ereff}$ are the effective intracellular and ER volumes, $\lambda = V_{ieff}/P_{ip3}$, and $\sigma = V_{ereff}/V_{ieff}$. The feedback controller $O_\infty$, the IP$_3$ receptor open probability, is modeled by

$$O_\infty = \frac{[Ca^{2+}]_i}{0.1 + [Ca^{2+}]_i} \frac{[IP_3]}{0.2 + [IP_3]} \frac{0.4}{0.4 + [Ca^{2+}]_i}. \tag{7.35}$$

The calcium flux $r_{serca}$ through SERCA is modeled by

$$r_{serca} = P_{ip3} R_{serca} \frac{[Ca^{2+}]_i^2}{(0.09)^2 + [Ca^{2+}]_i^2}. \tag{7.36}$$

The calcium flux $r_{pm}$ through the plasma membrane is modeled by

$$r_{pm} = -\alpha(I_{caf} + I_{cas}) - k_{ca}[Ca^{2+}]_i, \tag{7.37}$$

where $-\alpha(I_{caf} + I_{cas})$ denotes the influx through the voltage-gated calcium channel with $\alpha$ denoting the conversion factor from current to flux, and $k_{ca}[Ca^{2+}]_i$ denotes the efflux through the plasma membrane pump.

A numerical solution of the model is presented in Fig. 7.5 and the parameter values for the numerical solution are given in Table 7.2. The figure shows that the model can simulate the experimental action potentials and calcium oscillations (Figs. 7.1 and 7.2). However, the ER calcium concentration is too low as the resting free Ca$^{2+}$ in the ER is approximately 500 $\mu$M [16, 19].

## 7.3 The FTMP Feedback Control System

Fridlyand *et al* [5, 6] constructed a model that considers the plasma membrane potential, a variety of ionic channels and pumps, ATP/ADP-Ca$^{2+}$ interactions, as well as endoplasmic reticulum calcium sequestration. Using the current balance equation (6.149), the equation governing the plasma membrane potential $V$ is given by

$$C_{pm}\frac{dV}{dt} = -I_{ca} - I_{pm} - I_{naca} - I_{cran} - I_{na} - I_{nak} - I_{drk} - I_{kca} - I_{katp}, \tag{7.38}$$

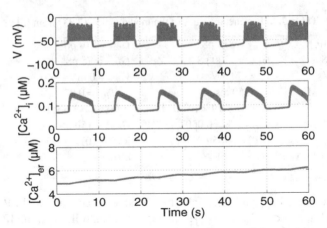

**Fig. 7.5.** Numerical simulation of action potentials and calcium oscillations by the model (7.15)-(7.37)

where $C_{pm}$ is the total membrane capacitance, $I_{ca}$ is a voltage-gated $Ca^{2+}$ current, $I_{pm}$ is a plasma membrane calcium pump current, $I_{naca}$ is a $Na^+/Ca^{2+}$ exchanger current, $I_{cran}$ is a $Ca^{2+}$-release-activated nonselective cation current, $I_{na}$ is an inward $Na^+$ current, $I_{nak}$ is a $Na^+/K^+$ pump current, $I_{drk}$ is a delayed rectifier $K^+$ current, $I_{kca}$ is a voltage-independent small conductance $Ca^{2+}$-activated $K^+$ current, and $I_{katp}$ is an ATP-sensitive $K^+$ current. The models of these currents are discussed below.

The voltage-gated $Ca^{2+}$ current was modeled by the function (6.70):

$$I_{ca} = G_{ca}P_{ca}\left(V - \frac{RT}{2F}\ln\left(\frac{[Ca^{2+}]_o}{[Ca^{2+}]_i}\right)\right), \tag{7.39}$$

where $V_{cah}$ and $K_{cah}$ are positive constants, $G_{ca}$ is the maximum whole cell conductance, R is the universal gas constant, T is absolute temperature, F is Faraday's constant, $[Ca^{2+}]_o$ denotes the extracellular calcium concentration, and the feedback controller $P_{ca}$ is given by

$$P_{ca} = \frac{1}{1+\exp[(V_{cah}-V)/K_{cah}]}.$$

The function (6.136) was used to model the $Ca^{2+}$ current through the plasma membrane calcium pump:

$$I_{pm} = \frac{P_{pm}[Ca^{2+}]_i^2}{K_{pm}^2 + [Ca^{2+}]_i^2}, \tag{7.40}$$

where $P_{pm}$ is a maximum current and $K_{pm}$ is the value for the half-activation calcium concentration.

The function (6.148) was used to model the electrogenic $3Na^+/Ca^{2+}$ exchanger current:

$$I_{naca} = G_{naca}P_{naca}\left(V - \frac{RT}{F}\left(3\ln\frac{[Na^+]_o}{[Na^+]_i} - \ln\frac{[Ca^{2+}]_o}{[Ca^{2+}]_i}\right)\right), \tag{7.41}$$

where $[Na^+]_o$ and $[Na^+]_i$ denote extracellular and intracellular $Na^+$ concentrations, respectively, $G_{naca}$ is the maximum whole cell conductance, and the feedback controller $P_{naca}$ is given by

$$P_{naca} = \frac{[Ca^{2+}]_i^5}{K_{naca}^5 + [Ca^{2+}]_i^5}.$$

In their model, Fridlyand et al [5] assumed that only $Na^+$ can penetrate into the cell via the nonselective cation channel, reflecting experimental results under physiological concentrations of cations [10, 16, 17]. This $Na^+$ inward current was modeled by

$$I_{cran} = -G_{cran}P_{cran}\left(V - \frac{RT}{F}\ln\frac{[Na^+]_o}{[Na^+]_i}\right), \qquad (7.42)$$

where $G_{cran}$ is the maximum whole cell conductance and $P_{cran}$ is a feedback controller to be designed. The current/voltage relationship of this current was roughly linear [10, 16, 17] in $V$. In addition, It was suggested that the regulation of the nonselective cation current depends on the endoplasmic reticulum $Ca^{2+}$ concentration $[Ca^{2+}]_{er}$ [2, 4, 12]. In mouse $\beta$-cells, it could be activated indirectly by endoplasmic reticulum $Ca^{2+}$ store depletion [16, 17, 18]. Taking this dependence into account, the feedback controller $P_{cran}$ was modeled by

$$P_{cran} = \frac{V - V_{cran}}{1 + \exp(([Ca^{2+}]_{er} - K_{er})/3)},$$

where $K_{er}$ is the half-activation $[Ca^{2+}]_{er}$ level, and $V_{cran}$ is the reversal potential.
The function (6.58) was used to model the inward $Na^+$ current:

$$I_{na} = G_{na}P_{na}\left(V - \frac{RT}{F}\ln\frac{[Na^+]_o}{[Na^+]_i}\right), \qquad (7.43)$$

where $G_{na}$ is the maximum whole cell conductance and the feedback controller $P_{na}$ is given by

$$P_{na} = \frac{1}{1 + \exp[(104 + V)/8]}.$$

The model (6.137) was used to describe the $Na^+/K^+$ pump current:

$$I_{nak} = \frac{P_{nak}(F_1 f_2 f_3 F_4 F_5 f_6 - b_1 B_2 B_3 B_4 b_5 B_6)}{D}, \qquad (7.44)$$

where

$$F_1 = f_1[Na^+]_i^3, \qquad (7.45)$$
$$f_5 = f_5^* \exp((VF/(2RT))), \qquad (7.46)$$
$$F_5 = f_5[ATP]_i, \qquad (7.47)$$
$$B_2 = b_2[ADP_f]_i, \qquad (7.48)$$
$$b_5 = b_5^* \exp(-(VF/(2RT))), \qquad (7.49)$$
$$D = f_2 f_3 F_4 F_5 f_6 + b_1 f_3 F_4 F_5 f_6 + b_1 B_2 F_4 F_5 f_6 + b_1 B_2 B_3 F_5 f_6$$
$$+ b_1 B_2 B_3 B_4 f_6 + b_1 B_2 B_3 B_4 b_5. \qquad (7.50)$$

The equation (6.23) was used to model the delayed rectifier $K^+$ current $I_{drk}$:

$$I_{drk} = G_{drk}P_{drk}\left(V - \frac{RT}{F}\ln\frac{[K^+]_o}{[K^+]_i}\right), \tag{7.51}$$

where $G_{drk}$ is the maximum whole cell conductance and the feedback controller $P_{drk}$ is given by

$$P_{drk} = n, \tag{7.52}$$

$$\frac{dn}{dt} = \frac{n_\infty - n}{\tau_n}, \tag{7.53}$$

$$n_\infty = 1.0/(1.0 + \exp(((V_n - V)/S_n))), \tag{7.54}$$

$$\tau_n = c/[\exp((V - V_\tau)/a) + \exp((V_\tau - V)/b)]. \tag{7.55}$$

The Hill-type function (6.35) was used to model the voltage-independent small conductance $Ca^{2+}$-activated $K^+$ current $I_{kca}$:

$$I_{kca} = G_{kca}P_{kca}\left(V - \frac{RT}{F}\ln\frac{[K^+]_o}{[K^+]_i}\right), \tag{7.56}$$

where $G_{kca}$ is the maximum whole cell conductance and the feedback controller $P_{kca}$ is given by

$$P_{kca} = \frac{[Ca^{2+}]_i^4}{K_{kca}^4 + [Ca^{2+}]_i^4}.$$

The equation (6.40) was used to model the ATP-sensitive $K^+$ current $I_{katp}$:

$$I_{katp} = G_{katp}P_{katp}\left(V - \frac{RT}{F}\ln\frac{[K^+]_o}{[K^+]_i}\right), \tag{7.57}$$

where the feedback controller $P_{katp}$ is given by

$$P_{katp} = \frac{0.08(1.0 + 2[MgADP]/k_{dd}) + 0.89[MgADP]^2/k_{dd}^2}{(1.0 + [MgADP]/k_{dd})^2(1.0 + 0.45[MgADP]/k_{td} + [ATP]_i/k_{tt})}, \tag{7.58}$$

The relation between MgADP and free ADP in the cytosol is

$$[MgADP] = 0.55[ADP_f]_i.$$

Calcium enters $\beta$-cells primarily through voltage-gated $Ca^{2+}$ channels by diffusion. The calcium homeostasis in the cytosol is regulated by a $Ca^{2+}$-extruding mechanism in the plasma membrane and $Ca^{2+}$ sequestration by intracellular organelles. At the plasma membrane, three processes are involved in transporting $Ca^{2+}$ out of the cell: a $Ca^{2+}$ pump, an $Na^+/Ca^{2+}$ exchanger, and removal of $Ca^{2+}$ sequestrated in insulin granules by exocytosis. In addition, both the ER and mitochondria can accumulate $Ca^{2+}$ via pumps. Even though $Ca^{2+}$ is critical for mitochondrial function, mitochondrial $Ca^{2+}$ is not considered since it appears that both the volume of mitochondria and its $Ca^{2+}$ concentration are small relative to the ER.

Ca$^{2+}$ are transported into the ER by sarco(endo) plasmic reticulum Ca$^{2+}$-ATPases (SERCAs) with an unusually low value for the half-activation calcium concentration ($\leq 0.4\ \mu M$) [11]. The equation (6.133) was used to describe the calcium uptake rate by SERCA:

$$r_{serca} = \frac{R_{serca}[Ca^{2+}]_i^2}{[Ca^{2+}]_i^2 + K_{serca}^2}. \tag{7.59}$$

Following the equation (6.79), a Ca$^{2+}$ efflux from the ER was modeled by

$$r_{ip3r} = (P_{leak} + P_{ip3}O_\infty)([Ca^{2+}]_{er} - [Ca^{2+}]_i), \tag{7.60}$$

where $P_{leak}$ is the rate of calcium leaking from the ER and the feedback controller $O_\infty$, the IP$_3$-activated channel open probability, was given by

$$O_\infty = \frac{[Ca^{2+}]_i[IP_3]^3}{([Ca^{2+}]_i + K_{rca})([IP_3]^3 + K_{ip3}^3)}. \tag{7.61}$$

The Ca$^{2+}$ leak from cells by insulin granule exocytosis was modeled previously [4] as a rate proportional to $[Ca^{2+}]_i$.

The cytosolic and ER calcium concentrations are determined by the ion fluxes across the plasma and ER membranes. However, calcium concentrations are strongly buffered in cells [15]. This was modeled using special coefficients for the fraction of free Ca$^{2+}$ in the cytoplasm and ER. On the basis of the foregoing consideration, the equations for Ca$^{2+}$ concentrations were written as

$$\frac{d[Ca^{2+}]_i}{dt} = f_i\left(\frac{-I_{ca} + 2I_{naca} - 2I_{pm}}{2FV_i} - r_{serca} + \frac{r_{ip3r}}{V_i}\right) - k_{sg}[Ca^{2+}]_i, \tag{7.62}$$

$$\frac{d[Ca^{2+}]_{er}}{dt} = \frac{f_{er}}{V_{er}}(r_{serca}V_i - r_{ip3r}), \tag{7.63}$$

where $f_i$ and $f_{er}$ are the fractions of free Ca$^{2+}$ in cytoplasm and ER, $V_{er}$ and $V_i$ are the effective volumes of the ER and cytosolic compartments, and $k_{sg}$ is a coefficient of the sequestration rate of Ca$^{2+}$ by the secretory granules.

Like Ca$^{2+}$, the Na$^+$ cytosolic concentration was determined by the ion fluxes across the plasma membrane. The equation for the concentration was written as

$$\frac{d[Na^+]_i}{dt} = \frac{-3I_{naca} - 3I_{nak} - I_{na} - I_{cran}}{FV_i}. \tag{7.64}$$

The kinetics of IP$_3$ in $\beta$-cells is unknown. It was assumed that [5, 9] IP$_3$ is degraded at a rate proportional to its concentration. Thus the kinetics of IP$_3$ may be modeled by

$$\frac{d[IP_3]_i}{dt} = R_{ip3}P_{ip3in} - k_{ipd}[IP_3]_i, \tag{7.65}$$

where $R_{ip3}$ is the maximal rate of IP$_3$ production and $P_{ip3in}$ is a feedback controller to be designed. Experiments by Mouillac et al [13] indicated that intracellular Ca$^{2+}$

modulates the production of IP$_3$. This suggests the possibility of a positive feedback mechanism of Ca$^{2+}$ on the production of IP$_3$. On the basis of this information, the feedback controller $P_{ip3in}$ was modeled by

$$P_{ip3in} = \frac{[Ca^{2+}]_i^2}{K_{ipca}^2 + [Ca^{2+}]_i^2}, \tag{7.66}$$

where $K_{ipca}$ is a positive constant.

Glucose is phosphorylated in a sigmoidal fashion, so the Hill equation was used to model this process. Therefore, an empirically derived rate expression for glucose phosphorylation by glucokinase was employed as follows

$$r_{glu} = R_{glu}P_{glu}\frac{[Glu]^{1.7}}{K_{glu}^{1.7} + [Glu]^{1.7}}, \tag{7.67}$$

where $P_{glu}$ is a feedback controller to be designed. The MgATP dependence of this reaction could be well fit to a Michaelis-Menten-type saturation equation. Therefore, the feedback controller $P_{glu}$ is modeled by

$$P_{glu} = \frac{[ATP]_i}{K_{atp} + [ATP]_i}. \tag{7.68}$$

Intermediate metabolites produced from glucose are used to produce ATP. Thus the consumption rate of these intermediate metabolites is determined by ATP production. ATP is synthesized through oxidative phosphorylation processes that use the intermediate metabolites and free cytosolic MgADP. The dependence of oxidative phosphorylation on free MgADP may be calculated using the Hill equation. Then, assuming the simplest linear dependence of reaction rate on the intermediate metabolites, an empirical equation for oxidative phosphorylation rate $r_{op}$ was written as

$$r_{op} = R_{op}[IM]_i\frac{[MgADP]^2}{[MgADP]^2 + K_{op}^2} \tag{7.69}$$

where $[IM]$ is the concentration of the intermediate metabolites. Then the concentration of these intermediate metabolites were described by the following equation

$$\frac{d[IM]_i}{dt} = k_{im}r_{glu} - r_{op}. \tag{7.70}$$

Since 95% of energy supply derives from mitochondrial oxidative phosphorylation, the rate of oxidative phosphorylation (7.69) was used as the ATP production rate. On the other hand, ATP is consumed by the plasma and ER Ca$^{2+}$ pumps and by the electrogenic Na$^+$/K$^+$-ATPase. In addition, there is some basal level of ATP consumption and there is considerable evidence that other ATP consumption processes are accelerated by an increase in Ca$^{2+}$. On the basis of this evidence, the balance equation for ATP was written as

$$\frac{d[ATP]_i}{dt} = r_{op} - \frac{I_{nak} + I_{pm}}{FV_i} - \frac{r_{serca}}{2} - \left(k_{atpca}[Ca^{2+}]_i + k_{atpd}\right)[ATP]_i. \tag{7.71}$$

**Table 7.3.** Parameter values of the model (7.38)-(7.73) from [5, 6]

| | | |
|---|---|---|
| $K_{cah}$= 9.5 mV | $k_{atpca}$= 8.0E-5 $\mu M^{-1}ms^{-1}$ | $V_{cran}$ = 0.0 mV |
| $k_{atpd}$ = 5.0E-5 $ms^{-1}$ | $G_{cran}$= 0.7 pS/mV | $f_5^*$= 0.0020 $\mu M^{-1}ms^{-1}$ |
| $G_{ca}$ = 670.0 pS | $k_{dd}$ = 17.0 $\mu M$ | $f_4$ = 1.5E-8 $\mu M^{-2}ms^{-1}$ |
| $[K^+]_o$= 8000.0 $\mu M$ | $F_4$= $f_4[K^+]_o^2$ | $G_{na}$= 1200.0 pS |
| $K_{ipca}$ = 0.4 $\mu M$ | $C_{pm}$= 6158.0 fF | $P_{leak}$ = 1.0E-4 pl/ms |
| $K_{atp}$ = 500.0 $\mu M$ | $K_{kca}$= 0.1 $\mu M$ | $[N^+]_o$= 140000.0 $\mu M$ |
| $V_{cah}$ = -19.0 mV | $R_{ip3}$= 3.0E-4 $\mu M/ms$ | $b_5^*$= 0.03 $ms^{-1}$ |
| $f_i$ = 0.01 | $K_{ip3}$ = 3.2 | $b_6$ = 6.0E-7 $\mu M^{-1}ms^{-1}$ |
| $[K^+]_i$ = 132400.0 $\mu M$ | $B_6 = b_6[K^+]_i^2$ | $b_4$ = 2.0E-4 $\mu M^{-1}ms^{-1}$ |
| $[P_i]$ = 4950.0 $\mu M$ | $B_4 = b4[P_i]$ | $b_3$= 1.72E-17 $\mu M^{-3}ms^{-1}$ |
| $B_3 = b_3[N^+]_o^3$ | $K_{op}$ = 20.0 $\mu M$ | $k_{ipd}$= 4.0E-5 $ms^{-1}$ |
| $k_{adpf}$ =2.0E-4 $ms^{-1}$ | $k_{adpb}$ = 2.0E-5 $ms^{-1}$ | $P_{ip3}$ = 0.0012 pl/ms |
| $P_{nak}$ = 600.0 fA | $G_{drk}$= 3000.0 pS | $f_{er}$= 0.03 |
| $f_6$ = 11.5 $ms^{-1}$ | $f_3$= 0.172 $ms^{-1}$ | $f_2$ = 10.0 $ms^{-1}$ |
| $f_1$ = 2.5E-10 $\mu M^{-3}ms^{-1}$ | $[Glu]$ = 8.0 mM | $K_{glu}$ = 7.0 mM |
| $K_{naca}$= 0.75 $\mu M$ | $G_{kca}$ = 130.0 pS | $\frac{F}{RT}$ = 26.73 mV |
| $V_\tau$ = -75.0 mV | $c$ = 20.0 ms | $V_{er}$ = 0.28 pL |
| $b$ =20.0 mV | $[Ca^{2+}]_o$ = 2600.0 $\mu M$ | $a$= 65.0 mV |
| $V_n$ = -14.0 mV | $V_i$ = 0.764 pL | $R_{serca}$ = 0.105 $\mu M/ms$ |
| $G_{naca}$= 271.0 pS | $P_{pm}$ = 2000.0 fA | $F$ = 96480.0 |
| $K_{rca}$ = 0.077 $\mu M$ | $k_{tt}$ = 50.0 $\mu M$ | $R_{glu}$ = 0.025 $\mu M/ms$ |
| $k_{td}$ = 26.0 $\mu M$ | $k_{im}$ = 31.0 | $b_2$ = 1.0E-4 $\mu M^{-1}ms^{-1}$ |
| $b_1$= 100.0 $ms^{-1}$ | $k_{sg}$ = 1.0E-4 $ms^{-1}$ | $G_{katp}$= 24000.0 pS |
| $R_{op}$= 2.0E-4 $ms^{-1}$ | $K_{er}$ = 200.0 $\mu M$ | $S_n$= 7.0 mV |
| $K_{pm}$ = 0.1 $\mu M$ | $K_{serca}$ = 0.5 $\mu M$ | $[Ca^{2+}]_i(0) = 0.107$ $\mu M$ |
| $[Ca^{2+}]_{er}(0) = 34.3$ $\mu M$ | $[ADP_b]_i(0) = 922$ $\mu M$ | $[ADP_f]_i(0) = 92.2$ $\mu M$ |
| $[ATP]_i(0) = 2985$ $\mu M$ | $[IP_3]_i(0) = 0.5$ $\mu M$ | $[Na^+]_i(0) = 7281$ $\mu M$ |
| $[IM]_i(0) = 1260$ $\mu M$ | $V(0) = -57$ mV | $n(0) = 0.0021$ |

In ATP hydrolysis, ATP is converted into ADP and ADP exists in a free form $ADP_f$ and a bound form $ADP_b$. Only a small fraction of total cellular ADP is free. Taking these facts into account and assuming that the interchange between $ADP_f$ and $ADP_b$ is linear, the balance equations for them were written as

$$\frac{d[ADP_f]_i}{dt} = \frac{I_{nak}+I_{pm}}{FV_i} + \frac{r_{serca}}{2} + \left(k_{atpca}[Ca^{2+}]_i + k_{atpd}\right)[ATP]_i$$
$$+k_{adpb}[ADP_b]_i - r_{op} - k_{adpf}[ADP_f]_i, \tag{7.72}$$

$$\frac{d[ADP_b]_i}{dt} = k_{adpf}[ADP_f]_i - k_{adpb}[ADP_b]_i. \tag{7.73}$$

A numerical solution of the model (7.38)-(7.73) is presented in Fig. 7.6 and the parameter values for the numerical solution are given in Table 7.3. The figure shows that the model can simulate the experimental action potentials and calcium oscillations (Figs. 7.1 and 7.2). However, the ER calcium concentration is low as the resting free $Ca^{2+}$ in the ER is approximately 500 $\mu M$ [16, 19].

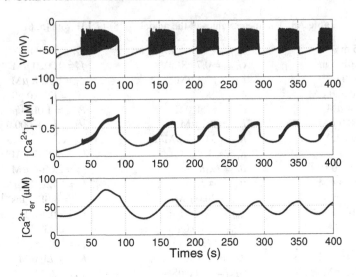

**Fig. 7.6.** Numerical simulation of action potentials and calcium oscillations by the model (7.38)-(7.73)

## Exercises

**7.1.** Modify the Chay-Keizer model (7.1)-(7.6) and (7.14) in the following different ways:

1. Replace $hm^3$ in the calcium current with the open probability (6.69) or (6.72) and keep others unchanged.
2. Replace $g_{kca}$ in the calcium-activated potassium current $g_{kca}(V - V_k)$ with the controller (6.35) and keep others unchanged.
3. Replace $n^4$ in the potassium current $G_k n^4 (V - V_k)$ with the activation model (6.23) or (6.29) and keep others unchanged.

Solve each modified model numerically to see whether it can simulate calcium oscillations.

**7.2.** Modify the BSSMAMS model (7.15)-(7.37) in the following different ways:

1. Replace $m_f$ in the fast inactivation calcium current $I_{caf}$ with the open probability (6.69) or (6.72) and keep others unchanged.
2. Replace $n$ in the potassium current $G_k n(V - V_k)$ with the activation model (6.29) and keep others unchanged.
3. Replace the ATP-sensitive potassium current $I_{katp}$ with the current equation (6.36) and (6.40) and keep others unchanged.
4. Replace the IP$_3$ receptor open probability $O_\infty$ with the probability $(0.1O + 0.9A)^4$ of the Sneyd *et al*'s model (6.91)-(6.95), the probability $x_1^4$ of Keizer *et al*'s model (6.97)-(6.100), or the probability (6.104) and keep others unchanged.

5. Replace the SERCA flux $r_{serca}$ with the flux (6.129) or (6.135) and keep others unchanged.
6. Replace $k_{ca}[Ca^{2+}]_i$ in the flux $r_{pm}$ with the flux (6.136) and keep others unchanged.

Solve each modified model numerically to see whether it can simulate calcium oscillations.

**7.3.** Solve the Fridlyand et el's model (7.38)-(7.73) numerically with different glucose inputs such as $[Glu] = 2, 5, 8, 11, 16$ mM to see how the glucose concentration affects the action potentials and calcium oscillations. Solve the model numerically with different values of $P_{leak}$ and $P_{ip3}$ to see how the ER affects the action potentials and calcium oscillations.

**7.4.** Modify the Fridlyand et el's model (7.38)-(7.73) in the following different ways:

1. Replace the voltage-dependent open probability in the calcium current $I_{ca}$ with the open probability (6.69) or (6.72) and keep others unchanged.
2. Replace the potassium current $I_{drk}$ with the current equation (6.12) and keep others unchanged.
3. Replace the sodium current $I_{na}$ with the current equation (6.41) and keep others unchanged.
4. Replace the IP$_3$ receptor open probability $O_\infty$ with the probability $(0.1O + 0.9A)^4$ of the Sneyd et al's model (6.91)-(6.95), the probability $x_1^4$ of Keizer et al's model (6.97)-(6.100), or the probability (6.104) and keep others unchanged.
5. Replace the SERCA flux $r_{serca}$ with the flux (6.129) or (6.135) and keep others unchanged.
6. Add the SOCE current (7.32) to the membrane potential equation (7.38) and the intracellular calcium equation (7.62) and keep others unchanged.

Solve each modified model numerically to see whether it can simulate calcium oscillations.

# References

1. Atwater I., Dawson C.M., Scott A., Eddlestone G., Rojas E.: The nature of the oscillatory behavior in electrical activity from pancreatic $\beta$-cell. J. Horm. Metabol. Res. **10** (suppl.), 100-107 (1980).
2. Bertram R., Smolen P., Sherman A., Mears D., Atwater I., Martin F., Soria B.: A role for calcium release-activated current (CRAC) in cholinergic modulation of electrical activity in pancreatic $\beta$-Cells. Biophysical J. **68**, 2323-2332 (1995).
3. Chay T.R., Keizer J.: Minimal model for membrane oscillations in the pancreatic $\beta$-cell. Biophysical J. **42**, 181-190 (1983).
4. Chay, T.R.: Effects of extracellular calcium on electrical bursting and intracellular and luminal calcium oscillations in insulin secreting pancreatic $\beta$-cells. Biophys. J. **73**, 1673-1688 (1997).
5. Fridlyand L.E., Tamarina N., Philipson L.H.: Modeling of Ca$^{2+}$ flux in pancreatic $\beta$-cells: role of the plasma membrane and intracellular stores. Am. J. Physiol. Endocrinol. Metab. **285**, E138-E154 (2003).

6. Fridlyand L.E., Ma L., Philipson L.H.: Adenine nucleotide regulation in pancreatic $\beta$-cells: modeling of ATP/ADP-$Ca^{2+}$ interactions. Am. J. Physiol. Endocrinol. Metab. **289**, E839-E848 (2005).

7. Frischauf I., Schindl R., Derler I., Bergsmann J., Fahrner M., Romanin C.: The STIM/Orai coupling machinery. Channels **2**, 1-8 (2008).

8. Keener J., Sneyd J.: Mathematical Physiology I: Cellular Physiology, II: Systems Physiology, Second Edition. Springer, New York (2009).

9. Keizer J., De Young G.W.: Two roles for $Ca^{2+}$ in agonist stimulated $Ca^{2+}$ oscillations. Biophys. J. **61**, 649-660 (1992).

10. Leech C.A., Habener J.F.: A role for $Ca^{2+}$ sensitive nonselective cation channels in regulating the membrane potential of pancreatic $\beta$-cells. Diabetes **47**, 1066-1073 (1998).

11. Lytton J., Westlin M., Burk S.E., Shull G.E., MacLennan D.H.: Functional comparisons between isoforms of the sarcoplasmic or endoplasmic reticulum family of calcium pumps. J. Biol. Chem. **267**, 14483-14489 (1992).

12. Mears D., Sheppard N.F. Jr, Atwater I., Rojas E., Bertram R., Sherman A.: Evidence that calcium release-activated current mediates transient glucose-induced electrical activity in the pancreatic $\beta$-cell. J. Membr. Biol. **155**, 47-55 (1997).

13. Mouillac B., Balestre M.N., Guillon G.: Positive feedback regulation of phospholipase C by vasopressin-induced calcium mobilization in WRK1 cells. Cellular Signaling **2**, 497-507 (1990).

14. Rorsman P., Trube G.: Calcium and delayed potassium currents in mouse pancreatic $\beta$-cells under voltage clamp conditions. J. Physiol. (Lond.) **375**, 531-550 (1986).

15. Rorsman P., Ammala C., Berggren P.O., Bokvist K., Larsson O.: Cytoplasmic calcium transients due to single action potentials and voltage-clamp depolarizations in mouse pancreatic $\beta$-cells. EMBO J. **11**, 2877-2884 (1992).

16. Straub S.G., Kornreich B., Oswald R.E., Nemeth E.F., Sharp G.W.: The calcimimetic R-467 potentiates insulin secretion in pancreatic $\beta$-cells by activation of a nonspecific cation channel. J. Biol. Chem. **275**, 18777-18784 (2000).

17. Worley J.K. III, McIntyre M.S., Spencer B., Dukes I.D.: Depletion of intracellular $Ca^{2+}$ stores activates a maitotoxin-sensitive nonselective cationic current in $\beta$-cells. J. Biol. Chem. **269**, 32055-32058 (1994).

18. Worley J.F. III, McIntyre M.S., Spencer B., Mertzm R.J., Roem M.W., Dukesm I.D.: Endoplasmic reticulum calcium store regulates membrane potential in mouse islet $\beta$-cells. J. Biol. Chem. **269**, 14359-14362 (1994).

19. Yu Y., Hinkle P.M.: Rapid turnover of calcium in the endoplasmic reticulum during signaling, J. Biol. Chem. **275**, 23648-23653 (2000).

20. Zimliki C.L., Chenault M.V., Mears D.: Glucose-dependent and -independent electrical activity in islets of Langerhans of Psammomys obesus, an animal model of nutritionally-induced obesity and diabetes. General and Comparative Endocrinology **161**, 193-201 (2009).

# Store-Operated Calcium Entry

Depletion of intracellular calcium stores such as the endoplasmic reticulum (ER) activates store-operated channels for $Ca^{2+}$ entry across the plasma membrane (PM), as demonstrated in Fig. 8.1. This process is called the *store-operated calcium entry* (*SOCE*), a common and ubiquitous mechanism of regulating $Ca^{2+}$ influx into cells [1, 7, 36, 38]. The best-studied store-operated channel (SOC) is the $Ca^{2+}$ release-activated $Ca^{2+}$ channel (CRAC) [5, 12, 13, 30, 38, 41, 53]. SOCE is a key feedback controller to stabilize ER $Ca^{2+}$ and has been proposed as the main $Ca^{2+}$ entry path-

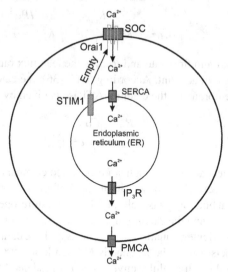

**Fig. 8.1.** A systematic sketch of an intracellular $Ca^{2+}$ regulatory system. Calcium ions $Ca^{2+}$ enter the cytosol through store-operated channels (SOCs). The sarcoplasmic or endoplasmic reticulum $Ca^{2+}$-ATPases (SERCA) pump $Ca^{2+}$ from the cytosol into ER and $Ca^{2+}$ in ER are released to the cytosol through the IP$_3$- and $Ca^{2+}$-mediated $Ca^{2+}$ channels. $Ca^{2+}$ exit the cytosol through the plasma membrane $Ca^{2+}$-ATPases (PMCA). Depletion of ER $Ca^{2+}$ stores causes STIM1 to move to ER-PM junctions, bind to Orai1, and activate SOCs for $Ca^{2+}$ entry [28]

Liu W.: Introduction to Modeling Biological Cellular Control Systems.
DOI 10.1007/978-88-470-2490-8_8, © Springer-Verlag Italia 2012

way in non-excitable cells [37, 38]. SOCE, originally known as capacitative calcium entry (CCE), was first proposed by Putney [42] and has been extensively studied later (for review, see Berridge [2, 3], Bird *et al* [4], Chakrabarti *et al* [6], Dirksen [9], Lewis [25], Parekh [37], Potier *et al* [39], Prakriya *et al* [40], Putney [43], and Shuttleworth *et al* [47]).

## 8.1 A Calcium Control System

To construct a mathematical model for the calcium dynamics described in Fig. 8.1, we need to model the calcium pumps or channels in this figure. Let $[Ca^{2+}]_i$ and $[Ca^{2+}]_{er}$ denote the concentrations of intracellular and ER calcium, respectively. According to (6.135), the rate of calcium transport by the sarcoplasmic or endoplasmic reticulum $Ca^{2+}$-ATPase (SERCA) can be modeled by

$$r_{serca} = \frac{R_{serca}[Ca^{2+}]_i}{(K_{serca} + [Ca^{2+}]_i)(1 + I_{serca}[Ca^{2+}]_{er})}, \tag{8.1}$$

where $R_{serca}, R_{serca}$ and $I_{serca}$ are positive constants. According to (6.79) and (6.104), the outward flux of $Ca^{2+}$ from ER through the IP$_3$ mediated channel is given by

$$r_{ipr} = (R_{ipr}P_{ipr} + R_{leak})([Ca^{2+}]_{er} - [Ca^{2+}]_i), \tag{8.2}$$

$$P_{ipr} = \frac{[Ca^{2+}]_i}{K_{rca} + [Ca^{2+}]_i} \frac{K_{inh}^3}{K_{inh}^3 + [Ca^{2+}]_i^3} \frac{[IP_3]^3}{K_{ip3}^3 + [IP_3]^3} \tag{8.3}$$

where $R_{ipr}$ is the maximum flow rate and $R_{leak}$ is the leak flux rate. In this model, an outward leak is taken into account. According to (6.136), the calcium extrusion rate across the plasma membrane by the calcium ATPase is given by

$$r_{pmca} = \frac{R_{pmca}[Ca^{2+}]_i^2}{K_{pmca}^2 + [Ca^{2+}]_i^2}. \tag{8.4}$$

The $Ca^{2+}$ entry mechanisms through store-operated channels have remained elusive. In yeast cells, experimental observations by Kellermayer *et al* [22] and Locke *et al* [29] indicated that budding yeast cells also have this store-operated calcium feedback control mechanism and the calcium uptake through the store-operated channels follows the Michaelis-Menten equation. On the basis of these data, we assume that the store-operated channels follow the Michaelis-Menten kinetics and then $Ca^{2+}$ entry into the cytosol from the extracellular environment is modeled by [28]

$$r_{soce} = R_{soce}p_{soce}\frac{[Ca^{2+}]_o}{K_{soce} + [Ca^{2+}]_o}, \tag{8.5}$$

where $R_{soce} > 0$ denotes the maximum SOCE rate, $K_{soce}$ is the Michaelis-Menten constant, $[Ca^{2+}]_o$ is the extracellular calcium concentration, and $p_{soce} = p_{soce}([Ca^{2+}]_{er}, [Ca^{2+}]_i)$ ranging from 0 to 1 is an output feedback controller to be designed.

Because the cytosolic calcium is strongly buffered in cells [44], we need to introduce the buffered $Ca^{2+}$ concentration $[Ca^{2+}]_b$ as a state variable. Taking this buffered $Ca^{2+}$ into account, we then derive from the law of mass balance that the $Ca^{2+}$ concentrations are governed by

$$\frac{d[Ca^{2+}]_i}{dt} = -r_{pmca} - r_{serca} + r_{ipr} + r_{soce}$$
$$+ k_{off}[Ca^{2+}]_b - k_{on}[Ca^{2+}]_i \left([Ca^{2+}]_{b,total} - [Ca^{2+}]_b\right), \quad (8.6)$$

$$\frac{d[Ca^{2+}]_{er}}{dt} = \gamma_{er}(r_{serca} - r_{ipr}) + r_{stim}, \quad (8.7)$$

$$\frac{d[Ca^{2+}]_b}{dt} = -k_{off}[Ca^{2+}]_b + k_{on}[Ca^{2+}]_i \left([Ca^{2+}]_{b,total} - [Ca^{2+}]_b\right), \quad (8.8)$$

where $k_{off}, k_{on}$ are positive constants, $\gamma_{er}$ is the cytoplasmic-to-ER volume ratio, and $r_{stim}$ is the net rate of $Ca^{2+}$ binding to and releasing from STIM1. According to (7.65), the IP$_3$ concentration can be modeled by

$$\frac{d[IP_3]}{dt} = R_{in}^{ip} p_{ip3}(t) + \frac{R_{ca}^{ip}[Ca^{2+}]_i}{K_m^{ca} + [Ca^{2+}]_i} + R_d^{ip}([\overline{IP_3}] - [IP_3]), \quad (8.9)$$

where $R_{in}^{ip}$ is the external IP$_3$ input rate, $p_{ip3}(t)$ is a pulse input caused by the extracellular stimuli, $R_{ca}^{ip}$ is the maximum $Ca^{2+}$ dependent IP$_3$ input rate, $R_d^{ip}$ is the IP$_3$ degradation rate, $K_m^{ca}$ is the activation constant that gives half of maximum rate of $R_{ca}^{ip}$, and $[\overline{IP_3}]$ is the IP$_3$ steady state.

## 8.2 Design of an Output Feedback Controller

In order to construct a feedback controller $p_{soce}$, we look at the SOCE molecular mechanism. Two major components of SOCE have been discovered: STIM1 (Stromal interaction molecule 1) and Orai1. STIM1 is a transmembrane protein residing primarily in the ER. STIM1 contains an EF-hand, an N-terminus directed towards the lumen of the ER, and a C-terminus facing the cytoplasmic side. STIM1 functions as an endoplasmic reticulum $Ca^{2+}$ sensor. Orai1 is a transmembrane protein present in the plasma membrane with intracellular N- and C-termini and is an essential component of the store-operated channel (for review, see Lewis [25], Potier *et al* [39], and Putney [42]).

The mechanism about how STIM1 senses the calcium in ER and communicates with Orai1 was discovered in the late 2000s (see, e.g., [8, 14, 17, 20, 25, 35, 38, 39, 43, 57]). Emptying of the calcium from ER changes the conformation of STIM1 and leads to oligomerization, which enables the polybasic region to target STIM1 to ER-PM junctions, and causes a conformational change in STIM1 to expose its CRAC-activation domain (CAD, amino acids 342-448 [38]; also called SOAR (STIM1 Orai activating region), amino acids 344-442 [57]; called CCb9 (coiled-coil domain containing fragment b9), amino acids 339-446 [20]. In the junctions, the exposed STIM1

CAD binds to Orai1. STIM1-Orai11 interactions change the conformation of Orai1, which drives the opening of CRAC channels and triggers calcium entry. Moreover, the channel opening is optimized by phosphatidylinositol 4-phosphate (PI4P) (see, e.g., [24, 52]) while the channel opening is disrupted by a large cell volume increase [27]. On the other hand, the CRAC channel is inactivated by calmodulin [35], annexin 6 [34], and protein kinase C [33]. Inhibition of mitochondrial $Ca^{2+}$ uptake decreases SOCE [11, 45].

The dynamic binding of the ER $Ca^{2+}$ to STIM1 and the ER $Ca^{2+}$ dissociation from STIM1 can be modeled by the differential equation

$$\frac{d[STIM1]}{dt} = -f_s[STIM1][Ca^{2+}]_{er}^{n_s} + b_s([TS] - [STIM1]),  \quad (8.10)$$

where $f_s$ is a binding rate, $b_s$ is a dissociation rate, $n_s$ is a positive exponent, and $[TS]$ is the concentration of total STIM1. Here we assume that as soon as the ER $Ca^{2+}$ is released from the luminal EF-hand of STIM1, the STIM1 is immediately in the ER-PM junctions. Thus $[STIM1]$ can be regarded as the concentration of the active cytosolic part of STIM1 in the ER-PM junctions. The net rate of $Ca^{2+}$ binding to and releasing from STIM1 is given by

$$r_{stim} = -f_s[STIM1][Ca^{2+}]_{er}^{n_s} + b_s([TS] - [STIM1]). \quad (8.11)$$

The dynamic binding of the active cytosolic part of STIM1 to Orai1 (the region of amino acids 70-91 [38]) and the dissociation from Orai1 can be described by

$$STIM1 + Orai1 \overset{f_o}{\underset{b_o}{\rightleftharpoons}} SO, \quad (8.12)$$

where SO denotes the complex of STIM1 and Orail1. It then follows that

$$\frac{d[SO]}{dt} = f_o(1 - [SO])[STIM1] - b_o[SO], \quad (8.13)$$

where $f_o$ is a binding rate, $b_o$ is a dissociation rate, and $[SO]$ is the fraction of STIM1-Orai1 complex among the total Orai1, that is, $[SO]$ = the concentration of STIM1-Orai1/ the total concentration of Orai1. The term $b_o[SO] - f_o(1 - [SO])[STIM1]$ is not included in the equation (8.10) because the binding of the cytosolic part of STIM1 to Orai1 does not affect the luminal part of STIM1. The equations (8.10) and (8.13) constitute a dynamical output feedback controller:

$$p_{soce} = [SO]. \quad (8.14)$$

Experimental data about the dependence of the SOCE current on the ER calcium concentration was obtained by Luik et al [30] (see Fig. 1C of [30]). These data can be used to determine the values of parameters in the equations (8.10) and (8.13). Since the data were obtained at equilibrium, we solve the steady state equations of (8.10) and (8.13) to obtain

$$\overline{[SO]} = \frac{f_o b_s [TS]}{b_o b_s + f_o b_s [TS] + b_o f_s [Ca^{2+}]_{er}^{n_s}}. \quad (8.15)$$

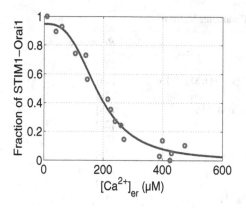

**Fig. 8.2.** Fit of the steady state $\overline{[SO]}$ of the equations (8.10) and (8.13) to the normalized data (red circles) of [30], which were read by the software Engauge Digitizer 4.1. Reproduced with permission from [28]

Fitting the steady state $\overline{[SO]}$ into the normalized data (Fig. 8.2), we obtain the values of these parameters as listed in Table 8.2. The fitting was done by using the MATLAB curve fitting toolbox. Note that different sets of parameter values can achieve the same best fitting since the minimum difference between the data and the fitting function can attain at the different sets of values. The set of parameter values is chosen such that our model can simulate experimental data.

## 8.3 SOCE Computer Simulation

Thapsigargin is used to deplete the ER calcium store by inhibiting the ER calcium pump SERCA in SOCE experiments [26, 55]. Cells grow initially in a medium without $Ca^{2+}$. An addition of thapsigargin to the medium raises cytosolic calcium concentration due to the calcium release from the ER. A consequent addition of calcium to the medium results in a sudden and bigger increase of cytosolic calcium concentration because the added extracellular $Ca^{2+}$ flood into the cells through the store-operated calcium channels, as shown in Fig. 8.3 (left). We now use the above SOCE model to simulate this experimental observation.

The model is solved by using the function ode15s of MATLAB, the MathWorks, Inc. The relative error tolerance and the absolute error tolerance were set to $10^{-6}$. The InitialStep is set to 0.01 and the MaxStep is set to 0.01. MATLAB does not recommend to reduce MaxStep for the accuracy of solutions since this can significantly slow down solution time. Instead the error tolerances can be used.

**Table 8.1.** Non-zero Initial Conditions

| Parameter | Value |
| --- | --- |
| $[Ca^{2+}]_i(0)$ | 0.1 ($\mu$M) |
| $[Ca^{2+}]_{er}(0)$ | 150 ($\mu$M) |
| $[IP_3](0)$ | 0.25 ($\mu$M) |

**Table 8.2.** Parameter Values

| Parameter | Value | Reference |
|---|---|---|
| $n_s$ | 3 | [28] |
| $f_s$ | $6.663 \cdot 10^{-6}$ (s$^{-1}$ · ($\mu$M)$^{-n_s}$) | [28] |
| $b_s$ | 2.535 (s$^{-1}$) | [28] |
| $f_o$ | 1.226 (s$^{-1}$ · ($\mu$M)$^{-1}$) | [28] |
| $b_o$ | 0.06774 (s$^{-1}$) | [28] |
| $[TS]$ | 1 ($\mu$M) | [28] |
| $R_{soce}$ | Varied (see text) | [28] |
| $K_{soce}$ | 500 ($\mu$M) | [29, 50] |
| $k_{off}$ | 500 (s$^{-1}$) | [19] |
| $k_{on}$ | 100 ($\mu$M$^{-1}$s$^{-1}$) | [19] |
| $[Ca^{2+}]_{b,total}$ | 660 ($\mu$M) | [46] |
| $R_{in}^{ip}$ | 20 ($\mu$M · s$^{-1}$) | [28] |
| $R_{ca}^{ip}$ | 2.8 ($\mu$M · s$^{-1}$) | [54] |
| $K_m^{ca}$ | 1.1 ($\mu$M) | [21, 54] |
| $R_d^{ip}$ | 1 (s$^{-1}$) | [54] |
| $[IP_3]$ | 0.25 ($\mu$M) | [54] |
| $R_{pmca}$ | 38 ($\mu$M · s$^{-1}$) | [48] |
| $K_{pmca}$ | 0.5 ($\mu$M) | [48] |
| $\gamma_{er}$ | 5.4 | [48] |
| $R_{ipr}$ | Varied (see text) | [28] |
| $R_{leak}$ | Varied (see text) | [28] |
| $R_{serca}$ | 100 ($\mu$M$^{-1}$ · s$^{-1}$) | [15] |
| $K_{serca}$ | 0.4 ($\mu$M) | [15, 31, 48] |
| $I_{serca}$ | Varied (see text) | [28] |
| $K_{rca}$ | 0.077 ($\mu$M) | [15, 32] |
| $K_{ip3}$ | 3.2 ($\mu$M) | [15, 18] |
| $K_{inh}$ | 5.2 ($\mu$M) | [32] |

Initial conditions are listed in Table 8.1. The values of parameters in the feedback control system are listed in Table 8.2. Since no data about the total STIM1 concentration are available, $[TS]$ is taken to be 1 $\mu$M. The parameters $n_s, f_s, b_s, f_o, b_o$ are determined by fitting the steady $\overline{[SO]}$ defined by (8.15) into the data of Luik *et al* [30] (Fig. 8.2). $R_{soce}$ and $I_{serca}$ are selected to result in the equilibrium calcium levels of $[Ca^{2+}]_i = 0.05$ $\mu$M and $[Ca^{2+}]_{er} = 450$ $\mu$M when $[Ca^{2+}]_o = 1500$ $\mu$M. The parameters $R_{ipr}$ and $R_{leak}$ are adjusted in simulation such that the simulations are close to experimental observations.

The resting free Ca$^{2+}$ in the cytosol is about 0.05 $\mu$M [56] and the resting free Ca$^{2+}$ in the ER is approximately 500 $\mu$M [16, 56]. Since the model solutions during the initial transient period have no biological meanings, we solve the model with the extracellular Ca$^{2+}$ concentration $[Ca^{2+}]_o$ of 1500 $\mu$M [10] for 1500 seconds such that the cytosolic calcium concentration $[Ca^{2+}]_i$ achieves a steady state of about 0.05 $\mu$M (Fig. 8.3(middle)) and the ER calcium concentration $[Ca^{2+}]_{er}$ achieves a steady

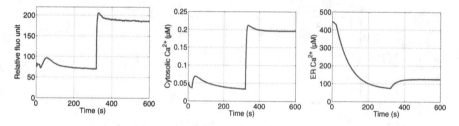

**Fig. 8.3.** SOCE simulations. *Left:* Reproduction from [55] (courtesy of S. Yang, J. Zhang, and X.-Y. Huang). *Middle and right:* model simulations. In this simulation, $p_{ip3}(t) \equiv 0, R_{ipr} = 0.18$ s$^{-1}$, $R_{leak} = 0.002$ s$^{-1}$, $I_{serca} = 0.025$ $\mu$M$^{-1}$, and $R_{soce} = 8.85$ $\mu$Ms$^{-1}$. Other parameter values are given in Table 8.2. Reproduced with permission from [28]

state of about 450 $\mu$M (Fig. 8.3 (right)), and then set this moment as the initial time ($t = 0$).

Following the experiments of Yang *et al* [55], before the 320th second, $[Ca^{2+}]_o$ is set to 10 $\mu$M, and after the 320th second, $[Ca^{2+}]_o$ is suddenly increased to 2000 $\mu$M. After the 20th second, the maximum velocity of the Ca$^{2+}$ pump SERCA, $R_{serca}$, is set to 5 $\mu$M/s to mimic the addition of thapsigargin. In this simulation, $p_{ip3}(t) \equiv 0$, $R_{ipr} = 0.18$ s$^{-1}$, $R_{leak} = 0.002$ s$^{-1}$, $I_{serca} = 0.025$ $\mu$M$^{-1}$, and $R_{soce} = 8.85$ $\mu$Ms$^{-1}$. Other parameter values are given in Table 8.2.

The simulated cytosolic calcium dynamics plotted in Fig. 8.3 (middle) agrees qualitatively with the experimental data of Yang *et al* [55] (Fig. 8.3 (left)). The model simulation also agrees qualitatively with the experimental Fig. 2Ai of Liao *et al* [26] and Fig. 7E of Park *et al* [38].

It was reported by Yu *et al* [56] that thapsigargin reduced $[Ca^{2+}]_{er}$ from about 500 $\mu$M to 50-100 $\mu$M over 10 minutes and the readdition of extracellular Ca$^{2+}$ led to a rapid increase in $[Ca^{2+}]_{er}$ to about 10 $\mu$M. The decrease in $[Ca^{2+}]_{er}$ was largely complete in the first minute after stimulation [56]. Fig. 8.3 (right) indicates that the simulated $[Ca^{2+}]_{er}$ approximately agrees with this observation.

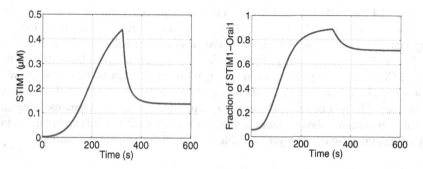

**Fig. 8.4.** Simulated dynamics of STIM1 and STIM1-bound Orai1. Reproduced with permission from [28]

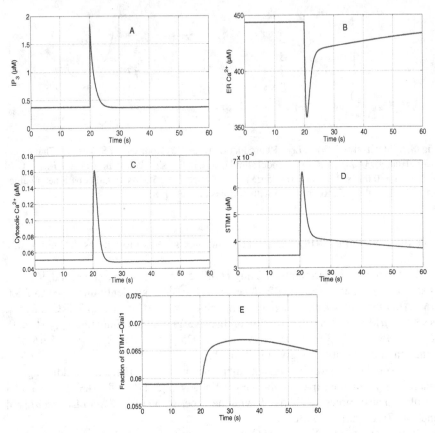

**Fig. 8.5.** Simulation of rejection of agonist disturbances. After disturbed by an agonist disturbance pulse, the ER Ca$^{2+}$ (B), cytosolic Ca$^{2+}$ (C), STIM1 (D), and STIM1-Orai1 (E) return
to their resting levels, respectively. In this simulation, $R_{in}^{ip} = 100 \ \mu M \cdot s^{-1}$, $R_{ipr} = 1.6 \ s^{-1}$,
$R_{leak} = 0.02 \ s^{-1}$, $I_{serca} = 0.00048 \ \mu M^{-1}$, and $R_{soce} = 8.85 \ \mu Ms^{-1}$. Other parameter values
are given in Table 8.2. Reproduced with permission from [28]

Fig. 8.4 shows that as the ER Ca$^{2+}$ decreases, the concentration of the cytosolic
portion of the active STIM1 in ER-PM junctions is increasing gradually, indicating that STIM1 is accumulating in ER-PM junctions. Simultaneously, the fraction
of STIM1-Orai1 is increasing gradually, indicating that the cytosolic portion of the
active STIM1 is binding to Orai1 and driving the opening of CRAC channels for Ca$^{2+}$
entry. This is consistent with the static observation of Park *et al* [38]. This simulated
dynamics of STIM1 and STIM1-Orai1 might provide a promising phenomenological
prediction since the above argument suggests that the model could capture the main
SOCE features. No data are yet available for testing this simulated dynamics.

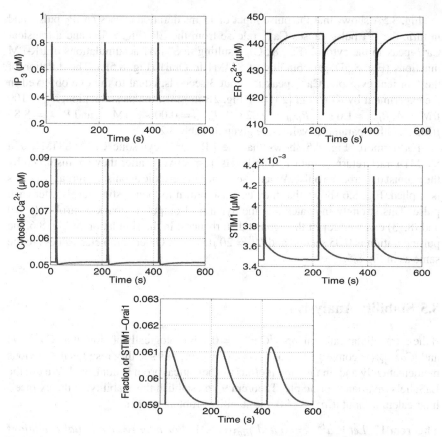

**Fig. 8.6.** Simulation of rejection of agonist disturbances. The ER $Ca^{2+}$, cytosolic $Ca^{2+}$, STIM1, and STIM1-Orai1 return to their resting levels, respectively, after each pulse of a sequence of periodic agonist disturbance pulses. In this simulation, the pulse duration is 0.05 second and $R_{in}^{ip} = 20 \ \mu M \cdot s^{-1}$. Other parameter values are the same as in Fig. 8.5 [28]

## 8.4 Simulation of Rejection of Agonist Disturbances

The action of an agonist on its specific receptor typically activates isoforms of the phosphoinositide-specific phospholipase C (PLC). PLC breaks down the phosphatidylinositol 4,5 bisphosphate ($PIP_2$) to generate two second messengers, the inositol 1,4,5 trisphosphate ($IP_3$) and diacylglycerol (DAG) [39]. Therefore, an agonist disturbance pulse results in an $IP_3$ pulse input. We assume that the pulse input is given by

$$p_{ip3}(t) = \begin{cases} 0 \ \text{if} \ t < 20 \ \text{s}, \\ 1 \ \text{if} \ 20 \ \text{s} \le t \le 20.015 \ \text{s}, \\ 0 \ \text{if} \ t > 20.015 \ \text{s}. \end{cases}$$

Fig. 8.5A shows that the pulse input of 15 ms duration causes an IP$_3$ pulse, and in turn, the IP$_3$ pulse causes Ca$^{2+}$ release from the ER (Fig. 8.5B) and a transient Ca$^{2+}$ peak in the cytosol (Fig. 8.5C), resulting in STIM1 accumulations in ER-PM junctions (Fig. 8.5D) and binding of STIM1 to Orai1 (Fig. 8.5E). The duration of the transient cytosolic Ca$^{2+}$ peak is about 3 seconds, equal to the one observed in the experiment by Kim *et al* [23] (see Fig. 2B in [23]). In this simulation, $R_{in}^{ip} = 100$ $\mu$M$\cdot$s$^{-1}$, $R_{ipr} = 1.6$ s$^{-1}$, $R_{leak} = 0.02$ s$^{-1}$, $I_{serca} = 0.00048$ $\mu$M$^{-1}$, and $R_{soce} = 8.85$ $\mu$Ms$^{-1}$. Other parameter values are given in Table 8.2.

Furthermore, Fig. 8.5 shows that the ER Ca$^{2+}$, cytosolic Ca$^{2+}$, STIM1, and STIM1-Orai1 return to their resting levels, respectively, after they are disturbed by the agonist disturbance pulse. When a sequence of periodic agonist disturbance pulses is applied, Fig. 8.6 shows that they can also return to their resting levels after each pulse. This strongly indicates that the dynamical output feedback controller (8.10) and (8.13) can well reject the agonist disturbances. In the simulation of Fig. 8.6, the pulse duration is 0.05 second and $R_{in}^{ip} = 20$ $\mu$M$\cdot$s$^{-1}$. Other parameter values are the same as in Fig.8.5.

## 8.5 Stability Analysis

If the extracellular calcium input $[Ca^{2+}]_o = 0$, it is biologically obvious that $[Ca^{2+}]_i(t)$ and $[Ca^{2+}]_{er}(t)$ converge to 0 as $t \to \infty$. However, this stability result is not obvious mathematically and, in fact, its rigorous mathematical proof is nontrivial. We use the LaSalle's invariance principle, Theorem 9, to establish this stability. If the extracellular calcium input $[Ca^{2+}]_o \neq 0$, the stability problem is open.

**Theorem 15.** *Let* $[Ca^{2+}]_o = 0$ *and* $p_{ip3}(t) \equiv 1$. *For nonnegative initial conditions with* $[Ca^{2+}]_b(0) \leq [Ca^{2+}]_{b,total}$, $[STIM1](0) \leq [TS]$, *and* $[SO](0) \leq 1$, *the solutions of the system (8.6), (8.7), (8.8), (8.9), (8.10), and (8.13) satisfy*

$$\lim_{t \to \infty}[Ca^{2+}]_i(t) = \lim_{t \to \infty}[Ca^{2+}]_{er}(t) = \lim_{t \to \infty}[Ca^{2+}]_b(t) = 0, \qquad (8.16)$$

$$\lim_{t \to \infty}[IP_3](t) = \overline{[IP_3]} + \frac{R_{in}^{ip}}{R_d^{ip}}, \qquad (8.17)$$

$$\lim_{t \to \infty}[STIM1](t) = [TS], \qquad (8.18)$$

$$\lim_{t \to \infty}[SO](t) = \frac{f_o b_s[TS]}{b_o b_s + f_o b_s[TS]}. \qquad (8.19)$$

In the following proof, we frequently use the result: If the real parts of all eigenvalues of a matrix **M** are negative and the vector function **f**(t) converges to **L** as $t \to \infty$, then the solution **x**(t) of the linear system

$$\frac{d\mathbf{x}}{dt} = \mathbf{Mx} + \mathbf{f} \qquad (8.20)$$

converges to $-\mathbf{M}^{-1}\mathbf{L}$.

Let

$$\mathbb{R}^n_+ = \{(x_1, x_2, \cdots, x_n) \in \mathbb{R}^n \mid x_1, x_2, \cdots, x_n \geq 0\},$$

$$\mathbb{B}^4_{+,Ca,K} = \{([Ca^{2+}]_i, [Ca^{2+}]_{er}, [Ca^{2+}]_b, [STIM1]) \in \mathbb{R}^4_+ \mid [Ca^{2+}]_b \leq [Ca^{2+}]_{b,total},$$
$$[STIM1] \leq [TS],$$
$$\gamma_{er} \left([Ca^{2+}]_i + [Ca^{2+}]_b\right) + [Ca^{2+}]_{er} + [TS] - [STIM1] \leq K\},$$

where $K > 0$ is selected such that $\gamma_{er} \left([Ca^{2+}]_i(0) + [Ca^{2+}]_b(0)\right) + [Ca^{2+}]_{er}(0) + [STIM1](0) \leq K$. We show that $\mathbb{B}^4_{+,Ca,K}$ is positively invariant and

$$[IP_3](t), [SO](t) \geq 0 \tag{8.21}$$

for all $t \geq 0$. Define

$$t_0 = \max\{T \mid [Ca^{2+}]_{er}(t) \geq 0 \text{ for all } 0 \leq t \leq T\},$$
$$t_1 = \max\{T \mid [Ca^{2+}]_i(t) \geq 0 \text{ for all } 0 \leq t \leq T\},$$
$$t_2 = \max\{T \mid 0 \leq [Ca^{2+}]_b(t) \leq [Ca^{2+}]_{b,total} \text{ for all } 0 \leq t \leq T\},$$
$$t_3 = \max\{T \mid [IP_3](t) \geq 0 \text{ for all } 0 \leq t \leq T\}\},$$
$$t_4 = \max\{T \mid 0 \leq [STIM1](t) \leq [TS] \text{ for all } 0 \leq t \leq T\},$$
$$t_5 = \max\{T \mid 0 \leq [SO](t) \leq 1 \text{ for all } 0 \leq t \leq T\}.$$

We claim that $t_0 = t_1 = t_2 = t_3 = t_4 = t_5 = \infty$. If it was not true, then $t^* = \min\{t_0, t_1, t_2, t_3, t_4, t_5\} < \infty$. We may as well assume that $t_0 = t^*$. Then $[Ca^{2+}]_{er}(t)$, $[Ca^{2+}]_i(t), [IP_3](t) \geq 0, 0 \leq [Ca^{2+}]_b(t) \leq [Ca^{2+}]_{b,total}, 0 \leq [STIM1](t) \leq [TS]$, and $0 \leq [SO](t) \leq 1$ for all $0 \leq t \leq t_0$ and $[Ca^{2+}]_{er}(t_0) = 0$. It then follows from (8.1), (8.2), and (8.7) that

$$\left.\frac{d[Ca^{2+}]_{er}}{dt}\right|_{t_0} = \gamma_{er} \frac{R_{serca}[Ca^{2+}]_i(t_0)}{(K_{serca} + [Ca^{2+}]_i(t_0))}$$
$$+ \gamma_{er}(R_{ipr}P_{ipr} + R_{leak})[Ca^{2+}]_i(t_0)$$
$$+ b_s([TS] - [STIM1](t_0))$$
$$\geq 0.$$

If either $[Ca^{2+}]_i(t_0) > 0$ or $0 \leq [STIM1](t_0) < [TS]$, then $\left.\frac{d[Ca^{2+}]_{er}}{dt}\right|_{t_0} > 0$. Thus $[Ca^{2+}]_{er}(t)$ is increasing near $t_0$ and then $[Ca^{2+}]_{er}(t) < [Ca^{2+}]_{er}(t_0) = 0$ for some $t < t_0$. This is a contradiction. Hence $[Ca^{2+}]_i(t_0) = 0$ and $[STIM1](t_0) = [TS]$. Then it follows from (8.6) that

$$\left.\frac{d[Ca^{2+}]_i}{dt}\right|_{t_0} = k_{off}[Ca^{2+}]_b(t_0).$$

If $[Ca^{2+}]_b(t_0) > 0$, then $\left.\frac{d[Ca^{2+}]_i}{dt}\right|_{t_0} > 0$. Thus $[Ca^{2+}]_i(t)$ is increasing near $t_0$ and then $[Ca^{2+}]_i(t) < [Ca^{2+}]_i(t_0) = 0$ for some $t < t_0$. This is a contradiction and so

$[Ca^{2+}]_b(t_0) = 0$. It then follows from (8.6), (8.7), (8.8), (8.9), (8.10), and (8.13) that $[Ca^{2+}]_i(t) = [Ca^{2+}]_{er}(t) = [Ca^{2+}]_b(t) = 0$, $[STIM1](t) = [TS]$, $[IP_3](t) \geq 0$, and $0 \leq [SO](t) \leq 1$ for $t \geq t_0$. This contradicts with $t^* = \min\{t_0, t_1, t_2, t_3, t_4, t_5\} < \infty$. Define

$$V = \gamma_{er}\left([Ca^{2+}]_i + [Ca^{2+}]_b\right) + [Ca^{2+}]_{er} + [TS] - [STIM1].$$

It follows from (8.6), (8.7), (8.8), and (8.10) that

$$\frac{dV}{dt} = -\frac{\gamma_{er} R_{pmca}[Ca^{2+}]_i^2}{K_{pmca}^2 + [Ca^{2+}]_i^2} \leq 0.$$

Thus $\mathbb{B}_{+,Ca,K}^4$ is positively invariant. Because

$$Z = \left\{([Ca^{2+}]_i, [Ca^{2+}]_{er}, [Ca^{2+}]_b, [STIM1]) \in \mathbb{B}_{+,Ca,K}^4 \,\middle|\, \frac{dV}{dt} = 0\right\}$$

$$= \left\{(0, [Ca^{2+}]_{er}, [Ca^{2+}]_b, [STIM1]) \in \mathbb{B}_{+,Ca,K}^4\right\},$$

it follows from the LaSalle's invariance principle, Theorem 9, that

$$\lim_{t \to \infty} [Ca^{2+}]_i(t) = 0.$$

For $\alpha > 1$, we can deduce from the equations (8.7) and (8.10) that

$$\frac{d}{dt}\left([Ca^{2+}]_{er} + \alpha\left([TS] - [STIM1]\right)\right) = -F(t, \alpha)\left([Ca^{2+}]_{er} + \alpha\left([TS] - [STIM1]\right)\right)$$
$$+ f(t),$$

where $\lim_{t \to \infty} f(t) = 0$ and $F(t, \alpha) \geq F_0 > 0$ for all $t \geq 0$ if $\alpha$ is sufficiently close to 1. It therefore follows that

$$\lim_{t \to \infty}\left([Ca^{2+}]_{er}(t) + \alpha\left([TS] - [STIM1](t)\right)\right) = 0.$$

In the same way, we deduce that

$$\lim_{t \to \infty}[Ca^{2+}]_b(t) = 0.$$

From the equation (8.9), we deduce that

$$\lim_{t \to \infty}[IP_3](t) = \overline{[IP_3]} + \frac{R_{in}^{ip}}{R_d^{ip}}.$$

## 8.6  Remarks

In this modeling, we assumed that as soon as the ER $Ca^{2+}$ is released from the luminal EF-hand of STIM1, the STIM1 is immediately in the ER-PM junctions. This assumption may be refined by introducing two state variables: $[STIM1]_{er}$ for the ER luminal portion of STIM1 and $[STIM1]_i$ for the cytosolic portion. These two states may be connected by the equation

$$\frac{d[STIM1]_i}{dt} = k([STIM1]_{er} - [STIM1]_i),$$

where the positive constant $k$ is a rate of conformational change from $[STIM1]_{er}$ to $[STIM1]_i$.

Stochastic factors were neglected in our modeling. Like telephone calls arriving at a switchboard, agonists arrive at the plasma membrane randomly and independently. Thus the agonist action on the plasma membrane, $p_{ip3}(t)$, should follow a Poisson process. In addition, the extracellular calcium $[Ca^{2+}]_o$ may be fluctuated randomly and then contain random "noises". These stochastic factors should be taken into account to further refine the model.

The distributions of $Ca^{2+}$ in the cytosol and ER are not uniform and the $Ca^{2+}$ concentrations near the ER are higher. The modeling of these spatial $Ca^{2+}$ distributions leads to a model consisting of partial differential equations [28].

Calcium ions play a central role in the process of insulin secretion. Release of calcium ions from intracellular stores is essential for the amplification of insulin secretion by promoting the replenishment of the readily releasable pool of secretory granules, while voltage-dependent calcium entry is directed to the sites of exocytosis via the binding of the L-type calcium channels to SNARE proteins [49, 51]. Therefore, the SOCE model will have potential applications in modeling insulin secretion.

# Exercises

**8.1.** Solve the steady state equations of (8.10) and (8.13).

**8.2.** Fit $[\overline{SO}]$ defined by (8.15) into the normalized data of Fig. 1C of [30]:

| $\left[Ca^{2+}\right]_{er}$ | 10 | 40 | 60 | 105 | 140 | 145 | 215 | 225 | 240 |
|---|---|---|---|---|---|---|---|---|---|
| $[SO]$ | | 1.0000 | 0.8958 | 0.9292 | 0.7417 | 0.7292 | 0.5625 | 0.4250 | 0.3542 0.2708 |
| $\left[Ca^{2+}\right]_{er}$ | 260 | 270 | 390 | 400 | 425 | 430 | 475 | | |
| $[SO]$ | | 0.2417 | 0.1458 | 0.0292 | 0.1375 | 0.0004 | 0.0500 0.1042 | | |

**8.3.** Assume that the thapsigargin inhibition effect on SERCA is modeled by

$$R_{serca} = \begin{cases} 5 & \text{if } t \le 25000 \text{ ms,} \\ (100 - 5)\exp(E_{tg}(25000 - t)) + 5 & \text{if } t > 25000 \text{ ms,} \end{cases}$$

where $E_{tg} = 0.00006$ ms$^{-1}$. Use the control system (8.6), (8.7), (8.8), (8.9), (8.10), and (8.13) to simulate SOCE with this thapsigargin inhibition model and explain how this thapsigargin inhibition model affects the calcium release from the ER.

**8.4.** Instead of the dynamic feedback controller defined by (8.10) and (8.13), use the static feedback controller defined by (8.15) to simulate SOCE and discuss their differences.

**8.5.** Solve the control system (8.6), (8.7), (8.8), and (8.9) numerically to find an equilibrium, linearize the system at the equilibrium, examine the controllability and observability of the linearized system, and then design an observer-based output feedback controller if possible.

**8.6.** Prove that if the real parts of all eigenvalues of a matrix $\mathbf{M}$ are negative and the vector function $\mathbf{f}(t)$ converges to $\mathbf{L}$ as $t \to \infty$, then the solution $\mathbf{x}(t)$ of the linear system

$$\frac{d\mathbf{x}}{dt} = \mathbf{M}\mathbf{x} + \mathbf{f}$$

converges to $-\mathbf{M}^{-1}\mathbf{L}$.

# References

1. Abdullaev I.F., Bisaillon J.M., Potier M., Gonzalez J.C., Motiani R.K., Trebak M.: Stim1 and Orai1 mediate CRAC currents and store-operated calcium entry important for endothelial cell proliferation. Circ Res. **103**, 1289-1299 (2008).
2. Berridge M J.: Elementary and global aspects of calcium signalling. J. Phyiol. **499**, 291-306 (1997).
3. Berridge M.J., Bootman M.D., Roderick H.L.: Calcium signalling: dynamics, homeostasis and remodelling. Nat. Rev. Mol. Cell Biol. **4**, 517-29 (2003).
4. Bird G.S., DeHaven W.I., Smyth J.T., Putney Jr., J.W.: Methods for studying store-operated calcium entry. Methods **46**, 204-212 (2008).
5. Brandman O., Liou J., Park W.S., Meyer T.: STIM2 is a feedback regulator that stabilizes basal cytosolic and endoplasmic reticulum $Ca^{2+}$ levels. Cell **131**, 1327-1339 (2007).
6. Chakrabarti R., Chakrabarti R.: Calcium signaling in non-excitable cells: $Ca^{2+}$ release and influx are independent events linked to two plasma membrane $Ca^{2+}$ entry channels. J. Cellular Biochem. **99**, 1503-1516 (2006).
7. DeHaven W., Jones B., Petranka J., Smyth J., Tomita T., Bird G., Putney J.: TRPC channels function independently of STIM1 and Orai1. J Physiol. **587**, 2275-2298 (2009).
8. Derler I., Fahrner M., Carugo O., Muik M., Bergsmann J., Schindl R., Frischauf I., Eshaghi S., Romanin C.: Increased hydrophobicity at the N-terminus/membrane interface impairs gating of the SCID-related ORAI1 mutant. J. Biol. Chem. **284** 15903-15915 (2009).
9. Dirksen R.T.: Checking your SOCCs and feet: the molecular mechanisms of $Ca^{2+}$ entry in skeletal muscle. J. Physiol. **587**, 3139-3147 (2009).
10. Dvorak M.M., Siddiqua A., War, D.T., Carter D.H., Dallas S.L., Nemeth E.F., Riccardi D.: Physiological changes in extracellular calcium concentration directly control osteoblast function in the absence of calciotropic hormones. Proc. Natl. Acad. Sci. USA **101** 5140-5145 (2004).
11. Feldman B., Fedida-Metula S., Nita J., Sekler I., Fishman D.: Coupling of mitochondria to store-operated $Ca^{2+}$-signaling sustains constitutive activation of protein kinase B/Akt and augments survival of malignant melanoma cells. Cell Calcium **47**, 525-537 (2010).
12. Feske S., Prakriya M., Rao A., Lewis R.S.: A severe defect in CRAC $Ca^{2+}$ channel activation and altered $K^+$ channel gating in T cells from immunodeficient patients. J. Experimental Medicine **202**, 651-662 (2005).
13. Feske S., Gwack Y., Prakriya M., Srikanth S., Puppel S.H., Tanasa B., Hogan P.G., Lewis R.S., Daly M., Rao A.: A mutation in Orai1 causes immune deficiency by abrogating CRAC channel function. Nature **441**, 179-185 (2006).

14. Feske S.: ORAI1 and STIM1 deficiency in human and mice: roles of store-operated $Ca^{2+}$ entry in the immune system and beyond. Immunological Reviews **231** 189-209 (2009).

15. Fridlyand L.E., Tamarina N., Philipson L. H.: Modeling of $Ca^{2+}$ flux in pancreatic $\beta$-cells: role of the plasma membrane and intracellular stores. Am. J. Physiol. Endocrinol. Metab. **285**, E138-E154 (2003).

16. Frischauf I., Schindl R., Derler I., Bergsmann J., Fahrne M., Romanin C.: The STIM/Orai coupling machinery. Channels **2** 1-8 (2008).

17. Frischauf I., Muik M., Derler I., Bergsmann J., Fahrner M., Schindl R., Groschner K., Romanin C.: Molecular determinants of the coupling between STIM1 and Orai channels, differential activation of Orail Orai1C3 channels by a STIM1 coiled-coil mutant. J. Biol. Chem. **284**, 21696-21706 (2009).

18. Hagar R.E., Ehrlich B.E.: Regulation of the type III InsP(3) receptor by InsP(3) and ATP. Biophys J. **79**, 271-278 (2000).

19. Hong D., Jaron D., Buerk D.G., Barbee K.A.: Transport-dependent calcium signaling in spatially segregated cellular caveolar domains. Am. J. Physiol. Cell Physiol. **294**, C856-C866 (2008).

20. Kawasaki T., Lange I., Feske S.: A minimal regulatory domain in the C terminus of STIM1 binds to and activates ORAI1 CRAC channels. Biochem. Biophys. Research Communications **385**, 49-54 (2009).

21. Keizer J., De Young G.W.: Two roles for $Ca^{2+}$ in agonist stimulated $Ca^{2+}$ oscillations. Biophys. J. **61**, 649-660 (1992).

22. Kellermayer, R., Aiello, D. P., Miseta, A., Bedwell, D.M.: Extracellular $Ca^{2+}$ sensing contributes to excess $Ca^{2+}$ accumulation and vacuolar fragmentation in a pmr1$\Delta$ mutant of S. cerevisiae. J. Cell Science **116**, 1637-1646 (2003).

23. Kim S.J., Jin Y., Kim J., Shin J.H., Worley P.F., Linden D.J.: Transient upregulation of postsynaptic $IP_3$-gated Ca release underlies short-term potentiation of mGluR1 signaling in cerebellar purkinje cells. J Neurosci. **28**, 4350-4355 (2008).

24. Korzeniowski M.K., Popovic M.A., Szentpetery Z., Varnai P., Stojilkovic S.S., Balla T.: Dependence of STIM1/Orai1-mediated calcium entry on plasma membrane phosphoinositides. J. Biol. Chem. **284**, 21027-21035 (2009).

25. Lewis R.S: The molecular choreography of a store-operated calcium channel. Nature **446**, 284-287 (2007).

26. Liao Y., Erxleben C., Yildirim E., Abramowitz J., Armstrong D.L., Birnbaumer L.: Orai proteins interact with TRPC channels and confer responsiveness to store depletion. PNAS **401**, 4682-4687 (2007).

27. Liu X., Ong H.L., Pani B., Johnson K., Swaim W.B., Singh B., Ambudkar I.: Effect of cell swelling on ER/PM junctional interactions and channel assembly involved in SOCE. Cell Calcium **47**, 491-499 (2010).

28. Liu W., Tang F., Chen J.: Designing dynamical output feedback controllers for store-operated $Ca^{2+}$ entry. Math. Biosci. **228**, 110-118 (2010).

29. Locke E.G., Bonilla M., Liang L., Takita Y., Cunningham K.W.: A Homolog of voltage-gated $Ca^{2+}$ channels stimulated by depletion of secretory $Ca^{2+}$ in yeast. Molecular and Cellular Biology **20**, 6686-6694 (2000).

30. Luik R.M., Wang B., Prakriya M., Wu M.M., Lewis R.S.: Oligomerization of STIM1 couples ER calcium depletion to CRAC channel activation. Nature **454**, 538-542 (2008).

31. Lytton J., Westlin M., Burk S.E., Shull G.E., MacLennan D.H.: Functional comparisons between isoforms of the sarcoplasmic or endoplasmic reticulum family of calcium pumps. J. Biol. Chem. **267**, 14483-14489 (1992).

32. Mak D.O., McBride S., Foskett J. K.: Regulation by $Ca^{2+}$ and inositol 1,4,5-trisphosphate (InsP$_3$) of single recombinant type 3 InsP$_3$ receptor channels. $Ca^{2+}$ activation uniquely distinguishes types 1 and 3 insp$_3$ receptors. J. Gen. Physiol. **117**, 435-446 (2001).

33. McElroy S.P., Drummond R.M., Gurney A.M.: Regulation of store-operated $Ca^{2+}$ entry in pulmonary artery smooth muscle cells. Cell Calcium **46**, 99-106 (2009).

34. Monastyrskaya K., Babiychuk E.B., Hostettler A., Wood P., Grewal T., Draeger A.: Plasma membrane-associated annexin A6 reduces $Ca^{2+}$ entry by stabilizing the cortical actin cytoskeleton. J. Biol. Chem. **284**, 17227-17242 (2009).

35. Mullins F.M., Park C.Y., Dolmetsch R.E., Lewis R.S.: STIM1 and calmodulin interact with Orai1 to induced $Ca^{2+}$-dependent interaction of CRAC channels. Proc. Natl. Acad. Sci. USA **106**, 15495-15500 (2009).

36. Parekh A.B., Putney Jr., J.W.: Store-operated calcium channels. Physiol. Rev. **85**, 757-810 (2005).

37. Parekh A.B.: On the activation mechanism of store-operated calcium channels. Pflugers Arch. – Eur. J. Physiol. **453**, 303-311 (2006).

38. Park C.Y., Hoover P.J., Mullins F.M., Bachhawat P., Covington E.D., Raunser S., Walz T., Garcia K.C., Dolmetsch R.E., Lewis R.S.: STIM1 clusters and activates CRAC channels via direct binding of a cytosolic domain to Orai1. Cell **136**, 876-890 (2009).

39. Potier M., Trebak M.: New developments in the signaling mechanisms of the store-operated calcium entry pathway. Pflugers Arch. **457**, 405-415 (2008).

40. Prakriya M., Lewis R.S.: CRAC channels: activation, permeation, and the search for a molecular identity. Cell Calcium **33**, 311-321 (2003).

41. Prakriya M., Lewis R.S.: Regulation of CRAC channel activity by recruitment of silent channels to a high open-probability gating mode. J. Gen. Physiol. **128**, 373-386 (2006).

42. Putney J.W. Jr.: A model for receptor-regulated calcium entry. Cell Calcium **7**, 1-12 (1986).

43. Putney J.W. Jr.: Recent breakthroughs in the molecular mechanism of capacitative calcium entry (with thoughts on how we got here), Cell Calcium **42**, 103-110 (2007).

44. Rorsman P., Ammala C., Berggren P.O., Bokvist K., Larsson O.: Cytoplasmic calcium transients due to single action potentials and voltage-clamp depolarizations in mouse pancreatic $\beta$-cells. EMBO J. **11**, 2877-2884 (1992).

45. Ryu S.Y., Peixoto P.M., Won J.H., Yule D.I., Kinnally K.W.: Extracellular ATP and P2Y2 receptors mediate intercellular $Ca^{2+}$ waves induced by mechanical stimulation in submandibular gland cells: Role of mitochondrial regulation of store operated $Ca^{2+}$ entry. Cell Calcium. **47**, 65-76 (2010).

46. Saftenku E.È.: Computational study of non-homogeneous distribution of $Ca^{2+}$ handling systems in cerebellar granule cells. J. Theoret. Biol. **257**, 228-244 (2009).

47. Shuttleworth T.J., Thompson J.L., Mignen O.: STIM1 and the noncapacitative ARC channels. Cell Calcium **42**, 183-191 (2007).

48. Sneyd J., Tsaneva-Atanasova K., Yule D.I., Thompson J.L., Shuttleworth T.J.: Control of calcium oscillations by membrane fluxes. PNAS. **101**, 1392-1396 (2004).

49. Tamarina NA., Kuznetso, A., Philipson L.H.: Reversible translocation of EYFP-tagged STIM1 is coupled to calcium infux in insulin secreting $\beta$-cells. Cell Calcium **44**, 533-544 (2008).

50. Tang F, Liu W.: An Age-dependent feedback control model for calcium in yeast cells. J. Math. Biol. **60**, 849-879 (2010).

51. Tengholm A., Gylfe E.: Oscillatory control of insulin secretion. Molecular and Cellular Endocrinology. **297**, 58-72 (2009).

52. Walsh C.M., Chvanov M., Haynes L.P., Petersen O.H., Tepikin A.V., Burgoyne R.D.: Role of phosphoinositides in STIM1 dynamics and store-operated calcium entry. Biochem. J. **425**, 159-168 (2010).
53. Wu M.M., Buchanan J., Luik R.M., Lewis R.S.: $Ca^{2+}$ store depletion causes STIM1 to accumulate in ER regions closely associated with the plasma membrane. J. Cell Biol. **174**, 803-813 (2006).
54. De Young G.W., Keizer J.: A single-pool inositol 1,4,5-trisphosphate-receptor-based model for agonist-stimulated oscillations in $Ca^{2+}$ concentration. Proc. Natl. Acad. Sci. USA **89**, 9895-9899 (1992).
55. Yang S., Zhang J.J., Huang X.-Y.: Orai1 and STIM1 are critical for breast tumor cell migration and metastasis. Cancer Cell **15**, 124-134 (2009).
56. Yu Y., Hinkle P.M.: Rapid turnover of calcium in the endoplasmic reticulum during signaling, J. Biol. Chem. **275**, 23648-23653 (2000).
57. Yuan J.P., Zeng W., Dorwart M.R., Choi, Y.-J., Worley P.F., Muallem, S.: SOAR and the polybasic STIM1 domains gate and regulate Orai channels. Nature Cell Biology **11** 337-343 (2009).

# Control of Mitochondrial Calcium

A mitochondrion is a membrane-enclosed organelle found in most eukaryotic cells. Mitochondria are sometimes described as "cellular power plants" because they generate most of the cell's supply of adenosine triphosphate (ATP), used as a source of chemical energy. A mitochondrion contains outer and inner membranes composed of phospholipid bilayers and proteins. The space between the outer and inner membranes is called the *inter-membrane space* and the space within the inner membrane is called the *matrix*.

Ions movements across the mitochondrial inner membrane are schematically described in Fig. 9.1. $H^+$ ions in mitochondria are ejected by the respiratory chain driven by the energy released from oxidation of NADP. The established electrochemical gradient drives the electrogenic transport of ions, including ATP and ADP by the adenine nucleotide translocator (ANT), $Ca^{2+}$ influx via the $Ca^{2+}$ uniporter, and $Ca^{2+}$ efflux via the $Na^{2+}/Ca^{2+}$ antiporter. The $H^+$ ions in the cytosol flow back to the mitochondrion through $F_1F_0$-ATPase to power the ATP synthesis.

Mitochondrial $Ca^{2+}$ uptake has profound consequences for physiological cell functions. Intramitochondrial $Ca^{2+}$ stimulates oxidative phosphorylation and controls the rate of ATP production. Mitochondrial $Ca^{2+}$ uptake modifies the shape of cytosolic $Ca^{2+}$ pulses or transients [4, 7, 8] and regulates store-operated calcium entry [5, 13]. Furthermore, for any repetitive physiological process dependent on intrami-

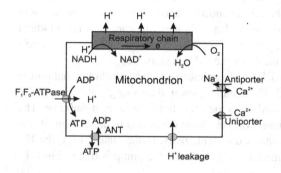

**Fig. 9.1.** Schematic description of ions movements across the mitochondrial inner membrane

Liu W.: Introduction to Modeling Biological Cellular Control Systems.
DOI 10.1007/978-88-470-2490-8_9, © Springer-Verlag Italia 2012

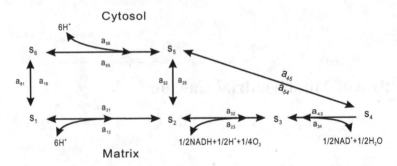

**Fig. 9.2.** Six state model of the respiratory chain proposed by Pietrobon and Caplan [11]. A counter-clock cycle of the diagram's perimeter corresponds to the transfer of 1 $e^-$ and the ejection of a maximum of 6 $H^+$, as driven by the oxidation of 1/2 NADH in the mitochondrial inner membrane

tochondrial free $Ca^{2+}$ concentration, a kind of intramitochondrial $Ca^{2+}$ homeostasis must exist and be controlled dynamically to avoid either $Ca^{2+}$ buildup or depletion in mitochondria [1, 7].

$Ca^{2+}$ transport both inward and outward across the mitochondrial inner membrane is controlled by an elaborate set of mechanisms and processes. Three mechanisms or modes of influx are the mitochondrial $Ca^{2+}$ uniporter, the rapid mode and the mitochondrial ryanodine receptor (mRyR). Two mechanisms of $Ca^{2+}$ efflux are the $Na^+$-dependent and the $Na^+$-independent mechanisms [8]. We present a minimal model for the mitochondrial $Ca^{2+}$ control system developed by Magnus and Keizer [9] and refined later by Cortassa et al [3].

## 9.1 Respiration-driven Proton Ejection

Most ATP is synthesized in mitochondria by the process of oxidative phosphorylation, as demonstrated in Fig. 9.1. In this process, energized electrons liberated from the oxidation of NADH and $FADH_2$ are delivered to $O_2$ via an electron transport chain (respiratory chain), which consists of four complexes. When passing through the chain, these electrons power proton $H^+$ pumps located in the chain. These powered pumps eject $H^+$ from the mitochondrial matrix to the outer compartment between the outer and inner mitochondrial membranes to establish a $H^+$ gradient across the inner mitochondrial membrane. The electrochemical energy of this gradient is then used to drive ATP synthesis by the $F_1F_0$-ATPase [10].

Pietrobon and Caplan [11] modeled the respiratory chain as a bidirectional proton pump, as demonstrated in Fig. 9.2. The pump is powered by a single electron, which is energized by the oxidation of NADH in the mitochondrial inner membrane. The powered pump then ejects a maximum of 6 $H^+$ from the matrix to the cytosol. It was assumed that the pump can stay in six different states. In the state 1 ($S_1$), the $H^+$ binding sites of the pump is oriented to the matrix and the pump is free to bind $H^+$

from the matrix. The pump is transited into the state 2 ($S_2$) when the mitochondrial $H^+$ binds to it. After energized by the oxidation of NADH, the pump is transited into the state 3 ($S_3$) and then the state 4 ($S_4$) with the production of $NAD^+$. The energized pump is able to undergo a conformational change reorienting its $H^+$ binding sites from the matrix to the cytosol and transited into the state 5 ($S_5$). After $H^+$ are released into the cytosol, the pump is transited into the state 6 ($S_6$) and then returns its original state 1 through a conformational change. The transition of the state 2 to the state 5 represents an outward $H^+$ leakage while the transition of the state 5 to the state 2 corresponds to a reaction slip that NADH oxidation occurs without $H^+$ ejection.

Rate constants may be functions of extrinsic quantities affecting pump turnover, such as membrane potential, ligand and metabolite concentrations. The rate constants $a_{21}$ and $a_{56}$ for the proton unbinding involve boundary potentials that are measured at each membrane surface with respect to the nearest bulk phase, and sum to a constant boundary voltage difference ($V_b = 50$ mV). Since the conformational changes between the state 1 and the state 6 occur within the membrane itself, the corresponding rate constants $a_{16}$ and $a_{61}$ depend on $V - V_b$, where $V$ is the membrane potential difference.

The $H^+$ ejection is driven by the chemical potential in the reaction

$$\frac{1}{2} \text{NADH} + \frac{1}{2} H^+ + \frac{1}{4} O_2 \rightleftharpoons \frac{1}{2} \text{NAD}^+ + \frac{1}{2} H_2O. \tag{9.1}$$

Assuming that the concentrations of $H^+$, $O_2$, and $H_2O$ are constant in the reaction (9.1), we derive from (2.56) that the chemical potential change is given by

$$\mu_{res} = -RT \ln \left( K_{res} \sqrt{\frac{[NADH]}{[NAD^+]}} \right), \tag{9.2}$$

where $K_{res}$ is the equilibrium constant. Mitochondrial $NAD^+$ is assumed to be conserved as follows

$$[NAD^+] = C_{pn} - [NADH]$$

with $C_{pn}$ as the total concentration of pyrimidine nucleotides.

On the other hand, the electrochemical gradient, or the proton motive force ($\Delta p$), drives mitochondria to uptake $H^+$. Let $\phi_o$ denote the electrical potential on the outside of the inner mitochondrial membrane, $\phi_i$ the electrical potential on the inside (the matrix side), and $V = \phi_o - \phi_i$ the potential difference. Let $\text{pH}_o$ denote the outer compartment pH, $\text{pH}_i$ the inside pH, and $\Delta\text{pH} = \text{pH}_o - \text{pH}_i$ the pH difference. It follows from (2.53) and (B.2) that the electrochemical gradient is given by

$$\begin{aligned}
\Delta p &= \phi_o - \phi_i + \frac{RT}{F} \left( \ln[H_o^+] - \ln[H_i^+] \right) \\
&= V + \frac{RT}{F \log e} \left( \log[H_o^+] - \log[H_i^+] \right) \\
&= V + 2.303 \frac{RT}{F} \Delta pH. 
\end{aligned} \tag{9.3}$$

The reactions in Fig. 9.2 can be written as follows:

$$S_1 + 6H_m^+ \underset{a_{21}}{\overset{a_{12}}{\rightleftharpoons}} S_2, \tag{9.4}$$

$$S_2 + \frac{1}{2}NADH + \frac{1}{2}H_m^+ + \frac{1}{4}O_2 \underset{a_{32}}{\overset{a_{23}}{\rightleftharpoons}} S_3, \tag{9.5}$$

$$S_3 \underset{a_{43}}{\overset{a_{34}}{\rightleftharpoons}} S_4 + \frac{1}{2}NAD^+ + \frac{1}{2}H_2O, \tag{9.6}$$

$$S_4 \underset{a_{54}}{\overset{a_{45}}{\rightleftharpoons}} S_5, \tag{9.7}$$

$$S_5 \underset{a_{25}}{\overset{a_{52}}{\rightleftharpoons}} S_2, \tag{9.8}$$

$$S_5 \underset{a_{65}}{\overset{a_{56}}{\rightleftharpoons}} S_6 + 6H_i^+, \tag{9.9}$$

$$S_6 \underset{a_{16}}{\overset{a_{61}}{\rightleftharpoons}} S_1, \tag{9.10}$$

where $H_m^+$ denotes the proton $H^+$ in the matrix of mitochondria and $H_i^+$ denotes the proton $H^+$ in the cytosol. At equilibrium, these six states satisfy the following equations

$$a_{12}[H^+]_m^6[S_1] + a_{16}[S_1] - a_{21}[S_2] - a_{61}[S_6] = 0, \tag{9.11}$$

$$a_{23}[H^+]_m^{1/2}[NADH]^{1/2}[O_2]^{1/4}[S_2]$$
$$+ (a_{21} + a_{25})[S_2] - a_{12}[H^+]_m^6[S_1] - a_{23}[S_3] - a_{52}[S_5] = 0, \tag{9.12}$$

$$a_{32}[S_3] + a_{34}[S_3] - a_{23}[H^+]_m^{1/2}[NADH]^{1/2}[O_2]^{1/4}[S_2]$$
$$- a_{43}[NAD^+]^{1/2}[H_2O]^{1/2}[S_4] = 0, \tag{9.13}$$

$$a_{43}[NAD^+]^{1/2}[H_2O]^{1/2}[S_4] + a_{45}[S_4] - a_{54}[S_5] - a_{34}[S_3] = 0, \tag{9.14}$$

$$(a_{54} + a_{52} + a_{56})[S_5] - a_{45}[S_4] - a_{25}[S_2] - a_{65}[H^+]_i^6[S_6] = 0, \tag{9.15}$$

$$[S_1] + [S_2] + [S_3] + [S_4] + [S_5] + [S_6] = [S_0], \tag{9.16}$$

where $[S_0]$ denotes the total concentration of the pump. Using a mathematical software such as the Maple, we can solve the above system to obtain $[S_1], [S_2], [S_3], [S_4], [S_5], [S_6]$, which are large fractions with denominators containing nearly 100 terms, each a product of 5 rate constants. Magnus and Keizer [9] approximated numerous terms and factors in the full expressions and greatly simplified them to obtain the following $H^+$ flux

$$r_{res} = \frac{6\rho_{res}\left[r_a \exp\left(\frac{-\mu_{res}}{RT}\right) - (r_a + r_b)\exp\left(\frac{6gF\Delta p}{RT}\right)\right]}{\left[1 + r_1 \exp\left(\frac{-\mu_{res}}{RT}\right)\right]\exp\left(\frac{6FV_b}{RT}\right) + \left[r_2 + r_3 \exp\left(\frac{-\mu_{res}}{RT}\right)\right]\exp\left(\frac{6gF\Delta p}{RT}\right)}$$
$$= \frac{6\rho_{res}\left[r_a K_{res}\sqrt{\frac{[NADH]}{[NAD^+]}} - (r_a + r_b)\exp\left(\frac{6gF\Delta p}{RT}\right)\right]}{D_{res}}, \tag{9.17}$$

where

$$D_{res} = \left[1 + r_1 K_{res}\sqrt{\frac{[NADH]}{[NAD^+]}}\right] \exp\left(\frac{6FV_b}{RT}\right)$$

$$+ \left[r_2 + r_3 K_{res}\sqrt{\frac{[NADH]}{[NAD^+]}}\right] \exp\left(\frac{6gF\Delta p}{RT}\right).$$

In the original $H^+$ flux model of Magnus and Keizer [9], $\Delta p$ was the potential difference $V$, which was changed to the current $\Delta p$ by Cortassa et al [3]. This equation indicates that if the ratio

$$\frac{[NADH]}{[NAD^+]} > \frac{(r_a + r_b)^2 \exp\left(\frac{12gF\Delta p}{RT}\right)}{r_a^2 K_{res}^2},$$

then $H^+$ are rejected from the matrix. Otherwise, $H^+$ flow back into the matrix due to the insufficient energy from NADH.

The $H^+$ ejection is also powered by the chemical potential

$$\mu_{res,f} = -RT \ln\left(K_{res,f}\sqrt{\frac{[FADH_2]}{[FAD]}}\right) \qquad (9.18)$$

from the complex II of the electron transport chain. The $H^+$ flux has the same form as $r_{res}$ except for a little adjustment as follows

$$r_{res,f} = \frac{6\rho_{res,f}\left[r_a \exp\left(\frac{-\mu_{res,f}}{RT}\right) - (r_a + r_b)\exp\left(\frac{4gF\Delta p}{RT}\right)\right]}{\left[1 + r_1 \exp\left(\frac{-\mu_{res,f}}{RT}\right)\right]\exp\left(\frac{4FV_B}{RT}\right) + \left[r_2 + r_3 \exp\left(\frac{-\mu_{res,f}}{RT}\right)\right]\exp\left(\frac{4gF\Delta p}{RT}\right)}$$

$$= \frac{6\rho_{res,f}\left[r_a K_{res,f}\sqrt{\frac{[FADH_2]}{[FAD]}} - (r_a + r_b)\exp\left(\frac{4gF\Delta p}{RT}\right)\right]}{D_{res,f}}, \qquad (9.19)$$

where

$$D_{res,f} = \left[1 + r_1 K_{res,f}\sqrt{\frac{[FADH_2]}{[FAD]}}\right] \exp\left(\frac{4FV_B}{RT}\right)$$

$$+ \left[r_2 + r_3 K_{res,f}\sqrt{\frac{[FADH_2]}{[FAD]}}\right] \exp\left(\frac{4gF\Delta p}{RT}\right).$$

The $H^+$ leakage from the outer compartment into the matrix is considered to be a linear function of the $\Delta p$ through a proportionality constant given by the $H^+$ conductance, $G_h$, described as follows:

$$r_{hleak} = G_h \Delta p. \qquad (9.20)$$

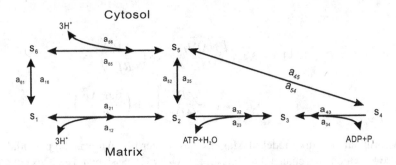

**Fig. 9.3.** Six state model of the $F_1F_0$-ATPase proposed by Pietrobon and Caplan [11]. A counter-clock cycle of the diagram's perimeter corresponds to the production of 1 ADP and the ejection of a maximum of 3 $H^+$, as driven by the hydrolysis of 1 ATP by the $F_1F_0$-ATPase in the mitochondrial inner membrane. The pump functions in the reverse direction in vivo

## 9.2 ATP Synthesis and Proton Uptake by the $F_1F_0$-ATPase

In analogy to the respiratory chain, the mitochondrial $F_1F_0$-ATPase was also modeled as a bidirectional proton pump [9, 11], as demonstrated in Fig. 9.3. If the respiratory chain is inhibited, ATP is hydrolyzed by the $F_1$-ATPase in the reaction

$$ATP + H_2O \rightleftharpoons ADP + H_2PO_4^-. \tag{9.21}$$

The energy from the ATP hydrolysis powers the proton pump to eject $H^+$ through the pore-like $F_0$ sector. This corresponds to the counter-clockwise cycle of the perimeter of Fig. 9.3. By (2.56), the chemical potential in the reaction (9.21) is given by

$$\mu_{f_1} = -RT \ln\left( K_{f_1} \sqrt{\frac{[ATP]_m}{[ADP]_m P_i}} \right), \tag{9.22}$$

where $K_{f_1}$ is the equilibrium constant for the reaction (9.21). Mitochondrial ATP, $[ATP]_m$, is assumed to be conserved as follows

$$[ATP]_m = C_m - [ADP]_m$$

with $C_m$ as the total concentration of adenine nucleotides and $[ADP]_m$ as the mitochondrial ADP concentration.

On the other hand, the system in vivo operates in the reverse direction. The electrochemical gradient $\Delta p$ provides the essential energy to drive ATP synthesis by complex V (ATP synthase) with the $H^+$ backflow into the matrix. As in the derivation of the proton ejection rate (9.17), it can be derived that the ATP production rate $r_{f1f0,atp}$ and the $H^+$ uptake rate $r_{f1f0,h}$ into the mitochondrial matrix are given by [3, 9]

$$r_{f1f0,atp} = -\rho_{f_1} \left\{ \left[ 10^2 p_a + p_{c1} \exp\left(\frac{3FV_B}{RT}\right) \right] \exp\left(\frac{-\mu_{f_1}}{RT}\right) \right.$$

$$- \left[ p_{c2} \exp\left(\frac{-\mu_{f_1}}{RT}\right) + p_a \right] \exp\left(\frac{3F\Delta p}{RT}\right) \left. \right\} \div$$

$$\left\{ \left[ 1 + p_1 \exp\left(\frac{-\mu_{f_1}}{RT}\right) \right] \exp\left(\frac{3FV_B}{RT}\right) \right.$$

$$+ \left[ p_2 + p_3 \exp\left(\frac{-\mu_{f_1}}{RT}\right) \right] \exp\left(\frac{3F\Delta p}{RT}\right) \left. \right\}$$

$$= -\rho_{f_1} \left\{ \left[ 10^2 p_a + p_{c1} \exp\left(\frac{3FV_B}{RT}\right) \right] K_{f_1} \sqrt{\frac{[ATP]_m}{[ADP]_m P_i}} \right.$$

$$- \left[ p_{c2} K_{f_1} \sqrt{\frac{[ATP]_m}{[ADP]_m P_i}} + p_a \right] \exp\left(\frac{3F\Delta p}{RT}\right) \left. \right\} \div$$

$$\left\{ \left[ 1 + p_1 K_{f_1} \sqrt{\frac{[ATP]_m}{[ADP]_m P_i}} \right] \exp\left(\frac{3FV_B}{RT}\right) \right.$$

$$+ \left[ p_2 + p_3 K_{f_1} \sqrt{\frac{[ATP]_m}{[ADP]_m P_i}} \right] \exp\left(\frac{3F\Delta p}{RT}\right) \left. \right\}, \qquad (9.23)$$

$$r_{f1f0,h} = \frac{-3\rho_{f_1} \left\{ 10^2 p_a \left[ 1 + \exp\left(\frac{-\mu_{f_1}}{RT}\right) \right] - (p_a + p_b) \exp\left(\frac{3F\Delta p}{RT}\right) \right\}}{\left[ 1 + p_1 \exp\left(\frac{-\mu_{f_1}}{RT}\right) \right] \exp\left(\frac{3FV_B}{RT}\right) + \left[ p_2 + p_3 \exp\left(\frac{-\mu_{f_1}}{RT}\right) \right] \exp\left(\frac{3F\Delta p}{RT}\right)}$$

$$= \frac{-3\rho_{f_1} \left\{ 10^2 p_a \left[ 1 + K_{f_1} \sqrt{\frac{[ATP]_m}{[ADP]_m P_i}} \right] - (p_a + p_b) \exp\left(\frac{3F\Delta p}{RT}\right) \right\}}{D_{f1f0,h}}, \qquad (9.24)$$

where

$$D_{f1f0,h} = \left[ 1 + p_1 K_{f_1} \sqrt{\frac{[ATP]_m}{[ADP]_m P_i}} \right] \exp\left(\frac{3FV_B}{RT}\right)$$

$$+ \left[ p_2 + p_3 K_{f_1} \sqrt{\frac{[ATP]_m}{[ADP]_m P_i}} \right] \exp\left(\frac{3F\Delta p}{RT}\right).$$

In the original rate model of Magnus and Keizer [9], $\Delta p$ was the potential difference $V$, which was changed to the current $\Delta p$ by Cortassa et al [3].

## 9.3 ATP and ADP Transport by the Adenine Nucleotide Translocator

Adenine nucleotide translocator (ANT), also known as the ADP/ATP transloca-tor, is a mitochondrial protein and functions as a monomer in mitochondrial mem-branes. ANT activity was modeled according to a sequential mechanism proposed by Bohnensack [2]. Oppositely oriented sites bind either $ATP^{4-}$ or $ADP^{3-}$, and both sites must be filled before the protein isomerization causes a ligand exchange across the inner membrane. These binding activities may be described by the following reac-tions:

$$ADP_i^{3-} + ANT + ADP_m^{3-} \rightleftharpoons [ADP_i^{3-}ANTADP_m^{3-}] \xrightarrow{k_{didm}} ADP_m^{3-} + ANT + ADP_i^{3-},$$

$$ADP_i^{3-} + ANT + ATP_m^{4-} \rightleftharpoons [ADP_i^{3-}ANTATP_m^{4-}] \xrightarrow{k_{ditm}} ADP_m^{3-} + ANT + ATP_i^{4-},$$

$$ATP_i^{4-} + ANT + ADP_m^{3-} \rightleftharpoons [ATP_i^{4-}ANTADP_m^{3-}] \xrightarrow{k_{tidm}} ATP_m^{4-} + ANT + ADP_i^{3-},$$

$$ATP_i^{4-} + ANT + ATP_m^{4-} \rightleftharpoons [ATP_i^{4-}ANTATP_m^{4-}] \xrightarrow{k_{titm}} ATP_m^{3-} + ANT + ATP_i^{4-}.$$

The subscripts i and m denote the cytosolic side and the matrix side of mitochondria, respectively. The total concentration $[ANT_0]$ of ANT is then given by

$$[ANT_0] = [ANT] + [ADP_i^{3-}ANTADP_m^{3-}] + [ADP_i^{3-}ANTATP_m^{4-}]$$
$$+[ATP_i^{4-}ANTADP_m^{3-}] + [ATP_i^{4-}ANTATP_m^{4-}]. \quad (9.25)$$

Assuming that each binding reaction is at equilibrium and defining the dissociation constants by

$$K_{didm} = \frac{[ADP^{3-}]_i[ANT][ADP^{3-}]_m}{[ADP_i^{3-}ANTADP_m^{3-}]},$$

$$K_{ditm} = \frac{[ADP^{3-}]_i[ANT][ATP^{4-}]_m}{[ADP_i^{3-}ANTATP_m^{4-}]},$$

$$K_{tidm} = \frac{[ATP^{4-}]_i[ANT][ADP^{3-}]_m}{[ATP_i^{4-}ANTADP_m^{3-}]},$$

$$K_{titm} = \frac{[ATP^{4-}]_i[ANT][ATP^{4-}]_m}{[ATP_i^{4-}ANTATP_m^{4-}]},$$

we derive from (9.25) that

$$[ANT_0] = [ANT] + \frac{[ADP^{3-}]_i[ANT][ADP^{3-}]_m}{K_{didm}} + \frac{[ADP^{3-}]_i[ANT][ATP^{4-}]_m}{K_{ditm}}$$
$$+ \frac{[ATP^{4-}]_i[ANT][ADP^{3-}]_m}{K_{tidm}} + \frac{[ATP^{4-}]_i[ANT][ATP^{4-}]_m}{K_{titm}},$$

and then
$$[ANT] = \frac{[ANT_0]}{D_{ant}},$$
(9.26)

where

$$D_{ant} = 1 + \frac{[ADP^{3-}]_i[ADP^{3-}]_m}{K_{didm}} + \frac{[ADP^{3-}]_i[ATP^{4-}]_m}{K_{ditm}}$$
$$+ \frac{[ATP^{4-}]_i[ADP^{3-}]_m}{K_{tidm}} + \frac{[ATP^{4-}]_i[ATP^{4-}]_m}{K_{titm}}.$$

It then follows that the rate equations for the four types of exchanges are

$$r_{didm} = k_{didm}[ADP_i^{3-} ANTADP_m^{3-}]$$
$$= \frac{k_{didm}[ANT_0][ADP^{3-}]_i[ADP^{3-}]_m}{K_{didm}D_{ant}},$$
(9.27)

$$r_{ditm} = k_{ditm}[ADP_i^{3-} ANTATP_m^{4-}]$$
$$= \frac{k_{ditm}[ANT_0][ADP^{3-}]_i[ATP^{4-}]_m}{K_{ditm}D_{ant}},$$
(9.28)

$$r_{tidm} = k_{tidm}[ATP_i^{4-} ANTADP_m^{3-}]$$
$$= \frac{k_{tidm}[ANT_0][ATP^{4-}]_i[ADP^{3-}]_m}{K_{tidm}D_{ant}},$$
(9.29)

$$r_{titm} = k_{titm}[ATP_i^{4-} ANTATP_m^{4-}]$$
$$= \frac{k_{titm}[ANT_0][ATP^{4-}]_i[ATP^{4-}]_m}{K_{titm}D_{ant}}.$$
(9.30)

Experimental observations showed that the dissociation constants and the first-order rate constants satisfy the following relations [2]

$$K_{ditm} = K_{didm}, \ K_{tidm} = K_{titm}, \ k_{ditm} = k_{titm} = k_{didm}.$$
(9.31)

The exchange between $ADP_i^{3-}$ and $ATP_m^{4-}$

$$ADP_i^{3-} + ATP_m^{4-} \rightleftharpoons ATP_i^{4-} + ADP_m^{3-}$$

is electrogenic due to the charge difference. It then follows from (2.54) and (B.2) that the change in electrochemical potential of this reaction is

$$\Delta\mu = \Delta\mu^\circ + RT \ln\left(\frac{[ATP^{4-}]_i[ADP^{3-}]_m}{[ADP^{3-}]_i[ATP^{4-}]_m}\right) - FV,$$

where $V = V_o - V_i$ is the transmembrane potential. At equilibrium, we must have $\Delta\mu = 0$. Assuming that the standard free energy for the reaction is the same on both sides of the membrane ($\Delta\mu^\circ = 0$), it then follows that

$$\frac{[ATP^{4-}]_i[ADP^{3-}]_m}{[ADP^{3-}]_i[ATP^{4-}]_m} = \exp\left(\frac{FV}{RT}\right).$$
(9.32)

Since, at equilibrium, the net flux $r_{ditm} - r_{tidm}$ is zero, it follows from (9.28) and (9.29) that

$$\frac{K_{tidm}k_{ditm}}{K_{ditm}k_{tidm}} = \frac{[ATP^{4-}]_i[ADP^{3-}]_m}{[ADP^{3-}]_i[ATP^{4-}]_m} = \exp\left(\frac{FV}{RT}\right). \tag{9.33}$$

The dependence of $K_{tidm}$ on the membrane potential was estimated experimentally as follows [2]

$$K_{tidm} = K_{didm}\exp\left(\frac{fFV}{RT}\right), \tag{9.34}$$

where the empirical factor $f$ expresses the fraction of the membrane potential producing the energy-dependent shift of the Michaelis-Menten constant for ATP. With these relations, we derive from (9.28) and (9.29) that the rate of the $ADP_i^{3-} - ATP_m^{4-}$ net exchange assumes the form [2, 3, 9]

$$r_{ant} = r_{ditm} - r_{tidm}$$

$$= \frac{k_{ditm}[ANT_0]\left([ADP^{3-}]_i[ATP^{4-}]_m - \frac{K_{ditm}k_{tidm}}{K_{tidm}k_{ditm}}[ATP^{4-}]_i[ADP^{3-}]_m\right)}{K_{ditm}D_{ant}}$$

$$= \left\{k_{didm}[ANT_0]\left([ADP^{3-}]_i[ATP^{4-}]_m - [ATP^{4-}]_i[ADP^{3-}]_m\exp\left(\frac{-FV}{RT}\right)\right)\right\} \div$$

$$\left\{K_{didm} + [ADP^{3-}]_i\left([ADP^{3-}]_m + [ATP^{4-}]_m\right)\right.$$

$$\left. + [ATP^{4-}]_i\exp\left(\frac{-fFV}{RT}\right)\left([ADP^{3-}]_m + [ATP^{4-}]_m\right)\right\}$$

$$= \frac{k_{didm}[ANT_0]\left([ADP^{3-}]_i[ATP^{4-}]_m - [ATP^{4-}]_i[ADP^{3-}]_m\exp\left(\frac{-FV}{RT}\right)\right)}{K_{didm} + \left([ADP^{3-}]_i + [ATP^{4-}]_i\exp\left(\frac{-fFV}{RT}\right)\right)\left([ADP^{3-}]_m + [ATP^{4-}]_m\right)}.$$

Since in general the concentration of the adenine nucleotides on both sides of the membrane is high enough, the constant $K_{didm}$ can be neglected in the denominator. The rate equation may then be approximated as

$$r_{ant} = R_{ant}\frac{1 - \frac{[ATP^{4-}]_i[ADP^{3-}]_m}{[ADP^{3-}]_i[ATP^{4-}]_m}\exp\left(\frac{-FV}{RT}\right)}{\left[1 + \frac{[ATP^{4-}]_i}{[ADP^{3-}]_i}\exp\left(\frac{-fFV}{RT}\right)\right]\left[1 + \frac{[ADP^{3-}]_m}{[ATP^{4-}]_m}\right]}, \tag{9.35}$$

where $R_{ant} = k_{didm}[ANT_0]$. The relations among $ADP^{3-}, ATP^{4-}, ADP$, and $ATP$ are given by [9]

$$[ATP^{4-}]_i = 0.05[ATP]_i, \tag{9.36}$$

$$[ADP^{3-}]_i = 0.45[ADP]_i, \tag{9.37}$$

$$[ATP^{4-}]_m = 0.05[ATP]_m, \tag{9.38}$$

$$[ADP^{3-}]_m = 0.45 \cdot 0.8[ADP]_m. \tag{9.39}$$

## 9.4 Calcium Uptake by the Uniporter

A uniporter is a mechanism which facilitates passive transport of $Ca^{2+}$ across the mitochondrial membrane without coupling $Ca^{2+}$ transport to the transport of another ion. The uniporter has both a transport site and a separate activation site. These sites have different binding affinities and the activation site can be activated by other divalent ions and also by trivalent lanthanides, such as $Pr^{3+}$, or by $Ca^{2+}$. These results suggest that the rate of $[Ca^{2+}]_i$ transport by the mitochondrial $Ca^{2+}$ uniporter is given by a second order Hill equation of the form (see Gunter *et al* [8])

$$r_{unip} = R^*_{unip} \frac{[Ca^{2+}]_i^2}{K_{0.5}^2 + [Ca^{2+}]_i^2},$$

where $R^*_{unip}$ is a function of membrane potential. The fastest reported $R^*_{unip}$ (1750 nmol/mg/min) was measured in dog heart mitochondria. The published values of $K_{0.5}$ were pretty diverse, ranging from 1 to 189 $\mu$M [8].

The membrane potential dependence of $Ca^{2+}$ uptake is also an important characteristic of uniporter behavior. Since a uniporter facilitates the transport of an ion down its electrochemical gradient without coupling it to the transport of any other ion, the membrane potential dependence should be consistent with the equations for electrochemical diffusion. This membrane potential dependence has been shown in liver mitochondria to fit the form [8]

$$\frac{bF(V - V_0)}{2RT} \left[ \exp \left( \frac{2bF(V - V_0)}{RT} \right) - 1 \right],$$

where $V$ is the mitochondrial membrane potential difference in millivolts, $F$, $R$, and $T$ are the Faraday constant, the gas constant and the Kelvin temperature, respectively, and $b = 1$ and $V_0 = 91$ mV are fitting parameters. Then the combined concentration and membrane potential dependence is then given by

$$r_{unip} = R_{unip} \frac{F(V - V_0)}{2RT} \left[ \exp \left( \frac{2F(V - V_0)}{RT} \right) - 1 \right] \frac{[Ca^{2+}]^2}{K_{0.5}^2 + [Ca^{2+}]^2}. \tag{9.40}$$

Another similar model developed by Magnus and Keizer [9] is as follows

$$r_{unip} = R_{unip} \frac{\frac{[Ca^{2+}]_i}{K_{trans}} \left( 1 + \frac{[Ca^{2+}]_i}{K_{trans}} \right)^3 \frac{2F(V - V_0)}{RT}}{\left[ \left( 1 + \frac{[Ca^{2+}]_i}{K_{trans}} \right)^4 + \frac{L}{\left( 1 + \frac{[Ca^{2+}]_i}{K_{act}} \right)^{na}} \right] \left[ 1 - \exp \left( \frac{-2F(V - V_0)}{RT} \right) \right]}. \tag{9.41}$$

## 9.5 Calcium Efflux via the Sodium/Calcium Exchanger

$Na^+/Ca^{2+}$ exchangers (NCX) are present in the mitochondrial inner membrane and are the primary mechanism of $Ca^{2+}$ efflux in brain, heart, skeletal muscle and many other types of mitochondria [8]. The kinetics of this mechanism in both heart and liver mitochondria have been determined to be first order in $Ca^{2+}$ and second order in $Na^+$. Thus, the ionic flux was described mathematically by (see, e.g., [8])

$$r_{ncx} = R_{ncx} \left( \frac{[Na^+]_i^2}{K_{na}^2 + [Na^+]_i^2} \right) \left( \frac{[Ca^{2+}]_m}{K_{ca} + [Ca^{2+}]_m} \right). \qquad (9.42)$$

For liver mitochondria, $R_{ncx}$ has been found to be $2.6 \pm 0.5$ nmol/(mg·min), $K_{ca}$ to be $8.1 \pm 1.4$ nmol/mg protein, and $K_{na}$ to be $9.4 \pm 0.6$ mM. For heart mitochondria, the corresponding parameters have been found to be 18 nmol/(mg·min), around 10 nmol/mg protein, and 7 - 12 mM $Na^+$, respectively, whereas for brain mitochondria, $R_{ncx}$ has been found to be around 30 nmol/(mg·min).

Another rate equation of the exchanger was established by Wingrove *et al* [14], and modified later by Magnus *et al* [9] and Cortassa *et al* [3]. Under normal conditions, the major $Ca^{2+}$ efflux channel through the inner mitochondrial membrane is the $Na^+/Ca^{2+}$ exchanger. However, the exchanger may import $Ca^{2+}$ while extruding $Na^+$ under ischemic conditions. Experimental observations showed that the exchanger is able to sense both $[Ca^{2+}]_m$ and $[Ca^{2+}]_i$ based on the reversal of its activity under pathological conditions [6]. Like the $Ca^{2+}$ influx uniporter, the exchanger is controlled by the inner membrane potential $V$ and its rate is

$$r_{ncx} = R_{ncx} \frac{\exp\left( \frac{bF(V-V_0)}{RT} \right)}{\left( 1 + \frac{K_{na}}{[Na^+]_i} \right)^n \left( 1 + \frac{K_{ca}}{[Ca^{2+}]_m} \right)}. \qquad (9.43)$$

Furthermore, a model of fifteen states for the $nNa^+/Ca^{2+}$ exchangers was proposed by Pradhan *et al* [12]. Because this model is complex, we do not discuss it here and refer to [12] for interested readers.

## 9.6 Governing Equations of Calcium Dynamics

Using the electric circuit model (6.149) for the membrane potential, we obtain the governing equations for the dynamics of mitochondrial calcium as follows

$$\frac{d[ADP]_m}{dt} = r_{ant} - r_{f1f0,atp}, \qquad (9.44)$$

$$C_{mito} \frac{dV}{dt} = r_{res} + r_{res,f} - r_{f1f0,h} - r_{ant} - r_{hleak} - r_{ncx} - 2r_{unip}, \qquad (9.45)$$

$$\frac{d[Ca^{2+}]_m}{dt} = f(r_{uni} - r_{ncx}). \qquad (9.46)$$

In these equations, $r_{res}$ is defined by the equations (9.17), $r_{res,f}$ is defined by the equations (9.19), $r_{hleak}$ is defined by the equations (9.20), $r_{f1f0,atp}$ is defined by the equations (9.23), $r_{f1f0,h}$ is defined by the equations (9.24), $r_{ant}$ is defined by the equations (9.35), $r_{unip}$ is defined by the equations (9.41), $r_{ncx}$ is defined by the equations

**Table 9.1.** Parameter values of the model (9.44)-(9.46) from [3]

| Symbol | Value | Description |
|---|---|---|
| $r_a$ | $6.394 \times 10^{-10}$ s$^{-1}$ | Sum of products of rate constants |
| $r_b$ | $1.762 \times 10^{-13}$ s$^{-1}$ | Sum of products of rate constants |
| $r_1$ | $2.077 \times 10^{-18}$ | Sum of products of rate constants |
| $r_2$ | $1.728 \times 10^{-9}$ | Sum of products of rate constants |
| $r_3$ | $1.059 \times 10^{-26}$ | Sum of products of rate constants |
| $\rho_{res}$ | 0.0006-0.05 mM | Concentration of electron carriers |
| $K_{res}$ | $1.35 \times 10^{18}$ | Equilibrium constant of respiration |
| $\rho_{res,f}$ | 0.0045 mM | Concentration of electron carriers |
| $V_b$ | 0.05 V | Phase boundary potential |
| $g$ | 0.85 | Correction factor for voltage |
| $K_{res,f}$ | $5.765 \times 10^{13}$ | Equilibrium constant of FADH$_2$ oxidation |
| $[FADH_2]$ | 1.24 mM | Concentration of FADH$_2$ (reduced) |
| $[FAD]$ | 0.01 mM | Concentration of FAD (oxidized) |
| $p_a$ | $1.656 \times 10^{-5}$ s$^{-1}$ | Sum of products of rate constants |
| $p_b$ | $3.373 \times 10^{-7}$ s$^{-1}$ | Sum of products of rate constants |
| $p_{c1}$ | $9.651 \times 10^{-14}$ s$^{-1}$ | Sum of products of rate constants |
| $p_{c2}$ | $4.585 \times 10^{-14}$ s$^{-1}$ | Sum of products of rate constants |
| $p_1$ | $1.346 \times 10^{-8}$ | Sum of products of rate constants |
| $p_2$ | $7.739 \times 10^{-7}$ | Sum of products of rate constants |
| $p_3$ | $6.65 \times 10^{-15}$ | Sum of products of rate constants |
| $\rho_{f1}$ | 0.06-1.8 mM | Concentration of F$_1$F$_0$-ATPase |
| $K_{F1}$ | $1.71 \times 10^6$ | Equilibrium constant of ATP hydrolysis |
| $R$ | 8.315 V C mol$^{-1}$ K$^{-1}$ | Gas constant |
| $T$ | 310.16 K$^{-1}$ | Mammalian body temperature |
| $F$ | 96480 C mol$^{-1}$ | Faraday constant |
| $P_i$ | 20.0 mM | Inorganic phosphate concentration |
| $C_m$ | 15.0 mM | Total sum of mitochondrial adenine nucleotides |
| $R_{ant}$ | 0.05-24.0 mM s$^{-1}$ | Maximal rate of the ANT |
| $[ADP]_i$ | 0.05-0.2 mM | Cytoplasmic ADP$_i$ concentration |
| $[ATP]_i$ | 6.5 mM | Cytoplasmic ATP$_i$ concentration |
| $G_h$ | 0.01 mM s$^{-1}$ V$^{-1}$ | Ionic conductance of the inner membrane |
| $\Delta pH$ | -0.6 pH | pH gradient across the inner membrane |
| $C_{pn}$ | 10.0 mM | Total sum of mito pyridine nucleotides |
| $C_{mito}$ | 1.812 mM V$^{-1}$ | Inner membrane capacitance |
| $R_{unip}$ | 0.625-1.25 $\mu$M s$^{-1}$ | Maximum rate of uniporter Ca$^{2+}$ transport |
| $[Ca^{2+}]_i$ | $2.0 \times 10^{-2}$-1.2 $\mu$M | Cytosolic Ca$^{2+}$ concentration |
| $V_0$ | 0.091 V | Offset membrane potential |
| $K_{act}$ | $3.8 \times 10^{-4}$ mM | Activation constant |
| $K_{trans}$ | 0.019 mM | Dissociation constant for translocated Ca$^{2+}$ |
| $L$ | 110.0 | Equilibrium constant for conformational transitions in uniporter |
| $n_a$ | 2.8 | Uniporter activation cooperativity |
| $R_{ncx}$ | 0.005-0.2 mM s$^{-1}$ | Maximum rate of Na$^+$/Ca$^{2+}$ exchanger |

**Table 9.1** continued

| Symbol | Value | Description |
|--------|-------|-------------|
| $b$ | 0.5 | V dependence of Na$^+$/Ca$^{2+}$ exchanger |
| $K_{na}$ | 9.4 mM | Exchanger Na$^+$ constant |
| $[Na^+]_i$ | 10.0 mM | Cytosolic Na$^+$ concentration |
| $K_{ca}$ | $3.75 \times 10^{-4}$ mM | Exchanger Ca$^{2+}$ constant |
| $n$ | 3 | Na$^+$/Ca$^{2+}$ exchanger cooperativity |
| $f$ | 0.0003 | Fraction of free $[Ca^{2+}]_m$ |

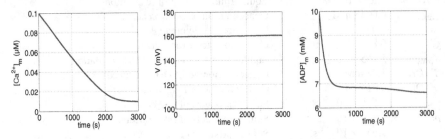

**Fig. 9.4.** Numerical solutions of the model (9.44)-(9.46)

(9.43), and $f$ denotes the fraction of free $[Ca^{2+}]_m$. This system is solved numerically with the parameter values given in Table 9.1. Fig. 9.4 shows that, for a constant intracellular calcium concentration $[Ca^{2+}]_i = 0.1\ \mu$M, all three states $[ADP]_m, V, [Ca^{2+}]_m$ converge to a steady state, which is close to experimental observations.

## Exercises

**9.1.** Solve the proton ejection system (9.11)-(9.16) with a mathematical software and find the proton ejection rate $r_{res} = a_{56}[S_5] - a_{65}[S_6][H^+]_i^6$.

**9.2.** Convert Fig. 9.3 to reaction equations, write the steady state equations for the six states $S_1, S_2, S_3, S_4, S_5, S_6$, solve them with a mathematical software, and find the proton uptake rate $r_{f1f0,h} = a_{21}[S_2] - a_{12}[S_1][H^+]_m^3$ and the ATP synthesis rate $r_{f1f0,atp} = a_{32}[S_3] - a_{23}[S_2][ATP]_m[H_2O]$.

## References

1. Balaban R.S.: Cardiac energy metabolism homeostasis: role of cytosolic calcium. J. Mol. Cell. Cardiol. **34**, 1259-1271 (2002).
2. Bohnensack R.: The role of the adenine nucleotide translocator in oxidative phosphorylation.Atheoretical investigationonthe basis of acomprehensive rate law of the translocator. J. Bioenerg. Biomem. **14**, 4561 (1982).

3. Cortassa S., Aon M.A., Marban E., Winslow R.L., O'Rourke, B.: An integrated model of cardiac mitochondrial energy metabolism and calcium dynamics., Biophysical J. **84**, 2734-2755 (2003).

4. Duchen M.R.: Mitochondria and calcium: from cell signalling to cell death. J. Physiol. **529**, 57-68 (2000).

5. Feldman B., Fedida-Metula S., Nita J., Sekler I., Fishman D.: Coupling of mitochondria to store-operated Ca2+-signaling sustains constitutive activation of protein kinase B/Akt and augments survival of malignant melanoma cells. Cell Calcium **47**, 525-537 (2010).

6. Griffiths E.J., Ocampo C.J., Savage J.S., Rutter G.A., Hansford R.G., Stern M.D., Silverman H.S.: Mitochondrial calcium transporting pathways during hypoxia and reoxygenation in single rat cardiomyocytes. Cardiovasc. Res. **39**, 423-433 (1998).

7. Gunter T.E., Yule D.I., Gunter K.K., Eliseev R.A., Salter J.D.: Calcium and mitochondria. FEBS Lett. **567**, 96-102 (2004).

8. Gunter T.E., Sheu, S.-S.: Characteristics and possible functions of mitochondrial $Ca^{2+}$ transport mechanisms. Biochim. Biophys. Acta. **1787**, 1291-1308 (2009).

9. Magnus G., Keizer J.: Minimal model of $\beta$-cell mitochondrial $Ca^{2+}$ handling. Am. J. Physiol. Cell Physiol. **273**, C717-C733 (1997).

10. Nicholls D.G., Ferguson S.J.: Bioenergetics 2. Academic, San Diego, CA (1992).

11. Pietrobon D., Caplan S.R.: Flow-force relationships for a six-state proton pump model: intrinsic uncoupling, kinetic equivalence of input and output forces, and domain of approximate linearity. Biochemistry **24**, 5764-5776 (1985).

12. Pradhan R.K., Beard D.A., Dash R.K.: A biophysically based mathematical model for the kinetics of mitochondrial $Na^+$-$Ca^{2+}$ antiporter. Biophys. J. **98**, 218-230 (2010).

13. Ryu S.Y., Peixoto P.M., Won J.H., Yule D.I., Kinnally K.W.: Extracellular ATP and P2Y2 receptors mediate intercellular $Ca^{2+}$ waves induced by mechanical stimulation in submandibular gland cells: Role of mitochondrial regulation of store operated $Ca^{2+}$ entry. Cell Calcium **47**, 65-76 (2010).

14. Wingrove D.E., Gunter T.E.: Kinetics of mitochondrial calcium transport. II. A kinetic description of the sodium-dependent calcium efflux mechanism of liver mitochondria and inhibition by ruthenium red and by tetraphenylphosphonium. J. Biol. Chem. **261**, 15166-15171 (1986).

# Control of Phosphoinositide Synthesis

Phosphatidylinositol 4,5-bisphosphate (PIP$_2$) is the predominant (99%) phospho-inositide in mammalian cells [7]. PIP$_2$ is synthesized from phosphatidylinositol-4-phosphate (PIP) by PIP$_2$ synthases while PIP is synthesized from Phosphatidylinositol (PI) by PIP synthases. PIP$_2$ in cells is normally hydrolyzed by phospholipase C (PLC) to generate inositol 1,4,5-trisphosphate (IP$_3$) and diacylglycerol (DAG), which serve as second messengers for intracellular Ca$^{2+}$ mobilization and PKC (protein kinase C) activation, respectively [4, 7]. Thus, PIP$_2$ plays important roles in PLC-mediated cellular processes, such as glucose-stimulated insulin secretion [1], store-operated calcium entry [2, 6], and sterol trafficking [5, 8]. Mathematical models for the process of phosphoinositide synthesis have been established (see, e.g., [3, 7]). In this chapter, we present the model developed by Xu $et$ $al$ [7] because of its simplicity.

## 10.1 PIP Synthesis from PI

The PIP synthesis from PI can be described by

$$PI \xrightarrow{k_{pip}} PIP. \tag{10.1}$$

The PIP synthesis is controlled by both PIP concentration and synthesis time. When the synthesized PIP reaches a basal level, the synthesis should stop to avoid over-production. Moreover, the synthesis rate could decrease in time. Taking this into account, Xu $et$ $al$ [7] expressed the PIP synthesis rate $k_{pip}$ as the sum of the basal rate of PIP synthesis and the stimulation rate of PIP synthesis due to PIP synthase activity:

$$k_{pip} = R_{pipb} + R_{pips}. \tag{10.2}$$

The basal rate of PIP synthesis depended on the PIP concentration and was given by

$$R_{pipb} = \begin{cases} 0.581 k_{pipb} \left( \exp\left( \frac{[PIP]_b - [PIP]}{[PIP]_b} \right) - 1 \right) & \text{if } [PIP] < [PIP]_b, \\ 0 & \text{if } [PIP] \geq [PIP]_b. \end{cases} \tag{10.3}$$

Liu W.: Introduction to Modeling Biological Cellular Control Systems.
DOI 10.1007/978-88-470-2490-8_10, © Springer-Verlag Italia 2012

The stimulation rate of PIP synthesis depended on time and was given by

$$R_{pips} = \begin{cases} 0 & \text{if } t \le \tau_{pip}, \\ k_{pips} \exp((\tau_{pip} - t)/D_{pip}) & \text{if } t > \tau_{pip}. \end{cases} \tag{10.4}$$

## 10.2 PIP$_2$ Synthesis from PIP

The PIP$_2$ synthesis from PIP can be described by

$$PIP \xrightarrow{k_{pip2}} PIP_2. \tag{10.5}$$

Like the PIP synthesis, the PIP$_2$ synthesis is also controlled by both PIP$_2$ concentration and synthesis time. Thus, the PIP$_2$ synthesis rate $k_{pip2}$ was expressed as the sum of the basal rate of PIP$_2$ synthesis and the stimulation rate of PIP$_2$ synthesis due to PIP$_2$ synthase activity [7]:

$$k_{pip2} = R_{pip2b} + R_{pip2s}. \tag{10.6}$$

The basal rate of PIP$_2$ synthesis depended on the PIP$_2$ concentration and was given by

$$R_{pip2b} = \begin{cases} 0.581 k_{pip2b} \left( \exp\left( \frac{[PIP_2]_b - [PIP_2]}{[PIP_2]_b} \right) - 1 \right) & \text{if } [PIP_2] < [PIP_2]_b, \\ 0 & \text{if } [PIP_2] \ge [PIP_2]_b. \end{cases} \tag{10.7}$$

The stimulation rate of PIP$_2$ synthesis depended on time and was given by

$$R_{pip2s} = \begin{cases} 0 & \text{if } t \le \tau_{pip2}, \\ k_{pip2s} \exp((\tau_{pip2} - t)/D_{pip2}) & \text{if } t > \tau_{pip2}. \end{cases} \tag{10.8}$$

## 10.3 PIP$_2$ Hydrolysis

PIP$_2$ in cells is normally hydrolyzed by phospholipase C (PLC) to generate inositol 1,4,5-trisphosphate (IP$_3$) and diacylglycerol (DAG), respectively [4, 7]. The hydrolysis can be described as follows:

$$PLC \underset{b_{plc}}{\overset{k_{plc}}{\rightleftharpoons}} PLC_a, \tag{10.9}$$

$$PIP_2 + PLC_a \xrightarrow{k_{pip2h}} IP_3 + DAG + PLC_a, \tag{10.10}$$

where $PLC_a$ denotes the active form of PLC. The activation rate of PLC depended on time and was constructed by Xu *et al* [7] as follows

$$k_{plc} = \begin{cases} 0 & \text{if } t \le \tau_0, \\ k_{plca} \exp((\tau_0 - t)/D_{plc}) & \text{if } t > \tau_0. \end{cases} \tag{10.11}$$

## 10.4  A Control System for Phosphoinositide Synthesis

Using the law of mass balance (2.1) and the law of mass action, we derive from the reaction equations (10.1), (10.5), (10.9), and (10.10) the control system governing phosphoinositide synthesis:

$$\frac{d[PI]}{dt} = -k_{pip}[PI], \tag{10.12}$$

$$\frac{d[PIP]}{dt} = k_{pip}[PI] - k_{pip2}[PIP], \tag{10.13}$$

$$\frac{d[PIP_2]}{dt} = k_{pip2}[PIP] - k_{pip2h}[PIP_2][PLC_a], \tag{10.14}$$

$$\frac{d[IP_3]}{dt} = k_{pmcyt}k_{mole}k_{pip2h}[PIP_2][PLC_a] - d_{ip3}([IP_3] - [IP3]_b), \tag{10.15}$$

$$\frac{d[DAG]}{dt} = k_{pip2h}[PIP_2][PLC_a], \tag{10.16}$$

$$\frac{d[PLC_a]}{dt} = k_{plc}([PLC]_T - [PLC_a]) - b_{plc}[PLC_a], \tag{10.17}$$

where $[PLC]_T$ is the concentration of total PLC. In the equation (10.15), it is assumed that the rate of IP3 degradation is proportional to its concentration ($d_{ip3}[IP_3]$) and there is a basal IP3 synthesis rate ($d_{ip3}[IP3]_b$).

The unit for the surface molecules, $[PI]$, $[PIP]$, $[PIP_2]$, $[DAG]$, $[PLC_a]$, is molecules/$\mu$m$^2$ and the unit for the cytosolic molecule $[IP_3]$ is $\mu$M. In the equation (10.15), the constants $k_{pmcyt}$ and $k_{mole}$ are used to convert the surface concentration unit, molecules/$\mu$m$^2$, to the cytosol volume concentration unit, $\mu$M. $k_{pmcyt}$ is the ratio of plasma membrane area to cytosol volume. Let $f_{nuc}$ denote the fraction of nucleus volume and define the ratio of plasma membrane area to cell volume by

$$R_{stv} = \frac{\text{area of plasma membrane}}{\text{volume of cell}}.$$

Then the cytosol volume is equal to (volume of cell)$\times(1 - f_{nuc})$ and then $k_{pmcyt}$ can be calculated by

$$k_{pmcyt} = \frac{\text{area of plasma membrane}}{(\text{volume of cell}) \times (1 - f_{nuc})} = \frac{R_{stv}}{1.0 - f_{nuc}}.$$

We note that the unit of $k_{pmcyt}k_{pip2h}[PIP_2][PLC_a]$ is molecules/$\mu$m$^3$/s. Thus the unit conversion ratio $k_{mole}$ is determined by

$$\frac{\text{molecule}}{\mu\text{m}^3} = k_{mole}\mu\text{M}.$$

**Table 10.1.** Parameter values for the system (10.12)-(10.17)

| Parameter | Value | Reference |
|---|---|---|
| $[IP3]_b$ | $0.16\ \mu M$ | [7] |
| $k_{pip2h}$ | $2.4\ (s \cdot molecules/\mu m^2)^{-1}$ | [7] |
| $k_{pipb}$ | $0.0055\ s^{-1}$ | [7] |
| $k_{pip2b}$ | $0.048\ s^{-1}$ | [7] |
| $k_{plca}$ | $5.0 \times 10^{-4}\ s^{-1}$ | [7] |
| $d_{ip3}$ | $0.08\ s^{-1}$ | [7] |
| $k_{mole}$ | $0.0017$ | [7] |
| $b_{plc}$ | $0.1\ s^{-1}$ | [7] |
| $k_{pips}$ | $0.019\ s^{-1}$ | [7] |
| $k_{pip2s}$ | $0.92\ s^{-1}$ | [7] |
| $[PIP_2]_b$ | $4000.0\ molecules/\mu m^2$ | [7] |
| $D_{pip2}$ | $1.0\ s$ | [7] |
| $[PIP]_b$ | $2857.0\ molecules/\mu m^2$ | [7] |
| $D_{pip}$ | $1.0\ s$ | [7] |
| $D_{plc}$ | $1.0\ s$ | [7] |
| $R_{stv}$ | $0.5\ \mu m^{-1}$ | [7] |
| $\tau_0$ | $0.05\ s$ | [7] |
| $\tau_{pip2}$ | $0.05\ s$ | [7] |
| $\tau_{pip}$ | $0.05\ s$ | [7] |
| $f_{nuc}$ | $0.1$ | [7] |
| $[PLC]_T$ | $100\ molecules/\mu m^2$ | [7] |
| $[PI](0)$ | $142857\ molecules/\mu m^2$ | Initial condition [7] |
| $[PIP](0)$ | $2857\ molecules/\mu m^2$ | Initial condition [7] |
| $[PIP_2](0)$ | $4000\ molecules/\mu m^2$ | Initial condition [7] |
| $[DAG](0)$ | $2000\ molecules/\mu m^2$ | Initial condition [7] |
| $[PLC_a](0)$ | $0\ molecules/\mu m^2$ | Initial condition [7] |
| $[IP_3](0)$ | $0.16\ \mu M$ | Initial condition [7] |

It then follows that

$$k_{mole} = \frac{molecule}{\mu m^3 \times \mu M}$$
$$= \frac{molecule}{\mu m^3 \times 10^{-6} \frac{mole}{liter}}$$
$$= \frac{molecule}{\mu m^3 \times 10^{-6} \frac{6.022 \times 10^{23} molecule}{10^{15} \mu m^3}}$$
$$= \frac{1}{6.022 \times 10^2}$$
$$\approx 0.0017.$$

The parameter values for the system (10.12)-(10.17) are listed in Table 10.1.

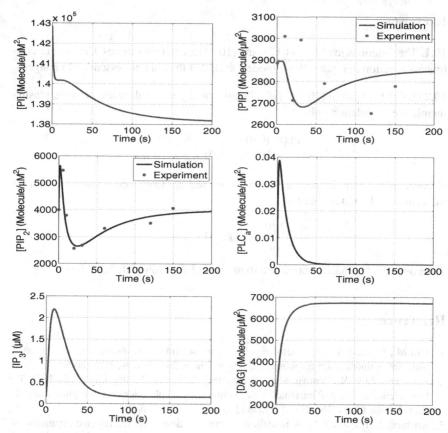

**Fig. 10.1.** A numerical solution of the system (10.12)-(10.17). The data are read from [7] using the software Engauge Digitizer 4.1. The original data in [7] is in percentage of change, which is converted into concentration with the basal concentration of 4000 molecules/$\mu$m$^2$ for $[PIP_2]$ and of 2857 molecules/$\mu$m$^2$ for $[PIP]$ (their initial concentrations) using the formula: concentration = basal $\cdot$(1+percentage of change)

The system (10.12)-(10.17) is solved numerically with MATLAB and the numerical solution is plotted in Fig. 10.1. The solution shows that a plasma membrane stimulation activates PLC and the activated PLC converts PIP$_2$ into IP$_3$ and DAG, resulting in a decrease in PIP$_2$ and an increase in IP$_3$ and DAG. Since PI is converted into PIP, which is, in turn, converted into PIP$_2$, the concentration of PI is decreasing while the concentrations of PIP and PIP$_2$ are increasing to a steady state.

## Exercises

**10.1.** The functions in (10.4), (10.8), and (10.11) are not continuous. Construct a continuous function to replace these models such that they are not essentially changed.

**10.2.** Solve the system (10.12)-(10.17) numerically with following periodic plasma membrane stimulation pulses:

$$k_{plc} = \begin{cases} 0.0005 & \text{if } 10n \leq t \leq 10n+0.01 \text{ s,} \\ 0 & \text{if } 10n+0.01 < t < 10(n+1) \text{ s.} \end{cases}$$

Because of the periodic pulses, $PIP_2$ will be used up. Introduce a control input $u$ in the equation (10.12) as follows

$$\frac{d[PI]}{dt} = -k_{pip}[PI] + u.$$

Then design a feedback controller $u$ to maintain $PIP_2$ around its basal level.

## References

1. Hao M., Bogan J.S.: Cholesterol regulates glucose-stimulated insulin secretion through phosphatidylinositol 4,5-bisphosphate. J. Biol. Chem. **284**, 29489-29498 (2009).
2. Korzeniowski M.K., Popovic M.A., Szentpetery Z., Varnai P., Stojilkovic S.S., Balla T.: Dependence of STIM1/Orai1-mediated calcium entry on plasma membrane phosphoinositides. J. Biol. Chem. **284**, 21027-21035 (2009).
3. Krishnan J., Iglesias P.A.: A modelling framework describing the enzyme regulation of membrane lipids underlying gradient perception in Dictyostelium cells II: InputCoutput analysis. J. Theor. Biol. **235** 504-520 (2005).
4. Potier M., Treba, M.: New developments in the signaling mechanisms of the store-operated calcium entry pathway. Pflugers Arch. **457**, 405-415 (2008).
5. Schulz T.A., Choi M.G., Raychaudhuri S., Mears J.A., Ghirlando R., Hinshaw J. E., Prinz W.A.: Lipid-regulated sterol transfer between closely apposed membranes by oxysterol-binding protein homologues. J. Cell. Biol. **187**, 889-903 (2009).
6. Walsh C.M., Chvanov M., Haynes L.P., Petersen O.H., Tepikin A.V., Burgoyne R.D.: Role of phosphoinositides in STIM1 dynamics and store-operated calcium entry. Biochem. J. **425**, 159-168 (2010).
7. Xu C., Watras J., Loew L.M.: Kinetic analysis of receptor-activated phosphoinositide turnover. J. Cell Biology **161**, 779-791 (2003).
8. Xu J., Dang Y., Ren Y.R., Liu J.O.: Cholesterol trafficking is required for mTOR activation in endothelial cells. Proc. Natl. Acad. Sci. USA **107**, 4764-4769 (2010).

# Appendix A

# Preliminary MATLAB

The name MATLAB stands for matrix laboratory. MATLAB was originally written to provide easy access to matrix software. So MATLAB is a high-level matrix/array language with control flow statements, functions, data structures, input/output, and object-oriented programming features. This brief introduction to MATLAB is adapted from the help documents of MATLAB, MathWorks, Inc., Natick, MA [2].

## A.1 MATLAB Desktop

### A.1.1 Command Window

This is the primary place where we interact with MATLAB. The prompt $>>$ is displayed in this window, and when this window is active, a blinking cursor appears to right of the prompt. This cursor and prompt signify that MATLAB is waiting to perform a mathematical operation. In this window, we can enter variables, mathematical expressions, MATLAB commands, and run them as demonstrated below:

Enter

>> 1+1

and hit the enter key:

ans =
        2

>> x = 1
x =
        1

>> y = 1
y =
        1

Liu W.: Introduction to Modeling Biological Cellular Control Systems.
DOI 10.1007/978-88-470-2490-8_A, © Springer-Verlag Italia 2012

```
>> z = x+y
z =
      2
>>
```

The expression $x = 1$ means that the value 1 is assigned to the variable $x$. Multiple commands are separated by commas or semicolons and they can be placed on one line:

```
>> x = 1, y = 1;
x =
      1
>>
```

Commas tell MATLAB to display results; semicolons suppress printing.

Performing mathematical operations one by one as above is not convenient for solving complex mathematical problems. Also the operations performed as above cannot be saved for later uses. Therefore, we usually write many MATLAB commands in one file, called an M-file, save it, and run it in this window. The next section will tell you how to do so.

## A.1.2  Help Browser

Use the Help browser to search and view documentation and demos for all your Math-Works products. The Help browser is a Web browser integrated into the MATLAB desktop that displays HTML documents.

To open the Help browser, click the help button in the toolbar, or type helpbrowser in the Command Window.

The Help browser consists of two panes, the Help Navigator, which we use to find information, and the display pane, where we view the information.

The following are useful commands for online help:

- help            lists topics on which help is available;
- helpwin         opens the interactive help window;
- helpdesk        opens the web browser based help facility;
- help *topic*    provides help on *topic*;
- lookfor *string*  lists help topics containing *string*.

## A.1.3  Editor / Debugger

Use the Editor/Debugger to create and debug M-files, which are programs we write to run MATLAB functions. The Editor/Debugger provides a graphical user interface for basic text editing, as well as for M-file debugging.

We can use any text editors to create M-files, such as Microsoft Word, and can use preferences (accessible from the desktop File menu) to specify that editor as the default. If we use another editor, we can still use the MATLAB Editor/Debugger for debugging, or we can use debugging functions, such as dbstop, which sets a breakpoint.

# A.2 Creating, Writing, and Saving a MATLAB File

To create a new M-file, click on the icon "New M-File". Then a blank m-file editor window is open. In this window, we will write MATLAB programs that consist of MATLAB statements and commands. The following is an example:

```
%%%%%%%%%%%%%%%%%%%%%%%%%%%%%%%%%%%%%%%%%%%%%%%%
%    Matlab code welcome.m
%%%%%%%%%%%%%%%%%%%%%%%%%%%%%%%%%%%%%%%%%%%%%%%

disp('Hello, Welcome to MATLAB class');

% disp is a built-in function to display a string.

% compute the area of a circle

r = 2;    %radius of a circle.
A = pi*r^2
```

The percent % is used for comments. Any things after it are ignored by MATLAB. If our comments cannot be finished in one line, we can use

```
%{
comments
%}
```

**disp** is a built-in function to display a string. A string is put between two single quotes. To set up the radius, we type

$$r = 2.$$

The semicolon here suppresses the display of variable. To compute the area, we type the MATLAB mathematical expression:

```
A = pi*r^2
```

where pi is built-in constant $\pi$.

To save this file, go to "File" and select "Save As". Use "welcome" (can be any other string) as the file name. The M-file extension is .m. Save it in your directory.

To execute this program, go to the directory where the file is saved. Then type the file name without the extension .m and hit "Enter". You should see

```
Hello, Welcome to MATLAB class
A =
   12.5664
```

After the execution of the file is complete, the variables $r$ and $A$ remain in the workspace.

## A.3 Simple Mathematics

### A.3.1 Variables

MATLAB does not require any type declarations or dimension statements. When MATLAB encounters a new variable name, it automatically creates the variable and allocates the appropriate amount of storage. If the variable already exists, MATLAB changes its contents and, if necessary, allocates new storage. For example,

>> numOfSudents = 25

creates a 1-by-1 matrix named numOfSudents and stores the value 25 in its single element.

Variable names consist of a letter, followed by any number of letters, digits, or underscores. MATLAB uses only the first 31 characters of a variable name. MATLAB is case sensitive; it distinguishes between uppercase and lowercase letters. A and a are not the same variable. To view the matrix assigned to any variable, simply enter the variable name. Variable names can contain up to 63 characters. Any characters beyond the 63rd are ignored.

### A.3.2 Operators

Expressions use familiar arithmetic operators and precedence rules:

- +:    Addition;
- −:    Subtraction;
- ∗:    Multiplication;
- /:    Division;
- ˆ:    Power;
- ( ):  Specify evaluation order.

### A.3.3 Built-in Functions

MATLAB provides a large number of standard elementary mathematical functions, such as abs(x), sqrt(x), exp(x), and sin(x). For example, we can enter

>> sqrt(2)

to compute $\sqrt{2}$. Taking the square root or logarithm of a negative number is not an error; the appropriate complex result is produced automatically. MATLAB also provides many more advanced mathematical functions, including Bessel and gamma functions. Most of these functions accept complex arguments. For a list of the elementary mathematical functions, type

>> help elfun

For a list of more advanced mathematical and matrix functions, type

```
>>help specfun
```

```
>>help elmat
```

Some of the functions, like sqrt and sin, are built in. They are part of the MAT-LAB core so they are very efficient, but the computational details are not readily accessible. Other functions, like gamma and sinh, are implemented in M-files. You can see the code and even modify it if you want.

Some useful constants are built in MATLAB:

- pi:      3.14159265...;
- i:       Imaginary unit, $\sqrt{-1}$;
- j:       Same as i;
- Inf:     Infinity;
- NaN:     Not-a-number.

Infinity is generated by dividing a nonzero value by zero, or by evaluating well defined mathematical expressions that overflow. Not-a-number is generated by trying to evaluate expressions like 0/0 or Inf-Inf that do not have well defined mathematical values.

The function or constant names are not reserved. It is possible to overwrite any of them with a new variable, such as

```
>> pi = 10
```

and then use that value in subsequent calculations. The original constant can be restored with

```
>> clear pi
```

## A.3.4 Mathematical Expressions

Like most other programming languages, MATLAB provides mathematical expressions, but unlike most programming languages, these expressions involve entire matrices. Here are a few examples and the resulting values.

```
>> rho = (1+sqrt(5))/2
rho =
        1.6180

>> a = abs(3+4i)
a =
        5
>>
```

Sometimes expressions or commands are so long that it is convenient to continue them onto additional lines. In MATLAB, statement continuation is denoted by three periods in succession, as shown in the following code:

```
>> b = 10/ ...
2
b =
        5
>>
```

## A.4 Vectors and Matrices

### A.4.1 Generating vectors

A vector can be generated in different ways:

- Enter an explicit list of elements

```
>> v1 = [16 3 2 7]

v1 =

    16      3      2      7

>>
```

- Use the colon operator

```
>> v2 = 1:10

v2 =

     1    2    3    4    5    6    7    8    9   10

>>
```

In this statement, 1 is the starting number, 10 is the ending number, and the step is 1. We can specify a different step by inserting the step between the starting and ending numbers: 1:step:10. For example,

```
>> v3   = 1:0.5:10

v3 =

    Columns 1 through 5

      1.0000    1.5000    2.0000    2.5000    3.0000

    Columns 6 through 10

      3.5000    4.0000    4.5000    5.0000    5.5000
```

```
Columns 11 through 15

   6.0000     6.5000     7.0000     7.5000     8.0000

Columns 16 through 19

   8.5000     9.0000     9.5000    10.0000
```

- Use MATLAB function linspace(a, b, n). This function creates a vector of *n* evenly-spaced elements with the starting element *a* and the ending element *b*. For example,

```
>> v4 = linspace(0,1,10)

v4 =

Columns 1 through 7

   0    0.1111    0.2222    0.3333    0.4444    0.5556    0.6667

Columns 8 through 10

   0.7778    0.8889    1.0000

>>
```

## A.4.2 *Generating matrices*

A Matrix can be generated by entering an explicit list of elements

```
>> A = [16 3 2 1; 3 5 8 9]

A =

   16     3     2     1
    3     5     8     9

>>
```

MATLAB provides four functions that generate basic matrices:

- **zeros**   all zeros;
- **ones**    all ones;
- **rand**    uniformly distributed random elements;
- **randn**   normally distributed random elements.

Here are some examples.

```
>> B = zeros(2,4)

B =

    0    0    0    0
    0    0    0    0

>>

> F = 5*ones(3,3)

F =

    5    5    5
    5    5    5
    5    5    5

>>
```

In the function zeros(m,n), the first argument m specifies the number of rows and the second argument n specifies the number of columns.

### A.4.3 Array Addressing or Indexing

The element in row i and column j of A is accessed by A(i,j).

```
>> A(1,3)

ans =

    2

>>
```

The element in row i and column j of B can be changed by

```
>> B(1,1) = 1

B =

    1    0    0    0
    0    0    0    0

>>
```

## A.4.4 Arithmetic Operations on Arrays

- Transpose: A'

```
>> C = A'

C =

    16    3
     3    5
     2    8
     1    9

>>
```

- Addition: A+B

```
>> A+B

ans =

    17    3    2    1
     3    5    8    9

>>
```

- Multiplication: *

```
>> A'*B

ans =

    16    0    0    0
     3    0    0    0
     2    0    0    0
     1    0    0    0

>>
```

- Element-by-element multiplication: .*

```
>> B.*A

ans =

    16    0    0    0
     0    0    0    0

>>
```

- Element-by-element division: ./

    ```
    >> B./A

    ans =

         0.0625         0         0         0
              0         0         0         0

    >>
    ```

- Operation between scalars and arrays: Addition, subtraction, multiplication, and division by a scalar simply apply the operation to all elements of the array.

    ```
    >> B./A+1

    ans =

         1.0625    1.0000    1.0000    1.0000
         1.0000    1.0000    1.0000    1.0000

    >>
    ```

- Change the first column of $B$ to 1:

    ```
    >> B(:,1) = 10

    B =

         10     0     0     0
         10     0     0     0

    >>
    ```

- Concatenation is the process of joining small matrices to make bigger ones. The pair of square brackets, [ ], is the concatenation operator. For example,

    ```
    >> [A B]

    ans =

         16     3     2     1    10     0     0     0
          3     5     8     9    10     0     0     0

    >>
    ```

- Deleting rows and columns: We can delete rows and columns from a matrix using just a pair of square brackets. To delete the second column of A, use

```
>> A(:,2) = []

A =

    16     2     1
     3     8     9

>>
```

# A.5 M-Files

MATLAB is a powerful programming language as well as an interactive computational environment. Files that contain codes in the MATLAB language are called M-files. There are two kinds of M-files:

- scripts, which do not accept input arguments or return output arguments. They operate on data in the workspace;
- functions, which can accept input arguments and return output arguments. Internal variables are local to the function.

## A.5.1 Scripts

The following is a script M-file named as circle_area.m.

```
%%%%%%%%%%%%%%%%%%%%%%%%%%%%%%%%%%%%%%%%%%%%%%%%
%    Matlab code circle_area.m
%    Compute the area of a circle
%%%%%%%%%%%%%%%%%%%%%%%%%%%%%%%%%%%%%%%%%%%%%%

r = 2;    %radius of a circle

% compute the area

A = pi*r^2
```

Typing the file name

>> circle_area

causes MATLAB to execute the statements in the file and compute the area of the circle. After the execution of the file is complete, the variables r and A remain in the workspace.

When we invoke a script, MATLAB simply executes the commands found in the file. Scripts can operate on existing data in the workspace, or they can create new data on which to operate. Although scripts do not return output arguments, any variables that they create remain in the workspace, to be used in subsequent computations.

## A.5.2 Functions

Functions are M-files that can accept input arguments and return output arguments. The name of the M-file and of the function should be the same. Functions operate on variables within their own workspace, separate from the workspace we access at the MATLAB command prompt. The following is an example of function M-files:

```
function y = square(x)
% Compute the square of x
% Comments from this line are not displayed
% when you use the lookfor
% command or request help on a directory

y = x^2;
```

The first line of a function M-file starts with the keyword function, followed by output arguments, the equal sign, the function name, and input arguments. The input arguments must be enclosed in the parenthesis. In the above example, x is the input argument, y is the output argument, and the function name is "square". The next several lines, up to the first blank or executable line, are comment lines that provide the help text. These lines are printed when we type

>> help square

The first line of the help text is the H1 line, which MATLAB displays when we use the lookfor command or request help on a directory. The rest of the file is the executable MATLAB codes defining the function.

The function can be used in several different ways:

```
>> square(10)
>> x = 10;
>> y=square(x)
```

To view the contents of an M-file, for example, square.m, use type square.m:

```
>> type square.m

function y = square(x)
% Compute the square of x
% Comments from this line are not displayed
% when we use the lookfor
% command or request help
```

```
y = x^2;
>>
```

If we duplicate function names, MATLAB executes the one that occurs first in the search path.

## A.6 Basic Plotting

To learn basic plotting, go to **MATLAB Help** → **Getting Started** → **Graphics**. The following example demonstrates how to use the plot function.

```
%%%%%%%%%%%%%%%%%%%%%%%%%%%%%%%%%%%%%%%%%%%%%%%
%    Matlab code basic_plotting.m
%    Plot the graph of a function
%%%%%%%%%%%%%%%%%%%%%%%%%%%%%%%%%%%%%%%%%%%%%%%

x = 0:pi/100:2*pi;
y = sin(x);
figure(1)
plot(x,y) % produce a graph of y versus x

xlabel('x = 0:2\pi') %label the axes
ylabel('Sine of x')
%add a title
title('Plot of the Sine Function','FontSize',12)

y2 = sin(x-.25);
y3 = sin(x-.5);
figure(2)
%create multiple graphs with a single call to plot
plot(x,y,x,y2,x,y3)
% provide an easy way to
% identify the individual plots.
legend('sin(x)','sin(x-.25)','sin(x-.5)')

x1 = 0:pi/100:2*pi;
x2 = 0:pi/10:2*pi;
figure(3)
% specify color, line styles, markers, and line width
plot(x1,sin(x1),'r:',x2,sin(x2),'r+', 'linewidth', 2)
```

Saving this program as basic_plotting.m and executing it by typing

>> basic_plotting

we obtain three figures, as shown in Fig. A.1.

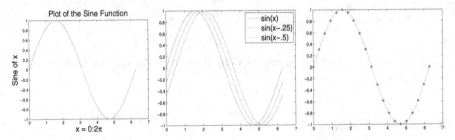

**Fig. A.1.** Figures obtained by executing the program basic_plotting.m

## A.7 Relational Operators

The relational operators are $<$ (less than), $>$ (greater than), $<=$ (less than or equal to), $>=$ (greater than or equal to), $==$ (equal to), and $\sim=$ (not equal to). Relational operators perform element-by-element comparisons between two arrays. They return a logical array of the same size, with elements set to logical 1 (true) where the relation is true, and elements set to logical 0 (false) where it is not. The operators $<$, $>$, $<=$, and $>=$ use only the real part of their operands for the comparison. The operators $==$ and $\sim=$ test real and imaginary parts. The following is an example:

```
>> a = [1 2]

a =

        1      2

>> b = [1, 3]

b =

        1      3

>> tf = a==b

tf =

        1      0
>>
```

Here 1 is returned because $a(1) = b(1) = 1$ and 0 is returned because $a(2) = 2 \neq b(2) = 3$. The following is another example:

```
b =

        1      3
```

```
>> c = [2 3]

c =

     2     3

>> tf = (c>b)

tf =

     1     0
>>
```

Here 1 is returned because $c(1) = 2 > b(1) = 1$ and 0 is returned because $c(2) = 3$ is not greater than $b(2) = 3$.

When an array is compared with a scalar, the scalar is compared with every element of the array. The following is an example:

```
>> a = 1

a =

     1
b =

     1     3

>> tf1 = (a==b)

tf1 =

     1     0

>>
```

# A.8 Flow Control

## A.8.1 *If-Else-End Constructions*

The if statement evaluates a logical expression and executes a group of statements when the expression is *true*. The optional elseif and else keywords provide for the execution of alternate groups of statements. An end keyword, which matches the if, terminates the last group of statements. The groups of statements are delineated by the four keywords–no braces or brackets are involved. The following is an example:

```
>> a = 1

a =

     1

>> if a ==1
       A = zeros(3,3)
    else
       B = ones(4,4)
    end

A =

     0     0     0
     0     0     0
     0     0     0

>> a = 0

a =

     0

>> if a ==1
       A = zeros(3,3)
    else
       B = ones(4,4)
    end

B =

     1     1     1     1
     1     1     1     1
     1     1     1     1
     1     1     1     1

>>
```

## A.8.2 *For Loops*

The for loop repeats a group of statements a fixed, predetermined number of times. A matching end delineates the statements.

The following program uses the for loop to compute the sum: $1 + 2 + \cdots + 9 + 10$.

```
>> sum = 0

sum =

    0

>> for n=1:10
       sum = sum +n;
   end
>> sum

sum =

    55

>>
```

## A.8.3 *While Loops*

The while loop repeats a group of statements an indefinite number of times under control of a logical condition. A matching end delineates the statements. The following program uses the while loop to compute the sum: $1 + 2 + \cdots + 9 + 10$.

```
>> sum = 0

sum =

    0

>> n = 1

n =

    1

>> while n<=10
       sum = sum + n;
       n = n +1;
   end
>> sum

sum =

    55

>>
```

## A.9 Logical Operators

The symbols &, |, and ~ are the logical array operators AND, OR, and NOT. These operators are commonly used in conditional statements, such as if and while, to determine whether or not to execute a particular block of codes. Logical operations return a logical array with elements set to 1 (true) or 0 (false), as appropriate.

"expression 1 & expression 2" represents a logical AND operation between values, arrays, or expressions "expression 1" and "expression 2". In an AND operation, if "expression 1" is true and "expression 2" is true, then the AND of those inputs is true. If either expression is false, the result is false. Here is an example of AND:

```
>> x = 1;
>> and = x==1 & x>0
and =
     1
>> and = x==1 & x<0
and =
     0
```

"expression 1 | expression 2" represents a logical OR operation between values, arrays, or expressions "expression 1" and "expression 2" . In an OR operation, if "expression 1" is true or "expression 2" is true, then the OR of those inputs is true. If both expressions are false, the result is false. Here is an example of OR:

```
>> x = 1;
>> or = x==1 | x>0
or =
     1
>> or = x==1 | x<0
or =
     1
>> or = x~=1 | x<0
or =
     0
```

"~expression" represents a logical NOT operation applied to expression "expression". In a NOT operation, if "expression" is false, then the result of the operation is true. If "expression" is true, the result is false. Here is an example of NOT:

```
>> x = 1;
>> not  = ~(x<0)
not =
     1
>> not = ~(x==1)
not =
     0
```

The expression operands for AND, OR, and NOT are often arrays. When this is the case, The MATLAB software performs the logical operation on each element of the arrays. The output is an array that is the same size as the input array or arrays.

If just one operand is an array and the other a scalar, then the scalar is matched against each element of the array. When the operands include two or more non-scalar arrays, the sizes of those arrays must be equal. Here are examples:

```
>> A = [1, 2, 0];
>> B = [0, 3, 4];
>> and = A&B
and =
       0     1     0
>> or = A|B
or =
       1     1     1
>> and = 2&A
and =
       1     1     0
>> C = [3, 4];
>> and = A&C
??? Error using ==> and
Inputs must have the same size.
```

## A.10 Solving Symbolic Equations

Using Symbolic Math Toolbox, we can solve different types of symbolic equations. The following is such an example:

```
%To declare variables as symbolic objects,
%use the syms command:
syms x y

%Equations to be solved
eq1 = 'a*x+b*y = f';
eq2 = 'c*x+d*y = g';

%Use the method "solve" to solve the
%system of equations for x and y
solution = solve(eq1, eq2, 'x', 'y')

%To see the solution, enter
x = solution.x
y = solution.y
```

After running the program, we obtain the symbolic solution

```
x =

   -(b*g - d*f)/(a*d - b*c)



y =

   (a*g - c*f)/(a*d - b*c)
```

## A.11  Solving Ordinary Differential Equations

In general, a system of nonlinear ordinary differential equations cannot be solved analytically. For example, the following system

$$\frac{dS}{dt} = -k_1 ES + k_{-1}(E_0 - E) + u, \qquad (A.1)$$

$$\frac{dE}{dt} = -k_1' ES + (k_{-1} + k_2)(E_0 - E) \qquad (A.2)$$

cannot be solved. Thus, we need to solve it numerically.

MATLAB has a number of built-in solvers to numerically solve ordinary differential equations. One of them is the **ode15s** solver. To use it, we first write a MATLAB function file to implement the system (A.1)-(A.2) as follows:

```
function f = reaction_kinetics(t, x)

global k1 k2 k_1 u E0;

f = zeros(length(x), 1);
S = x(1);
E = x(2);

f(1) = -k1*E*S +k_1*(E0-E) +u;
f(2) = -k1*E*S+(k_1+k2)*(E0-E);
```

We then write a program to solve the system:

```
clear all;
close all;

global k1 k2 k_1 u E0;
%specify the parameters
k1 = 0.5;
k2 = 10;
k_1= 0.1;
```

```
u = 0.2;
E0 = 3;

%specify the initial conditions
x_0=zeros(1,2);
x_0(1) = 0.0;
x_0(2) = 0.1;

%use ode15s to solve the system
tspan =0:0.02:5;
[t,y] = ode15s(@reaction_kinetics,tspan, x_0);

%plot the solution S
figure(1)
plot(t, y(:, 1),'k-', 'linewidth', 3);
grid on
box on
axis([0, max(t) 0 max(y(:, 1))*1.1]);
xlabel('t', 'fontsize', 20);
ylabel('S',  'fontsize', 20);
set(gca,'fontsize', 20);

%plot the solution E
figure(2)
plot(t, y(:, 2),'k-', 'linewidth', 3);
axis([0, max(t) 0 3.1]);
grid on
box on
xlabel('t', 'fontsize', 20);
ylabel('E',  'fontsize', 20);
set(gca,'fontsize', 20);

%plot the  phase portrait
figure(3)
plot(y(:, 2),y(:, 1),'k-', 'linewidth', 3);
grid on
box on
xlabel('E', 'fontsize', 20);
ylabel('S',  'fontsize', 20);
set(gca,'fontsize', 20);
```

Numerical solutions of the system (A.1)-(A.2) with $k_1 = 0.5, k_2 = 10, k_{-1} = 0.1, u = 0.2, E_0 = 3$ and the initial conditions $S(0) = 0.0, E(0) = 0.1$ are plotted in Fig. A.2. The plot of $S$ against $E$ in the right figure of Fig. A.2 is called a phase portrait and the $E - S$ plane is called the phase plane.

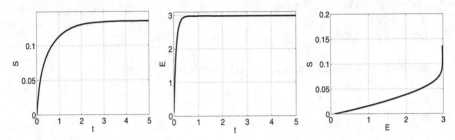

**Fig. A.2.** Numerical solutions of the system (A.1)-(A.2) with $k_1 = 0.5, k_2 = 10, k_{-1} = 0.1, u = 0.2, E_0 = 3$ and the initial conditions $S(0) = 0.0, E(0) = 0.1$

We can also combine the function file and the program file into one function file as follows:

```
function no_ouput = enzyme_reaction_simulation()
%Solve the enzymatic reaction system

clear all;
close all;

global k1 k2 k_1 u E0;

%specify the parameters
k1 = 0.5;
k2 = 10;
k_1= 0.1;
u = 0.2;
E0 = 3;

%specify the initial conditions
x_0=zeros(1,2);
x_0(1) = 0.0;
x_0(2) = 0.1;

%use ode15s to solve the system
tspan =0:0.02:5;
[t,y] = ode15s(@reaction_kinetics,tspan, x_0 );

%plot the solution S
figure(1)
plot(t, y(:, 1),'k-', 'linewidth', 3); % S
grid on
box on
axis([0, max(t) 0 max(y(:, 1))*1.1]);
xlabel('t', 'fontsize', 20);
```

```
ylabel('S', 'fontsize', 20);
set(gca,'fontsize', 20);


%plot the solution E
figure(2)
plot(t, y(:, 2),'k-', 'linewidth', 3); % E
axis([0, max(t) 0 3.1]);
grid on
box on
xlabel('t', 'fontsize', 20);
ylabel('E', 'fontsize', 20);
set(gca,'fontsize', 20);


%plot the  phase portrait
figure(3)
plot(y(:, 2),y(:, 1),'k-', 'linewidth', 3);
grid on
box on
xlabel('E', 'fontsize', 20);
ylabel('S', 'fontsize', 20);
set(gca,'fontsize', 20);


function f = reaction_kinetics(t,x)

global k1 k2 k_1 u E0;

f = zeros(length(x), 1);
S = x(1);
E = x(2);

f(1) = -k1*E*S +k_1*(E0-E) +u;
f(2) = -k1*E*S+(k_1+k2)*(E0-E);
```

## A.12 Data Fitting

Let data be as follows:

$x$ : $-109$ $-100$ $-88$ $-76$ $-63$ $-51$ $-38$ $-32$ $-26$ $-19$ $-10$ $-6$
$y$ : $0.915$ $0.866$ $0.748$ $0.61$ $0.524$ $0.419$ $0.31$ $0.241$ $0.192$ $0.15$ $0.095$ $0.085$

$$(A.3)$$

We want to fit the following function

$$y = \frac{a_1(a_2 + x)}{\exp\left(\frac{a_2 + x}{a_2}\right) - a_3}$$

$$(A.4)$$

into the data.

MATLAB has a curve fitting toolbox for fitting a function into data. The curve fitting toolbox software allows us to work in two different environments:

- an interactive environment with graphical user interfaces;
- a programmatic environment that allows us to write object-oriented MATLAB codes using curve fitting methods.

To work in the interactive environment, enter the following command:

```
>> cftool
```

A graphical user interface (GUI) named "Curve Fitting Tool" is opened.

Before we can import data into Curve Fitting Tool, the data variables must exist in the MATLAB workspace. Thus we enter:

```
>> x = [-109 -100 -88 -76 -63 -51 -38 -32 -26 -19 ...
        -10 -6];
>> y = [0.915 0.866 0.748 0.61 0.524 0.419 0.31 ...
        0.241 0.192 0.15 0.095 0.085];
```

We can now import the data into Curve Fitting Tool with the Data GUI. Open the Data GUI by clicking the **Data** button on Curve Fitting Tool. Then load $x$ and $y$ into Curve Fitting Tool as follows:

1. Select the variable names $x$ and $y$ from the **X Data** and **Y Data** lists. The data is displayed in the **Preview** window. If we do not import weights, then they are assumed to be 1 for all data points.
2. Specify the name of the data set.
3. Click the **Create data set** button to complete the data import process.
4. Click **Close**.

We then fit the data with the fitting GUI. We open the fitting GUI by clicking the **Fitting** button on Curve Fitting Tool. Then the data fitting procedure follows these steps:

1. From the **Fit Editor**, click **New Fit**. The new fit always defaults to a linear polynomial fit type. Use **New Fit** at the beginning of our curve fitting session, and when we are exploring different fit types for a given data set.
2. To use a customized equation for the fit, select **Custom Equations** from **Type of fit**. Edit the **Fit name** to myFunction.
3. Click the **New** button to open the **New Custom Equation** GUI. Select **General Equations**, type the function (A.4) in the **Equation** field, set **Lower** to 0, edit **Equation name**, and click **OK**.
4. Click the **Apply** button or select the **Immediate apply** check box. The function, fitted coefficients, and goodness of fit statistics are displayed in the **Results** area of the fitting GUI. Our new fit is plotted in Curve Fitting Tool. To export the figure, select **Print to Figure** from the **File** list.
5. To improve fitting, click **Fit options** from **Fit Editor**. Then a Fit Options GUI is opened. We can adjust the **StartPoint** field to improve our fitting.

Data fitting can be also done by writing programs that combine curve fitting functions with MATLAB functions and functions from other toolboxes. The following is such a program:

```
%Use the method "fit" to fit a function into data

%Data
x = [-109 -100 -88 -76 -63 -51 -38 -32 -26 -19 ...
       -10 -6]';
y = [0.915 0.866 0.748 0.61 0.524 0.419 0.31 ...
      0.241 0.192 0.15 0.095 0.085]';

%Use the method "fitoptions" to set up fit options
s = fitoptions('Method','NonlinearLeastSquares',...
               'Lower',[0,0, 0],...
               'Upper',[Inf,Inf,Inf],...
               'Startpoint',[0.005 9 1]);

%Use the method "fittype" to create a fit type
f = fittype('a1*(a2+x)/(exp((a2+x)/a2)-a3)', ...
            'options',s);

%Use the method "fit" to fit a function into data
[myfit,gof] = fit(x,y,f)

%Plot the fitted curve and the data
plot(myfit, 'b',  x,y,'ro')
```

In addition, if the curve fitting toolbox is not available, we can also write a simple program to do so by using the built-in function "fminsearch". The following is the program to do the fitting:

```
function output  = data_fitting()
% Fit a function into data

global x_data y_data;
% Data
x_data = [-109 -100 -88 -76 -63 -51 -38 -32 -26 -19
          ...   -10 -6 ];
y_data = [0.915 0.866 0.748 0.61 0.524 0.419 0.31 ...
           0.241 0.192 0.15 0.095 0.085 ];

% use the built-in function fminsearch to find
% the parameter values where the error between
% the data and the fitting function is minimized
a = fminsearch(@error, [0.01, 10, 1], ...
    optimset('TolX',1e-8, 'MaxFunEvals', 10000000));
```

```
output = a;
% Fitting function
x = -120:20;
y = a(1)*(a(2)+x)./(exp((a(2)+x)/a(2))-a(3));

figure(1)
plot(x, y, 'b-', x_data, y_data, 'r*', ...
        'linewidth', 3)
grid on
box on
xlabel('x',  'fontsize', 20);
ylabel('y', 'fontsize', 20);
legend('Fitting', 'Data');
set(gca,'fontsize', 20);

% calculate the error between the data and
% the fitting function
function f = error(a)
global x_data y_data;

y = a(1)*(a(2)+x_data)...
      ./(exp((a(2)+x_data)/a(2))-a(3));
f = (y-y_data)*(y-y_data)';
```

Fig. A.3 shows that the function is fitted well into the data with $a_1 = 0.0096$, $a_2 = 10.35$, and $a_3 = 1$.

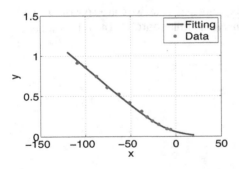

**Fig. A.3.** Fitting of the function (A.4) into the data (A.3)

## A.13 Parameter Estimation

In biological modeling problems, we often need to estimate parameters in differential equations such that the solutions of the differential equations agree with data. Consider the system of differential equations (A.1) and (A.2). Assume that we have collected the time course data of $S$ as follows:

$$t: 0 \quad 2 \quad 4 \quad 6 \quad 8 \quad 10 \ 12 \ 14 \ 16 \ 18 \ 20 \ 22 \ 24 \ 26 \ 28 \ 30 \ 32 \ 34 \ 36 \ 38 \ 40$$
$$S: 100 \ 98 \ 95 \ 91 \ 86 \ 80 \ 72 \ 62 \ 50 \ 46 \ 44 \ 38 \ 32 \ 28 \ 23 \ 19 \ 15 \ 10 \ 8 \quad 7 \quad 6$$

We need to estimate the parameters $k_1, k_{-1}$, and $k_2$ such that the solution $S$ of the system (A.1)-(A.2) is as close to the data as possible.

Functions of parameter estimation were implemented in the Systems Biology toolbox [1] and the SimBiology toolbox of MATLAB. The following program demonstrates how to use the function "SBparameterestimation" implemented in the Systems Biology toolbox to estimate parameters:

```
function no_output = parameter_estimation()
%%%%%%%%%%%%%%
% parameter estimation example
%%%%%%%%%%%%%%

model = SBmodel('enzymeReaction.txt');
data = SBdata('enzymeReactionMeasurements.xls');
parameters = [];
parameters.names = {'k1','k_1','k2'};
parameters.initialValues = [0.1 0.01 2];
parameters.signs = [1 1 1];
options = [];
options.optimizer ='fminsearch'; %'simannealingSB';
new_parameters = SBparameterestimation(model, ...
                parameters, data, options)

global k1 k2 k_1    E0;

%use the estimated parameters
k1 = new_parameters.parameterValues(1)
k_1= new_parameters.parameterValues(2)
k2 = new_parameters.parameterValues(3)
E0 = 3;

%specify the initial conditions
x_0=zeros(1,2);
x_0(1) = 100;
x_0(2) = 0;
```

```
%use ode15s to solve the system
tspan =0:0.02:40;
[t,y] = ode15s(@reaction_kinetics,tspan, x_0 );

%read the data from a file
read_data = importdata( ...
    'enzymeReactionMeasurements.xls',' ', 6);
x_data = read_data.data.Sheet1(4:end,1);
y_data = read_data.data.Sheet1(4:end,2);

%plot the solution S
figure(2)
plot(t, y(:, 1),'k-', x_data, y_data, ...
    'r o','linewidth', 3); % S
grid on
box on
axis([0, max(t) 0 max(y(:, 1))*1.1]);
xlabel('t', 'fontsize', 20);
ylabel('S',  'fontsize', 20);
legend('Solution','Data');
set(gca,'fontsize', 20);

function f = reaction_kinetics(t,x)

global k1 k2 k_1    E0;

f = zeros(length(x), 1);
S = x(1);
E = x(2);

f(1) = -k1*E*S +k_1*(E0-E)   ;
f(2) = -k1*E*S+(k_1+k2)*(E0-E);
```

To use this function, a text file for the differential equations needs to be created in the following format and saved as "enzymeReactoin.txt" with the extension ".txt":

```
********** MODEL NAME
Enzyme reaction Model

********** MODEL NOTES
Developed by Weijiu Liu in 2010

********** MODEL STATES
d/dt(S) = -k1*S*E+k_1*(3-E)
d/dt(E) =  -k1*S*E+(k_1+k2)*(3-E)
```

**Table A.1.** Make-up Data

| Name | enzyme reaction |
|---|---|
| Notes | make-up data |
| Components | S |
| StimulusFlag | 0 |
| NoiseOffset | 0 |
| NoiseVariance | 0 |
| 0 | 100 |
| 2 | 98 |
| 4 | 95 |
| 6 | 91 |
| 8 | 86 |
| 10 | 80 |
| 12 | 72 |
| 14 | 62 |
| 16 | 50 |
| 18 | 46 |
| 20 | 44 |
| 22 | 38 |
| 24 | 32 |
| 26 | 28 |
| 28 | 23 |
| 30 | 19 |
| 32 | 15 |
| 34 | 10 |
| 36 | 8 |
| 38 | 7 |
| 40 | 6 |

```
S(0) = 100
E(0) = 3


********** MODEL PARAMETERS

k1 = 0.5
k_1 = 0.02
k2 = 5


********** MODEL VARIABLES


********** MODEL REACTIONS
```

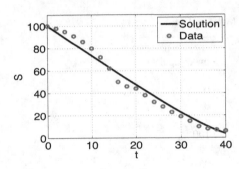

**Fig. A.4.** Comparison between the solution and the make-up data

```
********** MODEL FUNCTIONS




********** MODEL EVENTS




********** MODEL MATLAB FUNCTIONS
```

An Excel file for the data needs to be created in the format as shown in Table A.1 and saved as "enzymeReactionMeasurements.xls".

The functions "SBmodel" and "SBdata" are used to read the text file and the Excel data file, respectively. The function "SBparameterestimation" gives the estimation for the parameters as follows:

$$k_1 = 0.2144, \ k_{-1} = 0.0058, \ k_2 = 0.9344.$$

Fig. A.4 shows that the solution of the system (A.1)-(A.2) with these estimated parameter values is close to the make-up data.

## Exercises

**A.1.** Write the following expressions in MATLAB:

1. $\frac{3 \times 5 + 10^2}{2 \times 3}$.
2. $4^{2 \times 5}$.
3. $1.5 \times 10^{-4} + 2.5 \times \sqrt{2}$ (the square root of $x$ in MATLAB is sqrt(x)).
4. $\ln(e^2) - \log_{10}(10)$ ($\ln(x)$ in MATLAB is log(x), $\log_{10}(x)$ is log10(x), $e^x$ is exp(x)).
5. $\sin(\pi/4) + 2\frac{\cos(1)}{\tan(\pi/4)}$ ($\pi$ is pi in MATLAB).

**A.2.** Use MATLAB to evaluate the following expressions:

1. $\frac{1}{2\times 3}$.
2. $2^{2\times 3}$.
3. $1.5 \times 10^{-4} + 2.5 \times \sqrt{2}$.
4. $e^3 + \ln(e^3) - \log_{10}(10)$.
5. $\sin(\pi/3) + 2\frac{\cos(1)}{\tan(\pi/3)}$ ($\pi$ is pi in MATLAB).

**A.3.** Create the vector $v = [2\ 3\ -1]$ and then perform the following operations:

1. Add 1 to each element of the vector.
2. Multiply each element of the vector by 3.
3. Square each element of the vector.
4. Find the length of the vector.

**A.4.** Create a $3 \times 3$ matrix $A$ with all elements of zero and a $3 \times 3$ matrix $B$ with all elements of 10, and then perform the following operations:

1. Change the value of the element in row 2 and column 3 of $A$ to 5.
2. Change the values of all elements in row 3 of $A$ to 10.
3. Compute $A + B$, $A - 2B$, and $AB$.

**A.5.** Create a $2 \times 3$ matrix $A$ with all elements of zero and a $2 \times 3$ matrix $B$ with all elements of 10. Perform the following arithmetic operations:

1. Change the value of the element of $A$ in row 2 and column 3 to 100.
2. Change the values of all elements of $A$ in column 1 to 2.
3. Addition of $A$ and $B$.
4. Element-by-element multiplication of $A$ and $B$.
5. Transpose of $A$.
6. Matrix multiplication between the transpose of $A$ and $B$.

**A.6.** It has been suggested that the population of the United States may be modeled by the formula

$$P(t) = \frac{197273000}{1 + e^{-0.03134(t-1913.25)}}$$

where $t$ is the date in years.

1. Write a function M-file for the formula $P(t)$, where $t$ is allowed to be a vector. Name your function M-file as "population.m"
2. Write a script M-file to do the following:
   a) Call your function "population.m" to compute and display the population every ten years from 1790 to 2010.
   b) Plot your population function by calling it. Label the x-axis by "t" and y-axis by "population", and add a title and legend to the figure.

**A.7.** If $C$ and $F$ are Celsius and Fahrenheit temperatures, the formula

$$F = 9C/5 + 32$$

converts from Celsius to Fahrenheit. Write a MATLAB code to convert a temperature of $37°$ (normal human temperature) to Fahrenheit ($98.6°$).

**A.8.** Write a function M-file to compute $\sin(\pi t) + \cos(\pi t)$.

**A.9.** Write a function M-file that finds roots of

$$ax^2 + bx + c = 0.$$

Your function should take three input arguments $a$, $b$ and $c$, which are allowed to be vectors. You should test the sign of $b^2 - 4ac$. Then write a program to test your function with $a = 3$, $b = 1$, $c = 1$ and $a = [3\ 1\ 2]$, $b = [1\ -4\ 9]$, $c = [1\ 3\ -5]$.

**A.10.** Write MATLAB codes to compute the factorial $n!$.

**A.11.** Use Symbolic Math Toolbox to solve the following system for $E$ and $S$

$$-k_1 ES + k_{-1}(E_0 - E) + u = 0,$$
$$-k_1 ES + (k_{-1} + k_2)(E_0 - E) = 0.$$

**A.12.** Write MATLAB programs to numerically solve the Lorenz system

$$\frac{dx}{dt} = -\sigma x + \sigma y,$$
$$\frac{dy}{dt} = rx - y - xz,$$
$$\frac{dz}{dt} = xy - bz,$$
$$x(0) = 1,\ y(0) = 2,\ z(0) = 3,$$

where $\sigma = 2$, $r = 5$, and $b = 10$.

**A.13.** Fit the following function

$$y = b_1 \exp\left(\frac{V}{b_2}\right)$$

into the data from [3]:

$V : -109\ -100\ -88\ -76\ -63\ -51\ -38\ -32\ -26\ -19\ -10\ -6$
$y : \ 0.037\ 0.043\ 0.052\ 0.057\ 0.064\ 0.069\ 0.075\ 0.071\ 0.072\ 0.072\ 0.096\ 0.105$

## References

1. Systems Biology Toolbox for MATLAB. http://www.sbtoolbox.org/ (2011). Accessed 18 June 2011.
2. MATLAB, MathWorks, Inc., Natick, MA.
3. Hodgkin A.L., Huxley A.F.: A quantitativfe description of membrane current and its application to conduction and excitation in nerve. J. Physicol. **117**, 500-544 (1952).

# Appendix B

## Units, Physical Constants and Formulas

We collect frequently used units, physical constants and formulas in this appendix. They can be found in a general chemistry textbook, a physical chemistry textbook, or a mathematical biology textbook, such as [1, 2, 3].

## B.1 Physical Formulas

- 1 atm = 760 mmHg.
- $R = kN_A$.
- $F = qN_A$.
- $pH = -\log_{10}[H^+]$ with $[H^+]$ in M.
- $273.15 \, K = 0°C$.
- $T_{Kelvin} = T_{Celsius} + 273.15$.
- $T_{Farenheit} = \frac{9}{5}T_{Celsius} + 32$.
- Capacitance of the cell membrane $\approx 1 \, \mu F \cdot cm^{-2}$.
- 1 Liter = $10^{-3} \, m^3$.
- Dielectric constant for water = $80.4 \, \varepsilon_0$.
- Free energy of ideal dilute solutions:

$$G = G_0 + RT \ln[X], \tag{B.1}$$

where $G_0$ is the standard Gibbs free energy defined to be the free energy at a concentration of 1 mole per liter and the concentration $[X]$ is in unit of moles per liter.

- Relation between the cell potential $E$ and the free energy change $\Delta G$ for a reaction is given by

$$\Delta G = -nFE, \tag{B.2}$$

where $F$ is Faraday's constant and $n$ is the moles of electrons moved from the anode to the cathode.

Liu W.: Introduction to Modeling Biological Cellular Control Systems.
DOI 10.1007/978-88-470-2490-8_B, © Springer-Verlag Italia 2012

- Mole of ions:

$$1 \text{ mole of ions } = zF \text{ Coulombs,} \tag{B.3}$$

where $z$ is the charge of the ion.
- Voltage:

$$\text{Volt} = \frac{1 \text{ joule(J)}}{1 \text{ Coulomb(C)}}. \tag{B.4}$$

- Ampere: $1 \text{ A} = 1 \text{ C/s}$.
- The capacitance is given by

$$C = \frac{Q}{V}. \tag{B.5}$$

The SI unit of capacitance is the farad; 1 farad is 1 coulomb per volt.
- Electrical conductance is a measure of how easily electricity flows along a certain path through an electrical element. The SI derived unit of conductance is the siemens (also called the mho, because it is the reciprocal of electrical resistance, measured in ohms). Conductance is related to resistance by:

$$G = \frac{1}{R} = \frac{I}{V}, \tag{B.6}$$

where $R$ is the electrical resistance, $I$ is the electric current, and $V$ is the voltage.
- Energy Joule:

$$J = N \cdot m = C \cdot V. \tag{B.7}$$

- Ideal gas law:

$$pV = nRT, \tag{B.8}$$

where $p$ is the absolute pressure, $V$ is the volume of gas, $n$ is the number of moles of gas, and $T$ is absolute temperature.

## B.2 Units, Unit Scale Factors and Physical Constants

Units, unit scale factors and physical constants are listed in Tables B.1, B.2, and B.3, respectively.

**Table B.1.** Units

| Quantity | Unit | Symbol | Dimension |
|----------|------|--------|-----------|
| Amount | mole | mol | |
| Electric charge | coulomb | C | |
| Mass | gram | g | |
| Temperature | kelvin | K | |
| Time | second | s | |
| Length | meter | m | |
| Volume | liter | L | |
| Force | newton | N | $kg \cdot m \cdot s^{-2}$ |
| Energy | joule | J | $N \cdot m$ |
| Pressure | pascal | Pa | $N \cdot m^{-2}$ |
| Electric current | ampere | A | $C \cdot s^{-1}$ |
| Potential difference | volt | V | $N \cdot m \cdot C^{-1}$ |
| Capacitance | farad | F | $A \cdot s \cdot V^{-1}$ |
| Resistance | ohm | $\Omega$ | $V \cdot A^{-1}$ |
| Conductance | siemen | S | $A \cdot V^{-1}$ |
| Concentration | molar | M | $mol \cdot L^{-1}$ |
| Atomic mass | dalton | D | $g \cdot N_A^{-1}$ |

**Table B.2.** Unit scale factors

| Name | Prefix | Scale Factor |
|------|--------|--------------|
| femto | f | $\times 10^{-15}$ |
| pico | p | $\times 10^{-12}$ |
| nano | n | $\times 10^{-9}$ |
| micro | $\mu$ | $\times 10^{-6}$ |
| milli | m | $\times 10^{-3}$ |
| centi | c | $\times 10^{-2}$ |
| deci | d | $\times 10^{-1}$ |
| kilo | k | $\times 10^{3}$ |
| mega | M | $\times 10^{6}$ |
| giga | G | $\times 10^{9}$ |

**Table B.3.** Physical constants

| Physical Constants | Symbol | Value |
|---|---|---|
| Boltzmann's constant | $k$ | $1.381 \times 10^{-23}$ J·K$^{-1}$ |
| Planck's constant | $h$ | $6.626 \times 10^{-34}$ J · s |
| Avogadro's number | $N_A$ | $6.02257 \times 10^{23}$ mol$^{-1}$ |
| Unit charge | $q$ | $1.6 \times 10^{-19}$ C |
| Gravitational acceleration | $g$ | $9.78049$ m·s$^{-2}$ |
| Faraday's constant | $F$ | $9.649 \times 10^4$ C·mol$^{-1}$ |
| Universal gas constant | $R$ | $8.315$ J·mol$^{-1}$·K$^{-1}$ |
| Permittivity of free space | $\varepsilon_0$ | $8.854 \times 10^{-12}$ F·m$^{-1}$ |
| Atmosphere | atm | $1.01325 \times 10^5$ N·m$^{-2}$ |

# References

1. Atkins P.W., Beran J.A.: General Chemistry, Second Edition. Scientific American Books, New York (1992).
2. Ball D.W.: Physical Chemistry. Thomson Brooke/Cole, California (2003).
3. Keener J., Sneryd J.: Mathematical Physiology I: Cellular Physiology, II: Systems Physiology, 2nd ed. Springer, New York (2009).

# Index

$\omega$-limit point, 52
$\omega$-limit set, 52

Action potential, 174
Activation energy, 28
Adenine nucleotide translocator (ANT), 214
Algebraic multiplicity, 48
Arrhenius' formula, 28
Artificial pancreas, 69
Asymptotic tracking, 57
Asymptotically stable, 44
ATP-sensitive $K^+$ channel model, 133
ATP-sensitive potassium channel, 132
Autonomous, 44

Bendixson criterion, 88
Blood glucose, 69
Blood glucose control system, 2

Calcineurin, 95, 102
Calcium
    calcium-activated potassium channel, 131
    dynamics in ER, 101
    dynamics in the Golgi apparatus, 100
    dynamics in the vacuole, 99
    shock, 111
    uptake, 96
    uptake velocity, 97
Calmodulin, 95, 102
Capacitance, 160
Capacitative calcium entry, 97
Capacitative calcium entry (CCE), 190
Cell cycle model, 107

Cell cycle-dependent oscillation of calcium, 113
Characteristic equation, 41
Chemical potential, 27
COD1, 100
Cofactor, 40
Competitive inhibitor, 20
Compounding reaction rate, 98
Conductance, 127
Control system of blood glucose, 70
Control variables, 43
Controllability, 55
Cooperative effects, 4
Crz1p, 95, 102, 103

Depolarization, 127
Determinant, 40
Disturbance rejection, 57
Duality between controllability and observability, 57
Dynamic output feedback controller, 58
Dynamical state feedback controller, 58

Eigenvalue, 40
Eigenvector, 41
Electron transport chain, 208
Elementary reactions, 13
Endoplasmic reticulum (ER), 100
Enzyme, 11
    kinetics, 11
Equation of calcium in the cytosol, 101
Equilibrium
    constant, 13

point, 44
voltage, 127
Exogenous glucose input, 81
Exponentially stable, 44

Feedback controller
    for Pmc1p and Pmr1p, 107
    for Vcx1p, 103
    for Yvc1p, 107
    of calcium uptake, 103
Feedback gain matrix, 58
Feedback matrix, 58
Fick's law, 124
Frequency factor, 28

Gas constant, 27
Glucagon, 2, 69
    infusion rate, 71
    receptor, 72
    signaling pathway, 72
    transition, 71
Glucose, 2, 69
    transporter 2 (GLUT2), 2, 69
    transporter 3 (GLUT3), 69
    transporter 4 (GLUT4), 2, 69
GLUT4 transport, 78
Glycogen phosphorylase, 2, 16, 69
Glycogen synthase, 2, 69
Goldman-Hodgkin-Katz (GHK) current
    equation, 126
Golgi apparatus, 99

Half-life, 12
Hill
    equation, 26
    exponent, 4
    function, 4
    plot, 26
Hodgkin-Huxley model, 163

Insulin, 2, 69
    infusion rate, 71
    receptor, 74
    receptor substrate-1 (IRS-1), 75
    signaling pathway, 73
    transition, 71
Inter-membrane space, 207
Intracellular calcium control system, 2

Invariance principle, 52
$IP_3$ receptor, 144

Jacobian matrix, 51

Kalman observability matrix, 56

LaSalle's invariance principle, 52
Law of mass action, 4, 12
Law of mass balance, 3, 11
Linear
    current-voltage equation, 127
    dependence, 42
    voltage potential model, 161
Lineweaver-Burk plot, 16
Luenberger observe, 60
Lyapunov function, 46
Lyapunov's indirect method, 51
Lyapunov's Stability Theorem, 45

M (mitosis) phase promoting factor (MPF),
    107
M-file, 230
Matrix, 37, 207
    multiplication, 39
    sum, 38
Maximal velocity, 15
Membrane potential model, 160
Metabolic pathway, 19
Michaelis-Menten constant, 4, 15
Michaelis-Menten equation, 4, 15
Mitochondrion, 207
Model of protein X, 108

$Na^+/Ca^{2+}$ exchangers (NCX), 217
Negative cooperativity, 24
Nernst
    equation, 5, 125
    potential, 5, 125
Nernst-Planck equation, 125
NFAT model, 103
Non-cooperativity, 24, 25
Nonautonomous, 44
Noncompetitive inhibition, 23
Noncompetitive inhibitor, 23
Nonlinear voltage potential model, 161
Nuclear factor of activated T-cells, 103

Observable, 56
Observation system, 56

Observer gain matrix, 60
Open probability of IP$_3$ receptor, 148
Order, 12
Order of reaction, 13
Oscillation
    of glucose, 86
    of insulin, 86
Output, 43, 55, 58
    feedback control, 60
    feedback controller, 58
    injection matrix, 60

Pancreas, 2, 69
Phase
    plane, 249
    portrait, 249
Phosphatase homologous to tensin (PTEN), 77
Phosphatidylinositol 3,4,5-trisphosphate (PI(3,4,5)P$_3$), 77
Phosphatidylinositol 3,4-bisphosphate (PI(3,4)P$_2$), 77
Phosphatidylinositol 4,5-bisphosphate (PI(4,5)P$_2$), 77
PI 3-kinase, 75
Planck's equation, 124
Plasma membrane calcium ATPase, 156
PMC1, 95
Pmc1p, 95, 98
PMR1, 95
Pmr1p, 95, 99
Poisson's equation, 126
Positive cooperativity, 24, 26
Positively invariant, 44
Postreceptor signaling, 75
Potassium channel, 127
Power of a matrix, 39
Protein kinase B (PKB), 77
Protein kinase C (PKC), 77
Protein tyrosine phosphatases (PTP), 74

Quasi-steady state approximation, 15

Rank of matrix, 42
Rapid oscillation, 86
Rate constants, 12
Reaction
    rate, 12
    velocity, 12

Respiratory chain, 208
Resting
    calcium level in Golgi, 109
    calcium level in ER, 109
    voltage, 127
Routh-Hurwitz's Criterion, 50
Ryanodine receptors, 150

Sarcoplasmic or endoplasmic reticulum Ca$^{2+}$-ATPases, 2, 26, 154
Scalar multiple of matrix, 38
Sensitivity
    index, 62
    of glucose and insulin to parameters, 84
SH2-containing 5'-inositol phosphatase (SHIP2), 77
Small-signal finite-gain stable, 54
Small-signal stable, 54
SOCE output feedback controller, 192
Sodium/Calcium Exchanger, 158
Sodium/potassium ATPase, 157
Square matrix, 38
Stabilizable, 58
Stable, 44
State, 55, 58
    equation, 58
    feedback, 58
    feedback controller, 58
    observer, 60
    variables, 43
Steady state, 44
Store-operated calcium entry, 97, 189
Substrates, 11
Survival
    curve, 95
    rate, 95

Time constant, 130
Transport
    kinetics of Vcx1p, 98
    velocity of Pmc1p, 98
    velocity of Yvc1p, 99
Transpose of a matrix, 39

Ultradian oscillation, 86
Uncompetitive inhibitor, 22
Uniporter, 217
Unstable, 44

Upper limit of cytosolic calcium tolerance, 116

Vacuole, 95, 98
Vcx1p, 95, 98
Velocity of transport of calcium out of Golgi, 100

Velocity of transport of calcium out of ER, 101
Voltage-gated calcium channel, 2, 141
Voltage-gated sodium channel, 137

Yvc1p, 95

# MS&A – Modeling, Simulation and Applications

**Series Editors:**

Alfio Quarteroni
MOX – Politecnico di Milano (Italy)
and
École Polytechnique Fédérale
de Lausanne (Switzerland)

Tom Hou
California Institute of Technology
Pasadena, CA (USA)

Claude Le Bris
École des Ponts ParisTech
Paris (France)

Anthony T. Patera
Massachusetts Institute of Technology
Cambridge, MA (USA)

Enrique Zuazua
Basque Center for Applied
Mathematics
Bilbao (Spain)

**Editor at Springer:**

Francesca Bonadei
francesca.bonadei@springer.com

**THE ONLINE VERSION OF THE BOOKS PUBLISHED IN THE SERIES IS AVAILABLE ON SpringerLink**

1  L. Formaggia, A. Quarteroni, A. Veneziani (eds.)
Cardiovascular Mathematics
2009, XIV+522 pp, ISBN 978-88-470-1151-9

2.  M. Emmer, A. Quarteroni (eds.)
MATHKNOW
2009, XII+264 pp, ISBN 978-88-470-1121-2

3.  A. Quarteroni
Numerical Models for Differential Problems
2009, XVI+602 pp, ISBN 978-88-470-1070-3

4.  A. Alonso Rodríguez, A. Valli
Eddy Current Approximation of Maxwell Equations
2010, XIV+348 pp, ISBN 978-88-470-1934-8

5.  D. Ambrosi, A. Quarteroni, G. Rozza (eds.)
Modeling of Physiological Flows
2012, X+414 pp, ISBN 978-88-470-1934-8

6. W. Liu
   Introduction to Modeling Biological Cellular Control Systems
   2012, XII+268 pp, ISBN 978-88-470-2489-2

For further information, please visit the following link:
http://www.springer.com/series/8377